Arctic Geopolitics

Is the Arctic destined for conflict, or can cooperation prevail? This timely book explores the complex interplay of security, geography, and regions in the Arctic. Moving beyond simplistic narratives of geopolitical rivalry, it offers a nuanced, multilevel analysis of state security practices, foreign policies, and regional cooperation across distinct subregions. Challenging conventional notions of 'regions' and re-evaluating the role of geographic proximity, it provides fresh insights into how states engage with their neighbours. It also explores the enduring relevance of geography in international relations, demonstrating how the concept of an 'Arctic region' can be a powerful framework but also rests on some false assumptions. Essential for scholars, students, and policymakers, *Arctic Geopolitics* reshapes our understanding of security dynamics in the Arctic and beyond. This title is also available as Open Access on Cambridge Core.

ANDREAS ØSTHAGEN is the Research Director for Arctic and Ocean Politics at the Fridtjof Nansen Institute, Norway's leading foreign affairs institute concerned with Arctic issues. He comes from the Norwegian Arctic and has worked on explaining the complexities of Arctic geopolitics for more than a decade, across different continents. He is frequently used as a commentator on Arctic issues by the BBC, CNN, and the *Economist*. He holds a PhD from the University of British Columbia in Canada and a MSc from the London School of Economics in the UK, is also a Fellow at Harvard University, a Senior Advisor at the High North Center in Bodø, and teaches at Oslo New University.

Arctic Geopolitics

A Subregional Approach to Security and Geography

ANDREAS ØSTHAGEN
Fridtjof Nansen Institute

CAMBRIDGE
UNIVERSITY PRESS

CAMBRIDGE
UNIVERSITY PRESS

Shaftesbury Road, Cambridge CB2 8EA, United Kingdom

One Liberty Plaza, 20th Floor, New York, NY 10006, USA

477 Williamstown Road, Port Melbourne, VIC 3207, Australia

314–321, 3rd Floor, Plot 3, Splendor Forum, Jasola District Centre, New Delhi – 110025, India

Cambridge University Press is part of Cambridge University Press & Assessment, a department of the University of Cambridge.

We share the University's mission to contribute to society through the pursuit of education, learning and research at the highest international levels of excellence.

www.cambridge.org
Information on this title: www.cambridge.org/9781009743310
DOI: 10.1017/9781009743358

First published 2026

A catalogue record for this publication is available from the British Library

Library of Congress Cataloging-in-Publication Data
Names: Østhagen, Andreas author
Title: Arctic geopolitics : a subregional approach to security and geography / Andreas Østhagen, Fridtjof Nansen Institute, Norway.
Other titles: Subregional approach to security & geography
Description: First edition. | New York, NY : Cambridge University Press, 2026. | Includes bibliographical references and index.
Identifiers: LCCN 2026016748 (print) | LCCN 2026016749 (ebook) |
ISBN 9781009743341 Hardback | ISBN 9781009743310 Paperback |
ISBN 9781009743358 eBook
Subjects: LCSH: Geopolitics – Arctic regions | Security, International – Arctic regions | Arctic regions – Strategic aspects | Arctic regions – Government policy – United States | Arctic regions – Government policy – Norway | Arctic regions – Government policy – Russia (Federation) | Arctic regions – Government policy – China
Classification: LCC G615 .S74 2026 (print) | LCC G615 (ebook) |
DDC 320.1/209113–dc23/eng/20260608
LC record available at https://lccn.loc.gov/2026016748
LC ebook record available at https://lccn.loc.gov/2026016749

ISBN 978-1-009-74334-1 Hardback
ISBN 978-1-009-74331-0 Paperback

To all those who take time to reply to questions, answer emails, and give feedback and assistance along the way, seeing the world not as a zero-sum competition but as a collaborative expedition.

Contents

Figures

Preface

This book springs out of a simple and – unfortunately – increasingly relevant question: will there be conflict between states in the Arctic? When asked about my work, this is often the question I reference, as it has guided my research over the past fifteen years. Considering that question when I started doing Arctic-specific research, conflict was more or less unimaginable. Today it is not. Essentially, this book expands upon a decade of investigation into Arctic security issues, using this question as a foundational premise.

Having grown up in Bodø, North Norway, just above the Arctic Circle, I had no conception of hailing from 'the Arctic'. Despite the stunning landscapes and the best football team in Norway, I wanted to leave my hometown and the Arctic as quickly as possible, seeking warmer temperatures and bigger cities. As a relatively young master's student in London in 2009, I was told that it would be smart to choose a dissertation topic 'close to home' or one linked to future aspirations. This was the exact time that the Arctic – however you want to define it – started to appear on agendas and conferences around the world, due to climatic change, resource potential, and evolving political relations.

More than a decade and a half later, the Arctic's status has fluctuated between a climate change cautionary tale, a resource-rich frontier, and more recently, a symbol of great power rivalry. Today, the latter narrative predominates, although resource availability and the effects of climate change are more evident than ever.

The fundamental point in this book stems from the work I did after graduating, but before turning to academia, when I represented Northern Norway in Brussels from 2010 to 2013. My role was to inform EU policymakers that the Arctic, which was starting to figure in their strategic considerations, was a complex region. Saying, as some did and still do, that we need to 'Save the Arctic' just doesn't capture that nuance or the interests of the millions of people living in the Arctic.

That doesn't discount that parts of the Arctic, whether nature, wildlife, or peoples, might need saving at times. But simple one-liners about a space that comprises 4 per cent of the globe just doesn't cut it.

In 2014, at the Norwegian Institute for Defence Studies, I began researching how Norway and the USA, two Arctic states sharing a maritime boundary with Russia, responded to the Russian war in Ukraine starting in March that year. At the same time, the Arctic Coast Guard Forum came into being – something I had been briefly involved in proposing when working at the American think tank Center for Strategic and International Studies (CSIS) in the summer of 2011. However, it quickly became evident that discussions on pan-Arctic coast guard cooperation and broader narratives about Arctic geopolitics and security post 2014 often neglected the diverse regions within the Arctic and the varied geographical contexts in which Arctic coast guards operate. This realisation sparked the idea for this book, which seeks to explore the nuances and complexities of Arctic security.

In 2016, as a PhD student in Vancouver, I wrote a paper on regional security theories, which later became an academic article. With support from fellow PhD candidate Greg Sharp and after presenting the paper at a conference in Toronto in 2016, an initial interest in the linkages between security and regions was sparked. The article was only submitted in 2019, and published in 2021, delayed partly due to one reviewer's insistence on a more holistic approach to Arctic security. The article focused on state and military security – the traditional lens for security analysis in international affairs. I argued that, prior to 2022, there was less emphasis on circumpolar or pan-Arctic state security and more on other types of security, such as environmental, human, and Indigenous, with an increasing focus on governance and politics, albeit often excluding military aspects.

I argued that this emphasis, while not incorrect, led to a misconception among scholars, analysts, and the media that the Arctic is also a unified space for security interactions and military operations. Or, at a minimum, there was and is limited understanding about what security issue or issues in the Arctic *is* or *are* in focus, and where, when analysing geopolitical and security dynamics.

Though I would argue that the article accurately captured this fundamental point, it fell short in exploring the complexities of Arctic political dynamics and how national or international factors impact Arctic security. The application of the dominant regional security complex theory by Barry Buzan and Ole Wæver was also superficial, and the article failed to propose alternative frameworks for understanding Arctic security dynamics. This book attempts to address those shortcomings by building on recent scholarly work and, importantly, includes an analysis of Russia's full-scale invasion of Ukraine, the increased Arctic-related collaboration between China and Russia, particularly in the North Pacific, and even US President Trump's insistence on 'control' over Greenland.

In turn, this book is not merely a summary of my previous efforts; instead, it draws on the varied topics and writings I have explored since 2014.[1] It encompasses discussions from coast guard and Svalbard issues to broader Arctic geopolitics, reflecting a decade of scholarly engagement with Arctic security. I argue (and hope) that these writings also hold relevance for other parts of the world, when examining geopolitics and conflict potential. 'Geopolitics' has seldom been more in vogue than right now, while we simultaneously seem to increasingly rely on simple one-liners or 'quick takes' that promise to reveal the truth about complex issues in 400 words or less. Alas, the world is complex, as are regional security relations, and empirics matter. This book is, in other words, a call for acknowledging, embracing, and diving into that complexity.

There might be some obvious biases in here too. As I touch on at the end of this book, I've found myself labelled both a 'hawk' and a 'dove'

[1] As such, Chapter 2, in outlining security regions theories leans on 'The Arctic Security Region: Misconceptions and Contradictions' (2021b). Chapter 3 leans on, in the part of the Arctic Coast Guard Forum, *Coast Guards and Ocean Politics in the Arctic* (2020a); Arctic regional dynamics, 'Security Dynamics In, Through, and Over the Arctic "Region"' (with Østhagen, Andreas and Whitney Lackenbauer, 2023) and 'Geopolitics and Increased Tension?' (2023, with Svein Vigeland Rottem). Chapter 4 leans on some of the latter works, and, for the NATO and Canada discussion, 'At Opposite Poles' (2018, with Greg Sharp and Paal Sigurd Hilde). Chapter 5 leans on, for the Norway part, *Norway's Arctic Policy: Still High North, Low Tension?* (2021a), 'Ine Eriksen Søreide: Nordområdene, USA og Balansekunst' (2023b), and 'The Myths of Svalbard Geopolitics' (2024). Chapter 6 leans on 'Reluctant No More? The Complexities of US Arctic Security Strategy' (*Handbook of the Politics of the Arctic*, with Pavel Devyatkin, Lillian Hussong, and Andreas Østhagen, pp. 283–300, Cheltenham, Edward Elgar Publishing, 2026). Chapter 7 leans on, for Svalbard-discussions, 'The Myths of Svalbard Geopolitics' (2024), and for the China–Russia discussion, 'China, Russia, and the U.S. in the Bering Sea: Military Exercises and Great Power Politics' (*Arctic Yearbook, 1, 2025*, with Erdem Lamazhapov and Andreas Østhagen, pp. 1–21: https://arcticyearbook.com/images/yearbook/2025/Scholarly_Papers/2SP_AY2025_Lamazhapov_Osthagen.pdf, 2025). Finally, Chapter 9 leans on elements from 'High North, Low Politics' (2016) when discussing low-level cooperation.

in academic circles, depending on the context. While working on this book, I stumbled upon some older writings from my time in Brussels between 2010 and 2013, where I argued – much as I do here – for the necessity for differentiation and nuance. Maybe I've always been drawn to the subtleties and distinctions when we discuss the Arctic, not only because it's my home but also because I don't quite fit with how the Arctic is typically portrayed.

My experience representing a subnational level in Brussels to a supranational audience might explain my focus on the multilevel structure highlighted in this book. Ultimately, however, this work emphasises the importance of taking state security (i.e., security practices) seriously, while also recognising the various dynamics at play on other 'levels'. This is indeed a Norwegian outlook, although I have lived at various intervals in two Arctic states – Canada and the USA – as well as worked in the 'capital' of the EU. Engaging in Arctic debates in those contexts might be exactly why I am, in this book, emphasising distinction and nuance.

Finally, if you are looking for the 'one ring to rule them all', this book probably won't meet your needs. It doesn't set out to define a comprehensive theory of international relations, nor does it propose an all-encompassing framework. What it does argue is that our common understanding of the 'regional' is often too broad, lacking the explanatory depth and specificity needed to truly grasp the reality (or as close as we can get) of state security practices. With the rise of quick analyses promising to decode the 'geopolitics' of particular regions or even continents, it's worth taking a moment to rethink what geopolitics really entails and how we can enhance the accuracy of our descriptions. That should, in theory, also serve a purpose beyond enlightenment, hopefully influencing the prospects of conflict management in the Arctic and beyond.

Acknowledgements

Have you ever tried writing a book while on paternity leave with a strong-willed boss-baby named Cornelia? Or reading up on incomprehensible academic theory written half a century ago while a three-year-old Theodora is planning her next scheme to acquire chocolate? I think most academics base their careers on their ability to work at odd hours, having oceans of time to dive into theory and/or nitty gritty empirics, and being in relative solitude when doing so. Alas, this book has been a struggle – perhaps more so than writing my doctoral thesis or any other project I've had in academia. Not because I don't love the ups and downs of fatherhood, but because concentration and plenty of time have become elusive perks. Therefore, I must first and foremost thank my wife, Victoria, not only for support but also for managing the wildcats.

As also outlined in the Preface, this is a book based on work done over a decade as a researcher at the Fridtjof Nansen Institute, the Norwegian Institute for Defence Studies, the High North Center at Nord University, during my PhD work at the University of British Columbia, and my brief stint as a Fulbright grantee at Harvard University Belfer Center's Arctic Initiative. I could thus thank a range of scholars, many of whom are referred to in this book.

For feedback and critique concerning this book and the emphasis on geography, security, regions, region-building, and geopolitics, I must especially thank Arild Moe, Paal Sigurd Hilde, Svein Vigeland Rottem, P. Whitney Lackenbauer, Gunhild Hoogensen Gjørv, Jennifer Spence, Ole Wæver, Olav Schram Stokke, and Klaus Dodds. They might not agree with my assumptions or conclusions, but they have all given invaluable input and feedback to various aspects of this book. I hope I am able to do the same, not only to senior colleagues such as these, but also for the next generation of scholars engaged with Arctic geopolitics. I must also thank the very constructive feedback I received from the reviewers, as well as from Cambridge University Press editor

John Haslam, who provided guidance with this book project as it progressed over several years.

As I am working at an applied research institute concerned with international relations completely dependent on external research funding, I must also acknowledge and thank those who actually funded all the research hours and months spent writing this book. This is primarily the Norwegian Research Council, but also the Norwegian Ministry of Defence and the Norwegian Ministry of Defence. At a time when questions are increasingly being asked about the value of facts, knowledge generation, and academic studies, having the opportunity to dedicate considerable time to connect the various dots of Arctic security and geopolitics is not a given – and very much appreciated. Funding from the Norwegian Research Council (grant number: 343154 / Arctic3D and 345131 / Norwegian Centre for Geopolitics) made it possible for this book to be published open access, making the digital version freely available for anyone to read and reuse under a Creative Commons licence.

1 *The Return of Geopolitics and the Emergence of the Arctic*

Vodka, Trump, and Adjusting the Geopolitical Scope

In June 2008, Russia's foreign minister, Sergey Lavrov, was on board the Russian nuclear icebreaker *Lenin* in the Arctic port of Murmansk together with the Norwegian foreign minister, Jonas Gahr Støre. The two ministers had met more than a dozen times in the previous months, and on this occasion, they watched the Euro 2008 football match between Russia and Spain while drinking vodka.[1] Two years later, in April 2010, the ministers announced a breakthrough in Norway–Russia relations and a 'new era' was declared, as the two countries agreed to settle a four-decade-old dispute in the Arctic (V. L. Pettersen 2010).

Geographic proximity was the foundation for this dispute and subsequent agreement. Norway and Russia share a land border near the northernmost point of the European mainland, initially agreed upon in 1326 and finalised in 1826. A disagreement emerged in 1977 over how to divide the maritime space *north* of the European mainland, extending into the Arctic Ocean all the way to the North Pole. Geographic proximity between Norway and Russia was also the reason for the breakthrough agreement of 2010: by finally resolving this longstanding dispute, a thorn in Norway–Russia relations was removed, paving the way for joint offshore oil and gas development in the Arctic. As sea ice was receding and prospects of new shipping routes via the 'shortcut' to Asia became a reality, the two countries' shared geographic space became a source of cooperation and new relations.

In the Arctic, geography looms large. The Arctic is one of the places on the planet where climate change is most apparent. Large land and

[1] This story is described in *Seierherren*, an unauthorised biography of Gahr Støre by Ståle Wig (2014).

sea areas that were previously covered by ice are changing character. Resources previously inaccessible are not anymore. The millennia-old dream of a shortcut between continents via the North Pole is suddenly a reality.

Geography is also increasingly a cause for concern. In 2022, Russia launched a full-scale invasion of Ukraine in breach of international law, having already engaged in conflict in Ukraine since 2014. All the other Arctic states joined a Western response through sanction regimes scaled-up from those put in place in 2014, while they cut most forms of political and economic ties with Russia. All Arctic states apart from Russia have joined NATO. Both Russia and NATO member states signal their military capabilities in the Arctic, and fears of escalation and conflict abound.

Russia has turned towards some of the BRICS countries in order to fulfil its economic and political ambitions in the Arctic. Russia has even engaged in deeper military cooperation with China in parts of the Arctic to enhance its security vis-à-vis the other seven 'Western' Arctic states – something thought unthinkable a decade earlier. Because the Arctic is one of the geographic spaces where Russia borders Western states along two fronts (in both Europe and in the North Pacific), the Arctic is touted by many as the next arena for direct conflict, even war, between Russia and the West.

A clear illustration of this materialised in 2024 on 22 December, as incoming and returning US President Donald Trump stated that 'the United States of America feels that the ownership and control of Greenland is an absolute necessity' (Trump 2024). Trump's interest in acquiring control over Greenland, articulated for the first time publicly in 2019, reflected a securitised economic and strategic interest in Arctic geopolitics partly in response to what is perceived as Russia's northern dominance and China's increasing Arctic interest. Greenland's vast natural resources, including rare earths and hydrocarbons, combined with its location between North America and Europe, renders it a site of so-called geopolitical significance. Trump framed a potential purchase as enhancing US control over Arctic access routes and resource security, seeking to pre-empt rival powers' influence (Vock 2025).

The counter-reaction, from Danish authorities, Greenlandic leadership, and international observers, highlighted norms of sovereignty, self-determination, and NATO cohesion. Governments in both Copenhagen and Nuuk rejected the notion as an affront to constitutional order and a misreading of contemporary statecraft, emphasising Greenland's autonomous governance and aspirations.

The episode also catalysed renewed attention to Arctic security policy, prompting a range of stories in both media and amongst analysts about the Arctic as an arena for geopolitics and conflict between great powers (i.e., states with relatively more recognised capacity than their peers).[2] The problem, however, is that although the Trump administration's attention to the Arctic was perhaps overdue due to the USA's relative lack of interest in the rapidly changing political, economic, and climatic conditions in the North, the emphasis was placed on the wrong *part* of the Arctic.

If geopolitical concern over China truly was the motivation behind the statements and ensuing visits by top Trump officials to Greenland, then Alaska would have been a better place to focus policy and efforts. If Russia, on the other hand, was the motivation, Alaska, again, but also the European parts of the Arctic are more relevant to managing security threats that emanate from Russia's Arctic (Østhagen 2025b, 2025c).

The way we describe, understand, and analyse the Arctic is a clear example of the pitfalls of conflating security, geography, and state practices. Or, to put it another way, the Arctic is prone to quick 'geopolitical' analyses that make bold claims about the region, lacking specificity, nuance, or even a comprehension of what geographical space we are talking about. Examples of this abound. Descriptions such as a 'Great Game' (Borgerson 2009), 'Great Race' (E. Buchanan 2023b), and 'Great Power Competition' (Fridtjof Nansen Institute 2024b) are common. Others claim that 'War will start with Russia in the Arctic' (Geri 2024) or that 'a battle for the Arctic is underway' (K. R. Rosen 2022).[3]

Indeed, geography is a key, if not *the* key, factor in understanding the security concerns of a state. The premise of (traditional) international

[2] See Iver B. Neumann (2008) for an extensive discussion on different ways of understanding what a great power is: 'Weber's realist criteria – material resources and the ability to project power … Durkheim's moral criteria – setting an example for how a state should be in the world', and – what Neumann argues for – a 'governance' account which 'takes cognizance of the materiality of the discourse' and 'is predicated on the intersubjective character of recognition'.

[3] This author is also prone to leaning on such simplified headlines in order to draw attention to seminars, conferences, or research proposals. See, for example, the 'Arctic Security Conference': www.arcticsecuritynet.com/asc.

relations theory concerning how states behave and interact with each other is that power can to a degree help explain outcomes in the international domain. And power, whether based on economic, military, or population attributes, is contingent on the territory in question.

At the same time, the political developments in the Arctic amongst the eight Arctic states (those with territory above the Arctic Circle at 66.34 degrees latitude) – Canada, the Kingdom of Denmark, Finland, Iceland, Russia, Sweden, and the USA – in the period 2005–2014 actually attempted to *transcend* the constraints of geography and construct a 'region' defined by cooperative traits. Russia and the other Arctic states, alongside a range of non-Arctic actors such as China, India, Japan, the UK, and the EU, have more or less avoided a 'race for resources' or a 'scramble' for northern territory,[4] instead engaging in what is typically conceived of as region-building (Neumann 1994).

Over the last three decades, instead of racing towards conflict, Arctic states alongside other actors have been setting up governance mechanisms and forums for the Arctic, while creating a whole discourse and practice around Arctic peacefulness, an Arctic 'exceptionalism', joint sustainable development, and the region's inherent cooperative elements. State leaders everywhere from Russia to the USA, China, and India have attempted to politically define the Arctic as a region that they either belong to or have strong connections with. Emphasising everything *but* state security (or traditional military security) concerns, the 'Arctic region' emerged as a constructed object at the turn of the millennium and is now accepted not only as the geographic space above the Arctic Circle but also as a political entity where states engage and interact in both adversarial and cooperative ways.

However, if you – as an analyst of security in international affairs – retain a focus on the state, and its concerns with securing its own survival, the Arctic region becomes fuzzy. It seems doubtful that immediate security concerns and state practices covers the whole Arctic region, that is, across the still-frozen Arctic Ocean and this vast space that occupies 4 per cent of our planet. And yet we continue to talk about political, climatic, and economic developments in 'the Arctic'. And states highlight their Arctic credentials, while emphasising the

[4] As predicted by, for example, Borgerson 2013; Sale and Potapov 2010; Rainwater 2013.

role of pan-Arctic bodies such as the Arctic Council in dealing with 'Arctic issues' (whatever they might be).

Officials from the various Arctic states seem to recognise this discrepancy when they highlight the importance of immediate geography in explaining investments in military forces in the 'North'. As Daniel Riviere, the Canadian chief of Joint Task Force North,[5] put it in 2024: 'I challenge anyone to have a land invasion of the Canadian North … it would end up being a search and rescue operation' (Fridtjof Nansen Institute 2024a: pt. 51 min).

Other Arctic states see it differently, depending on geography. For Finland, also an Arctic country, a land invasion by Russia is indeed a plausible scenario – as the USSR did exactly that in 1939. After Russia's war in Ukraine, fears of Russian aggressive activity along the extensive Finnish and Norwegian border in the Arctic and the maritime boundary stretching towards the North Pole have become much more immediate. Russia simulates attacks on essential infrastructure across the border, while continuously jamming GPS signals in both Finland and Norway, disrupting everything from emergency services to civilian aviation.

The discrepancy between the politically defined Arctic 'region' and how Arctic states actually practice their security interests and concerns leads to a fundamental question in international affairs. Is state security 'trapped' by the confines of geography? Or can geographical parameters be transcended by political ingenuity and constructs, such as governance mechanisms and cooperation, as the Arctic states attempted to do at the start of the 2000s?

The answer naturally lies somewhere in between these two counterpoints, the former being particularly reductionist and deterministic, and the latter being rather utopian. But *how* do we understand the linkages between geography and security in a given territorial space? How can examining these linkages help us unpack the increasingly complex security dynamics in a part of the world where geography itself is also changing due to more fundamental planetary changes? In essence, *will* the changing climate and thus also geography of the Arctic lead to resource wars and interstate conflict in the North?

More fundamentally, are there certain inherent traits in the Arctic regional context that makes it prone to conflict, or cooperation?

[5] The military body responsible for protecting the Canadian North.

Or is there perhaps something about regions in general that by virtue of geographic proximity tend to exacerbate underlying security concerns and dynamics, that is, so-called geopolitics?

Here the concept of regions come into play: (1) as a level of analysis (i.e., something less than global but more than national and domestic), (2) as a distinct political process creating the 'region' (i.e., regionalism), and (3) as a so-called security complex defined by geographical limits in states' power or immediate reach. All of this is framed by the re-emergence of geopolitics as a fuzzy, albeit increasingly used, concept to understand international relations. Although there is no doubt that geopolitics is a relevant frame from which to analyse Arctic policies and security developments, that frame requires that we get our level of analysis right and adjust the scope. That holds true not only for the Arctic but also for geopolitical analyses in general – which increasingly seem to be employed by a range of state leaders around the world.

The Re-emergence of Regions and Geopolitics

In 2023, more than a decade after drinking vodka and highlighting the Arctic's economic potential, Gahr Støre – now prime minister of Norway – reminded an Arctic conference crowd about the perils of forgetting the rules of geography: 'I would warn against what we see in part of the debate in Europe now, where people are cutting Russia out of the map, as if it is no longer there. It is. We do not choose our geography' (Gahr Støre 2023b). In other words, Norway cannot escape the 'destiny' of its geography.[6] Despite strong objections to the Russian war in Ukraine, Norway had to, and continuously must, find constructive ways to manage issues in the North with its larger north-eastern neighbour because of geography.[7]

[6] Representing a small state at the edges of Europe (and the world), bordering a much larger unpredictable Russia, Norwegian leaders have indeed been fond of emphasising the necessities and restraints that come with geographical attributes and a state's physical position on the map (Tamnes 2019; Steinveg 2024).

[7] Before becoming prime minister, Gahr Støre underscored this by referring directly to Robert D. Kaplan's *The Revenge of Geography* (2012) in op-eds (Gahr Støre 2019).

The same point about the destiny of geography seems to underline some of the Trump administration's obsession with control over Greenland (albeit with a different policy approach emphasising control, not cooperation): that the island is geographically part of the North American continent, and that in turn makes it crucial for the USA to have 'control' over it. In fact, the same can be said about Russia's insistence of control over Ukraine, or China's adamant position regarding Taiwan: the simplified notion that the geography of these large states demands some form of control over adjacent territories.

The emphasis on the role of geography in defining international relations is, however, nothing new. Since the nineteenth century,[8] scholars, politicians, and pundits have made claims linking *how* states act and relate to each other in the international system and geographic attributes. The word geopolitics derives from the German term *Geopolitik* and was first introduced by Rudolf Kjellén in 1899 and later popularised by Halford Mackinder and other scholars (Mackinder 1904). The roots of the term can be traced to Kjellén's vision of a framework that integrates geography, political science, and sociology to analyse the spatial dynamics of power (R. Kjellén 1924).

Traditionally, geopolitics focuses on the connections between power politics and geographic characteristics, such as location, topography, population numbers, and territorial size, all of which are sensitive to expansionist inclinations and interstate rivalries over finite territories and resources (Sloan and Gray 1999; Stokke 2017; Dodds 2019). Today, 'geopolitics' has become shorthand for rivalry and potential conflict between states. The critique of traditional geopolitical analyses has often been that they are too narrowly focused on resource drives and tend to be deterministic in their analytical capacity.

The prevalence of geopolitical analyses has varied over time, often in sync with fluctuations in the levels of conflict seen amongst leading states (Mead 2014). The notion of geopolitics has also been tainted by its use by expansionist states engaging in territorial expansion and – in the case of Nazi Germany – deterministic race-based justification for

[8] Or before – see, for example, writing from Thucydides in the fifth century BC (1910) and Niccolo Machiavelli in the fifteenth to sixteenth centuries (1950). However, the advent of geopolitics as an analytical approach is reckoned to have started in the late nineteenth century. See more in Chapter 2.

those actions. In recent decades, however, more critical assessments of geopolitics – linking it to everything from language to popular imageries (M. Müller 2013) – have expanded on the concept, attempting to break what is often a somewhat deterministic link between geography and explaining outcomes in world politics.

Despite the overuse or misuse of the term, over the last decade 'geopolitics' has seen a revival amongst both media and scholarly work (Topalidis et al. 2024). What these works have in common is that they often explicitly or implicitly lean on more academic concepts of geography, security, and state behaviour within a *geographically bounded space* – or, to put it differently, 'regions' as both a political and social concept and as a defined material entity observable to the analyst frame our understanding of world affairs.

An increasing interest in the regional as a counterweight to the global (or at times national) derives from several political and economic trends apparent since the start of the 2010s. The financial crisis from 2008 and the decline of the USA as a global hegemon and the subsequent (possible) unravelling of a unipolar international system has prompted renewed interest in regional or anything-less-than-global power relations and security concerns.

Similarly, as the simplistic belief from the 1990s and early 2000s that all states were on a one-way path towards a liberal democratic system has been dispelled with the decline in democratic states and the growth and/or success of non-democratic states, a reappreciation for the complexities of states and their immediate neighbourhood has emerged. The disruption of economic 'globalism' – ranging from renationalisation in response to domestic income gaps and populism and an increase in trade barriers (e.g., the trade war between the USA and China) – has led to an emphasis on regional trade blocs and market integration. The COVID pandemic from 2020 also seems to have led to a more insular or protective approach to international politics by states. This comes alongside a policy of tariffs and 'regionalisation' by the world's largest economy – the USA – led by both the first Trump administration and the Biden administration (The Economist 2024), and also the second Trump administration.

In response to these trends, the focus in international affairs has somewhat shifted from the global to the *national* and the *regional*. In part, the practical response by states (i.e., their bureaucrats and politicians) has been an emphasis on securing regional cooperation with their near abroad. Simultaneously, in a related albeit somewhat separate intellectual space, 'regionalism' – the study of the social-political processes where actors construct and reinterpret regions – has again become trendy in international relations scholarship. Regionalism, however, in contrast to geography or security-oriented approaches, is conceived as more of a constructed or sociological phenomenon dependent on the *relations* amongst the states (or other actors) involved.

Another approach altogether, in order to 'objectively' unpack the geopolitical security concerns of states in the twenty-first century, is to focus solely on geography. Tim Marshall, for example, in the bestseller *Prisoners of Geography: Ten Maps That Tell You Everything You Need to Know about Global Politics* (2015), promises to unlock international politics. If one goes beyond the immediate critique of such relatively simple and reductive assertions,[9] the underlying concept framing such an examination of, say, the South China Sea, Eastern Europe, or the Arctic is that security relations amongst states in a *regional space* overlap and interplay. What, for example, Vietnam does in terms of military investments or security doctrines will inherently impact security considerations in, for example, Malaysia and China, because these states share the same geographical space. Moreover, the totality of these relations and interactions defines security dynamics within that geographically bounded space, or region.

In other words, several trends are occurring at the same time in the study of international relations. Geopolitics and an emphasis on the linkages between security and geography – especially within a confined geographical space (i.e., region) – is becoming a popular topic for both politicians and pundits. Even some state leaders actively refer to the relevance of Robert Kaplan's *Revenge of Geography* (2012) when considering the geopolitical and security implications of a changing Arctic (Gahr Støre 2019). Everyone is a 'geopolitician' these days.

Regions, as a level of analysis or as a social-political-economic process perpetuated by actors, are also again attracting academic attention – after previous 'waves' of interest in the 1970s and 1990s. And the fundamental driver of these issues is an increasing concern with state-oriented notions of security, linked to developments in the

[9] See, for example, Immerwahr 2022.

2020s from an emerging multipolar system to Russia's initiating of war in Europe, or fears of conflict erupting over Taiwan.

Security, geopolitics, and regional studies undoubtedly overlap, although these concepts are often not linked or explored in tandem. Moreover, they tend to either ignore elements of geography or have blind spots – at least when it comes to the study of states and their (traditional) security interests. At the same time, the emphasis on geopolitics and geographic attributes tends to – still – produce wide, sweeping conclusions that do not hold up under scrutiny. As put by Klaus Dodds (2019: 35): 'Geography is either greatly underappreciated or perversely overexaggerated.' Similarly, the regionalist focus tends to ignore core aspects of geography, and the barriers posed by distance, at least when it comes to security and military affairs.

The in-between framework often employed to merge the study of security and the regional level – regional security complex theory or variants thereof – proposes a relatively neat solution to the study of regional security interactions. Such approaches merge the fundamental restrictions levied on state interactions by geographic distance *and* the socially constructed element emerging from 'securitising' acts amongst states – acts that tend to have a regional flair.

Yet even this framework does not accurately reflect or explain region- and state-specific actions and security interests, as it arguably does not place enough emphasis on the geographic dimensions inherent in a 'security region' – at least from a state perspective. Drawing lines around most of Europe and arguing that Greece and Finland, or around Southeast Asia and arguing that Papua New Guinea and Myanmar, or around the Arctic and arguing that Canada and Norway belong to the same regional security complexes might hold true for theoretical and/or system-level analytical purposes. But for a military planner or foreign minister the scale is arguably too extensive to hold direct relevance to the task of ensuring the security of the state – a task, however, that seems increasingly to occupy the political agenda.

Instead, as argued in this book, it is time to take a closer look at the linkages between state security, the role of geography as perpetuated by geopolitics, and regional security complexes. This is especially true for regions that are undergoing rapid change with the potential to alter and even upend security relations amongst states in that very space.

The Arctic: An Ideal Case of Geopolitics, Regionalism, and Security

Certain regions are experiencing change more rapidly than others. Regional change involving states can, for example, be climatic or environmental (e.g., desertification in the Sahel), economic (e.g., trade agreement concerning market access in North America), or political (e.g., the security concerns derived from Russia's invasion of Ukraine). These dynamics overlap and can reinforce each other.

In no place is regional change more apparent than in the Arctic. If you find yourself driving, flying, sailing, walking, or even sledding northwards in any of the eight countries that surround the Arctic Ocean, you will at some point cross the invisible line that is the Arctic Circle (66°33′49.0″ north of the Equator). This Circle marks the point where the sun will not rise all day on the shortest day of the year and will not set on the longest day of the year. Apart from that, you can initially find no marked difference as you make your way from the south to the north. Except that after a while, nature changes, the population becomes *even* sparser, and summer days can feel chilly.[10] Thus, you have ventured into the Arctic.

For clarity, Figures 1.1 and 1.2 set the geographical parameters for this book. This book operates under the most common political definition of the Arctic as the region located above, or north of, the Arctic Circle. Another common definition often used by natural scientists is the 10°C isotherm that is defined as the northern area of the globe, where the average temperature in the warmest month, July, is below 10°C/50°F. Whole books have been written about defining the regional boundaries of the Arctic. However, this book is about the political dynamics in the Arctic, not water temperatures or animal habitation. This implies that what matters for our enquiries is not how *we* define the Arctic but how the dominant actors that we study – *states* – define the Arctic. The chapters of this book will also explore how a *security* region is defined and constituted as more than just geography through its social processes involving actors and interests. This has a bearing on security debates and Arctic geopolitics too.

[10] In Greenland or Iceland, particularly, this is perhaps an exaggeration, given that the whole of Greenland can by most definitions be deemed 'Arctic', and that almost all of Iceland is located just south of the Arctic Circle but exhibits Arctic climatic conditions.

Figure 1.1 Map of the Arctic. Source: CIA World Fact Book, public domain, www.cia.gov/the-world-factbook/static/97828fc01510ce36884bf7e07a508436/arctic_phy.pdf.

None of the eight Arctic states are located solely above the Arctic Circle. Iceland comes closest. Greenland, if it were to become independent at some point, would almost qualify, although the southernmost part of the island is actually below the Arctic Circle (not that many would question that all of Greenland is Arctic in terms of climate or identity (Jacobsen 2015)). Note the emphasis on the *Kingdom*

Figure 1.2 Map of the Arctic in a wider geographical context. Made by the author using Google Earth.

of Denmark here. Denmark, in and of itself, is not even close to the Arctic Circle.[11]

In turn, this book makes use of the Arctic spatial domain that has repeatedly been portrayed as an 'emerging' *security region* to question some of the fundamentals that constitute the concept of security regions and – fundamentally – international relations. This builds on

[11] When Norway, Denmark, and Sweden entered into the Kalmar Union in 1397, the extended Norwegian 'Viking Kingdom' included tentative control over Greenland, Iceland, and the Faroe Islands (Stenersen and Libæk 2007). After 417 years, in 1814, when Norway gained independence from Denmark with the Kiel Treaty after the first Napoleonic war, Denmark retained the these territories. Iceland later gained full independence during World War II, in 1944, by using the US military presence and political interest as leverage (Bergmann 2014). However, Greenland remains part of the Kingdom of Denmark to date, although discussions continue over what possible forms of increased autonomy are possible for Greenland (Jacobsen 2015; Kristensen and Rahbek-Clemmensen 2017). These discussions are heavily influenced by topics on which this book focuses: the increased political and economic interests in the Arctic and the related security questions and issues. The Faroe Islands are also part of the Kingdom of Denmark. Thus, when discussing Arctic affairs, Danish officials distinguish between the Kingdom at large and Denmark specifically.

developments in, or about, the Arctic for more than two decades now. During 2006–2007, reports of increasingly receding Arctic sea ice began to appear in headlines. Around approximately the same time, under the leadership of Vladimir Putin, Russia set out to reassert itself militarily, both regionally and globally. The symbolic pinnacle of this was on 2 August 2007, when a Russian expedition used a submersible to plant a Russian flag on the Arctic seabed beneath the North Pole.

However, since then, *all* eight Arctic states have held several high-level gatherings both separately and in the context of the regional cooperative science-oriented forum the Arctic Council where they emphasised the importance of international law, governance, and cooperation amongst Arctic states and peoples. The Arctic was even deemed an 'exceptional' region and used by some as the ultimate example of the virtues of region-building and transcending a state security-oriented approach to international affairs (Heininen 2013; Heininen, Exner-Pirot, and Barnes 2019; Devyatkin 2023; Olesen 2020).

Accordingly, some argued that a regional security complex was emerging, albeit due to environmental and human security concerns, not state or military security issues (Exner-Pirot 2013; Padrtová 2015; Lanteigne 2016; Sadurski 2024). Others have, however, questioned this – for example, Østhagen 2021b; Jacobsen, Wæver, and Pram Gad 2024c – albeit without delving into an in-depth study of *security dynamics*, *geography*, and notions of *security regions*. Some even mused on a future security community emerging around the North Pole (Keil 2013: 252; Åtland 2014: 162; Schaller 2015).

Russia's partial invasion of Ukraine in 2014 and its subsequent full-scale invasion in February 2022 marked watersheds in relations between the West and Russia, including in the Arctic, as cooperation was halted and sanctions on Russia were put in place. Some of the ideas about an Arctic exceptionalism and a shared community with Russia were disbanded (Bennett 2023; Koivurova and Shibata 2023). Other analysts emphasise that the Arctic is *still* a region characterised by cooperation, hoping that despite the invasion of Ukraine, low-level forms of collaboration with Russia can still be possible in the future (Spence et al. 2023; Oma Vyvial 2023).

The Arctic Council – briefly suspended – continues at the time of writing to function with Russian participation, albeit at a low level (Andreeva and Rottem 2024). At the same time, Russian politicians and military leaders actively seek the partnership of non-Arctic actors,

China first and foremost (Stensdal and Heggelund 2024), in order to counter all the other seven Arctic states being NATO members (after Finland and Sweden joined in 2023 and 2024 respectively).

There are, in other words, a range of different political and security dynamics at play in the field of security and geopolitics in the Arctic, both before and after 2014 – some might even seem contradictory. All the while analysts, media, politicians, and academics refer to the Arctic as one 'region' – as it has been more than a dozen times already in this chapter. Especially since 2014, as rhetoric and statements concerning Arctic security threats have intensified, there has been a general tendency to describe security challenges as coherent across the region. Scholars interested in this 'new' region

often describe the Arctic as one geographical space where states' various types of security interests are inherently intertwined.[12]

This Book and the 'Regional' Level

This book is about the intersection of security dynamics *in* the Arctic, and about a critical examination of the labels often used to describe those dynamics in the Arctic and in other spatial domains around the globe. On the one hand, this book is about the Arctic. More accurately, it is about the political-legal-security construction of the Arctic region and the respective Arctic states' security interests, concerns, and entanglements; that is, their *security practices*. The fundamental questions that this book aims to explore and eventually answer are: What is 'Arctic security' as seen from a traditional state or military security perspective? Is the Arctic a 'security region'? Do the Arctic states' security interests overlap? And, perhaps most importantly, why does this matter?

These questions are crucial to answer because, increasingly, we are talking about Arctic security and geopolitics, but with a lack of nuance or understanding of the complex causal mechanisms that may lead to interstate conflict. Statements such as 'Melting ice in the Arctic is unlocking a valuable natural resource base ... pushing countries to jostle over territorial claims and control of the region's shipping routes and navigation rights' (Rozpedowski 2024) are emblematic of how the Arctic is perceived and described. Others claim that 'Whoever controls the Arctic has a clear strategic advantage' (G. Johnsen 2024).

[12] See Borgerson 2008; Weber 2015; Huebert 2013a; and Åtland 2014.

The US Chamber of Commerce (Sapsford 2024) states that 'Russia and the northern members of US-led NATO are manoeuvring amid the receding ice packs for strategic advantage. ... Among the expected rewards are the region's abundant resources of fish, minerals and energy, and quick maritime passage from one ocean to another.'

The main trend in these descriptions, and a range of others,[13] is that everything is mixed with everything else. Chinese interests in the Arctic region are combined with the Russian presence, the mere fact that seven states in the Arctic are NATO members, and that parts of the Arctic hold resource potential that may or may not be commercially relevant. Yet how accurate are such descriptions and narratives when it comes to the Arctic? Which part of the Arctic is China investing in? Where is Russia increasing its presence, and for what purpose and with what goal?

This book questions the assumptions underpinning such descriptions and the concerns over security and geopolitics that derive from them. There is no doubt that there are a range of relevant security issues in the North and that geopolitics defines many Arctic issues. However, one goal here is to provide better accuracy and introduce nuance to debates about 'Arctic geopolitics' and the role of the Arctic space in global affairs.

On the other hand, by using the Arctic as a case study, we can better understand the notions of geopolitics, security, and regions more broadly, merging all this into the academic concept of security regions (e.g., Kelly 2007). An interesting concept in and of itself, the idea of 'security' issues and interests bounded by a region (either defined by geography, by political constructs, or a combination of both) – security regions or 'regional security complexes' – have oscillated in and out of fashion since they were made popular by primarily Buzan and Wæver in the late 1990s. The focus here is not only on evaluating the relevance of theories concerned with security regions but also examining their empirical accuracy.

To do so, it is helpful to lean on separating into 'levels of analysis', a basic concept in the studies of international relations as formulated by Kenneth Waltz in 1959. In his book *Man, the State, and War*, he classified theories of the causes of war into three categories, or levels of analysis. Waltz refers to these levels as 'images'. The first image is

[13] See, for example, Hopper 2014; Martinez 2015; Mandraud 2014.

the individual level of analysis, whereas the second is the state itself. A spotlight on the state is about understanding states' foreign policy decisions and their specific security strategies. The third image is the framework of world politics. Whereas the first and second images are the determining forces that create the policies of a state, it is primarily the third image where the cause of war is found; namely, that the anarchic structure of the international system is the root cause of war.[14]

However, we cannot get very far into unpacking security concerns and practices with only an emphasis on the individual, the state, or the global system of states. How can we explain that the same actor (state) might behave differently across different geographic parts of the world? How do we theorise and explain the intensity of interests and actions within certain spaces close to the state in question, but with effects or practices that dissipate the further removed we are from that very state? Hence, a range of scholars (as explored further in the Chapter 2) have argued for and examined the *regional* level in studies of security policy. The decisive factor in such studies is geographical proximity, that is, states that are geographically close to each other have more intense interactions (positive and negative) than those located on different continents.

The question driving this book is fundamentally what role and limitations geographic factors – primarily revolving around proximity to other states – pose for states, as they orient themselves in the immediate geographic vicinity. The focus is on unpacking and sorting the different political and security dynamics that take place across the Arctic today and have been taking place over the last two decades since the emergence of the Arctic on the international (geo)political agenda. Leaning on a basic distinction between levels of analysis, separating between the national, regional, and global levels, but also – crucially – the 'subregional', allows for a more granular understanding of Arctic

[14] David Singer (1961, 80–2) around the same time argued for a division between the 'international system' and 'nation state' to better distinguish between events in international politics that occurred at one level but not the other, with a particular focus on security. It has since become commonplace in studies of international relations to use some variant of level analysis, implicit or explicit. Joseph Grieco, John Ikenberry, and Michael Mastanduno (2022: 15) argue that 'a writer who puts forward any theory or explanation about international relations has had to make choices about which actors and causal processes to emphasize', referring to these three levels as posited by Waltz.

security dynamics, but also for a more accurate assessment of what are actually *regional* dynamics, and how these intertwine with state security concerns, geography, and geopolitics.

The book thus positions itself somewhere between a desire to explain the nuances and empirics of one particular part of the world – the Arctic – in terms of state behaviour and geopolitics (to the extent that is objectively possible), and the theorisation of international relations and security studies – directly answering the challenge put forth in 2003 by Buzan and Wæver when they developed a rather complex and comprehensive theory to attempt to understand all security regions.[15] This approach builds on international security studies, political geography, and legal studies. However, the scope of this book is the state and its security practices, thus this book is closer to foreign policy analysis approaches than to more broad-ranging international relations theories.

Ultimately, the argument here is that if we want to understand a state's security actions, interests, and concerns (or at least an approximation of these), the typical way of analysing regional security complexes is too broad and limited in its explanatory leverage. A fundamental premise here is that we must take the role of, and the constraints defined by, geography seriously, and scale down what 'regions' states operate in. Most states are not great powers or superpowers and do not operate or have security entanglements in several geographic regions: they are bound to their immediate vicinity. However, that vicinity is arguably smaller than the typical regional level often employed by scholars and analysts alike.

Therefore, to understand states' security concerns and related policies and actions, we must take the regional level seriously but also scale down to a subregional level – that is, less than regional (at least in the

[15] Although the approach in this book does not completely overlap with that of Buzan and Wæver, it shares their underlying sentiment expressed in 2003:

> 'Traditional geopolitics are too materialist and mechanical … while "critical geopolitics" on the other hand seem too absolutist in studying only the social construction of space … We believe that geography as such matters but that it has to be analysed in a political framework. In the policy literature, titles with 'The Geopolitics of …' are common (of Caspian oil, of the Yugoslav wars, etc.) but this usually means an atheoretical survey of some power politics. RSCT is a theory of security in which geographical variables are central' (Buzan and Wæver 2003, 69–70).

way region is commonly used), but still more than just the national. 'Western Europe', 'North America', or 'Southeast Asia' become too extensive and nondescript (Friis 2021).[16] These geographic domains are undoubtedly relevant, but they are too broad and extensive when trying to distil the linkages between state security practices and the regional aspect whether linked to geography or the construction of political regions. China might be a constant security factor across Southeast Asia for all other related states, but China's policies and actions differ in their scope and intensity within that very space. The same goes for the USA in North America. Despite still being the most powerful country in world measured according to traditional power characteristics, US security policies and practices are not uniform from the Gulf of Mexico to the Hudson Bay, nor is it possible for them to be due to constraints linked to traditional geographic traits such as sheer distance and topography.

However, this book is not arguing that all that matters for understanding state security practices is the regional, or more limited subregional, domain. The subregional level of analysis must naturally be conceived in conjunction with the systemic and national levels of analysis, as studying only the regional or subregional provides a lopsided picture of the importance of *that* level of analysis. The key argument, however, is that we utilise this bottom-up approach alongside an understanding of how global dynamics shape specific regions and subregions and how national and domestic politics in states that are located in that geographic space conceive of and act according to their security interests.

In turn, this book argues we need to combine an analysis of four levels in order to understand the geopolitics of a specific part of the world: *the global* (the system level), *the regional* (in this instance, the Arctic, level), *the national* (the state level), and finally the *subregional* (less-than-regional, more-than-national level). The focus is on mapping the regional, namely, the Arctic. But this book also tries to combine different approaches to regions. Conflating different dynamics and issues at various levels is often the root cause of many erroneous descriptions of the Arctic as an arena for geopolitical issues or security

[16] Karsten Friis states that 'security subregions are a level of analysis in their own right, with their own material, ideational, economic, and political dynamics' (Friis 2021: 1).

concerns; although these do exist, we must distinguish between different types of issues and concerns.

It is the interconnectedness and co-constitution of these levels that can help us better understand why states act as they do, when it comes to security practices, in a particular geographic domain. In turn, the Arctic as a case study exhibits the importance of geographic factors and a need to scale down the regional into something less than regional but more than national. Security dynamics linked to the Arctic and Arctic states take place in subregional security complexes where the most relevant – from a security point of view – are the Barents subregional security complex (Northern Fennoscandian region) and the Beringia subregional security complex (North Pacific region). Moreover, in contrast to a core tenet of regional security complex theory, which states that complexes do not overlap, this examination of the Arctic with state security as the primary concern shows how the subregions do indeed overlap. This becomes particularly apparent when studying the Arctic-specific security approaches by relevant states – here with case studies of Norway and the USA.

At the same time, the Arctic as a regional construct *has* been developing for decades, spurring circumpolar cooperation and even actions, including in the realm of state security. This in turn means we cannot ignore a broader regional scope and related region-building as both an arena for state interactions and its possible role in shaping state security agendas. This should in turn inspire additional studies of other scaled-down subregional security complexes and related security dynamics across different levels and geographies. How about further performing the same empirical analysis of security practices in, say, the Caucasus region (Oskanian 2013), or around the East China Sea (Midford and Østhagen 2024)?

The forthcoming chapters of this book explore various dimensions of Arctic geopolitics, security, and regional dynamics. First, Chapter 2 delves into the relationships between security, geography, and geopolitics, outlining the significance of the regional dimension and so-called regionalism in understanding state security practices and dynamics in given spatial domains.

Chapter 3 starts with the level analysis of security dynamics and practices by looking at the *regional* level – what does 'the Arctic' mean in terms of the ideational construction of a region and subsequent state actions? This chapter looks at efforts to construct an Arctic region and notions of Arctic exceptionalism and cooperation by tracing the evolution of regional governance post the Cold War.

Chapter 4 examines *global* factors impacting Arctic security practices, including climate change, naval developments, and the rising influence of China, alongside historic and contemporary responses from Russia and NATO. What are the international issues (or systemic developments if you will) that impact the Arctic states' security practices? To what extent are they pan-Arctic in their impact?

Chapters 5 and 6 shift to *national* perspectives, contrasting the unique security approaches among Arctic states. This means asking: What are the security practices of particular Arctic states and how do they relate to the previous levels explored? These two chapters place particular weight on Norway and the USA and how they manage relations with Russia and China, as examples of two very different, but also – in an Arctic context – somewhat similar states.

Chapter 7 draws some preliminary conclusions based on the three-level analysis performed in Chapters 3–6 and then concludes we must place weight on a fourth level – the subregional. It goes on to analyse the two prominent subregional security complexes in the Arctic, namely the Barents and Bering Seas, to reveal intricate security interactions.

Chapter 8 expands on the findings from the previous chapters and attempts to answer whether the Arctic is a 'region', or even a 'security region', and why that matters to both academic studies and those interested in the Arctic in general. This chapter also analyses the various ways in which we can understand the Arctic and what this means for our use of the region label in geopolitics more generally.

Finally, Chapter 9 concludes by reflecting on the implications of developments in these subregions on Arctic security dynamics, suggesting that recognising the region's complexities while also upholding the artificial Arctic construct can yield advantages amidst looming tensions. That in turn holds relevance not only for those who want to move beyond theorising international affairs and impact the very state practices we study, but also for other subregional parts of the world in need of fewer one-liners and more nuanced and lengthy descriptions.

2 *Security, Geopolitics, and Regions*

A question has repeatedly been asked in studies of state security policies and interests: What determines the security outlook of a state? An underlying assumption is that the security concerns and priorities of states located within a region are interlinked and overlapping. States located next to each other – sharing a border or boundary (at sea) – are, naturally, statistically more inclined to end up in conflict with each other (Bremer 1992). In other words, the geographic location of a state matters to understand its security concerns and priorities. As with the larger body of international relations theory, regional approaches are grounded in different epistemological foundations and emphasise different variables.

This chapter outlines approaches to studying security that deal with the following topics: (1) security amongst states, (2) the role of geography in defining that security, or so-called geopolitics, and (3) the concept of regional security. This lays the foundation for an in-depth study of Arctic security dynamics. In essence, security relations amongst states inherently entail a regional dimension (states of geographic proximity being more concerned with similar or the same security issues), although various approaches have grappled with this dimension in different ways.

Security of States

The topic of security lies at the core of international studies. It encompasses issues such as *war*, *survival*, and *power*. The first writings on international relations – written before the advent of 'modern' statehood – were all concerned with these issues in an attempt to secure the survival of their community through power expansion and warfare.[1] Several scholars have utilised this drive for security to explain the

[1] Such as Thucydides (1910) and Machiavelli (1950).

formation of a state and attributes related to its structure (Tilly 1990; Fukuyama 2011; Vu 2010). Similarly, realist scholars have argued that survival (and thus security) is the ultimate goal for any state in the current international system due to its anarchic structure (Waltz 1959; Mearsheimer 2001).

However, security is not a uniform term. As David Baldwin (1997) argues, this term has experienced a proliferation of usage and meaning after the end of the Cold War. He asked some relevant questions related to security: Security for whom? For which values? How much? From which threats? By what means? At what cost? And in which time period? While recognising the whole range of security studies and the interlinkages between the different types of security and safety, this chapter, and the book at large, focuses on what is often referred to as state (or military) security (Hoogensen Gjørv and Goloviznina 2014: 2–4). This means a state-centric point of view, where the survival of the state as a community, structure, and institution has primacy (Waltz 1979: 126; Tripp 2013), a point returned to in the final section of this chapter.

Then another question arises: What determines the security outlook of a *state*? In other words, in terms of security, how can the interests and behaviours of states be conceptualised in the international system? Well-trodden realist approaches hold that anarchy defines the international structure and states are primarily concerned with survival. In a neorealist view of international relations, states ultimately take advantage of each other in a relative power balance, as states are rational actors seeking the optimal course of action within the constraints and determinants of the anarchical international system. Thus, states primarily differ in terms of relative capabilities (Waltz 1979: 97).

While Hans Morgenthau (1954) and classical realists before him place emphasis on human behaviour and leaders' interests, Waltz's (1979) neorealism argues for the structural constraints under which states operate and influence their security actions. Robert Jervis (1976) in turn argues that security interests are defined by perception and misperception, which can be used to explain the outbreak of hostilities. Robert Gilpin (1981) sees the structure as an equilibrium, where changes in states' power forces a 'redistribution' of power in the system. In addition, states are security-maximising actors seeking security *and* welfare on an indifference curve (Gilpin 1981: 18). Mearsheimer (2001: 29) argues that great powers 'are always searching for opportunities to gain power over their rivals, with hegemony as their final goal'. Power (as latent power and military power) equals material capabilities. In sum, a state's security interests revolve around its survival through acquisition of power and maintaining a constant structural and *relative* balance with other states.

So-called constructivist approaches challenge some of the previously mentioned assumptions. As Alexander Wendt (1999) argues, anarchy is what you make of it. Thus, the focus remains on *how* state interests are formed, instead of taking them as given. The type of anarchy present amongst states depends on the ideas that drive relations, the internalisation of the anarchic culture, and the concept of 'othering' (Wendt 1999: 299–312). In other words, '[i]deas thus define the meaning of material power' (Tannenwald 2005: 19). The argument is that, in contrast to only material capabilities, ideas and states' conceptions of each other matter. This goes beyond perceiving the international system as only a system of similar entities that vary on certain parameters, and instead allows a more nuanced conception of how different states interact with each other in different contexts and settings; that is, '[t]he answer to the question of who are enemies and friends begins at home' (Hopf 2002: 294). Others have expanded on this to examine how states' interests and behaviours are influenced by norms, identity, and conceptions of 'morality' (Finnemore and Sikkink 1998; Price 2008).

Understanding a state's security interests is not just a consequence of material constraints and attributes but also that of states' perception of *each other*, their historical relationships, and their identities. Where does geography come into play? Often, mainstream international relations has not paid sufficient attention to geography and its role in shaping outcomes in international affairs (Agnew 1994; Sjoberg 2008).[2] One can speculate whether this is due to its historical focus on state-centric models and territorial boundaries, which homogenise global spaces and obscure the complexities of socio-spatial phenomena such as transnational migration (Karayiğit 2021). The focus has been on a narrower set of questions related to the role of territory in conflict/war between states and as a source of friction (e.g., Hensel 1999; Hensel et al. 2008; Hensel 2017; Huth 1998; J. K. Choi and Eun 2017). The neglect of wider theorising of

[2] See also, for example, Buzan and Wæver (2003) in their justification of a regional approach to IR studies.

the relations between geography/territory and international relations could perhaps also be explained by the decline of academic (political) geography in the USA post World War II after the concept of geopolitics became tainted (Dodds 2019: 28–30).

Geography and Geopolitics

Examining the linkages between security and international relations, it is apparent that geographic proximity serves as a cornerstone in shaping the security relations amongst states. As a consequence of European state formation and finite territorial space, the concepts of territorial sovereignty and boundaries have come to define the modern state and its relations to others (Krasner 1999; Zacher 2001; A. Buchanan and Moore 2003). As states formed, developed, and expanded, the need to define and uphold territorial boundaries became increasingly relevant (Tilly 1990: 131). Friedrich Kratochwil (1986: 33) holds that the 1659 Treaty of the Pyrenees between France and Spain established the first modern state boundary. In turn, because of European state formation and the finite territorial space in this part of the world, the concept of territorial sovereignty and boundaries has come to define the modern state and its relations to other states across the globe (Elden 2013; Agnew 1994; Sack 1986).[3]

When the emphasis was placed on delimitation of, or establishing jurisdiction in, all territory in the nineteenth and twentieth centuries, 'frontier' regions became a source of interstate friction as they lacked clear demarcation. Disputes have further emerged as states have sought, and continue to seek, to expand their territory and define their

[3] For a lengthy examination of the concept of territoriality and fixed territory, see, for example, Elden 2013; Agnew 1994; Storey 2012. Territoriality can be defined as the process whereby territory is claimed by individuals or groups. 'Territoriality can be seen as the spatial expression of power and the processes of control and contestation over portions of geographic space are central concerns of political geography' (Storey 2012: 8). Studies of territory and territoriality are primarily concerned with land and the human need/desire to inhabit and control land. However, the idea of 'socialised territoriality' is relevant also as it enables the role of territory to be conceived more broadly. Sack sees territoriality as a 'device to create and maintain much of the geographic context through which we experience the world and give it meaning' (1986: 219). In turn, once 'territories have been produced, they become spatial containers within which people are socialized' (Storey 2012: 20; see also Paasi and Prokkola 2008).

borders. The link between territory, sovereignty, and conflict has been extensively proven (Holsti 1991; Goertz and Diehl 1992; Vasquez 1993; Forsberg 1996; Huth 1998; Wiegand 2011; J. K. Choi and Eun 2017; Carter 2010). John Vasquez (1995), for example, shows how at least 79 per cent of all wars between 1648 and 1990 were fought over *territory*-related issues.

Disputing economic and political interests amongst states fixated on, or shaped by, geographical traits – sometimes emerging into outright conflict – is an inherent feature of international politics. This brings attention to the notion of *geopolitics*. The term geopolitics was born out of the confluence of geography and politics, trying to encapsulate an approach to understanding the interplay between geographical space, political power, and international relations. The word derives from the German term *Geopolitik* and was first introduced by Rudolf Kjellén in 1899 and later popularised by Halford Mackinder and other scholars (Mackinder 1904). The roots of the term can be traced to Kjellén's vision of a framework that integrates geography, political science, and sociology to analyse the spatial dynamics of power (R. Kjellén 1924).[4]

The critique of traditional geopolitical analyses has often been that they are too narrowly focused on resource drives and tend to be deterministic in their analytical capacity. Moreover, the term has also been tainted by its usage especially by Nazi Germany but also other regimes and state leaders to legitimise expansionist state actions (Dodds 2019: 20–27).[5]

During the Cold War, however, George Kennan (1947: 575), the architect of US containment policy, emphasised a strategic approach to counter the USSR, advocating for 'patient but firm and vigilant

[4] Mackinder expanded on this concept, introducing the idea of a geopolitical chessboard where control of strategic land masses, or the 'World Island', would dictate global dominance (Mackinder 1904). Also Ratzel's idea from 1901 of a *Lebensraum* ('living space', or room for manoeuvre) and Spykman's Rimland Theory (1944) exemplify different geopolitical perspectives, each emphasising the central role of specific regions in determining global power dynamics. Classical geopolitical writers were either informed by racial and environmental determinism or strategic consideration of imperial rivalries and great power projection (Raspotnik 2018: 10–13).

[5] Karl Haushofer, a German military officer and geography professor, integrated Mackinder's strategic perspective with Ratzel's ideas on racial and environmental determinism and called for Germany's expansion post World War I.

containment'. In the 1970s, US National Security Adviser Henry Kissinger used the term geopolitics to describe the power struggle between the superpowers (Ó Tuathail 2010: 256). The end of the Cold War in the early 1990s marked a shift in international politics. To fill the resulting geopolitical void, intellectuals proposed new visions, through, for example, Fukuyama's (1989, 2006) 'The End of History', Huntington's (1993, 2003) 'The Clash of Civilizations', and the shift from geopolitics to geoeconomics (Sparke 1998). New geopolitical perspectives emerged, focusing on environmental issues and transnational challenges such as sustainable development and inequality. Kaplan, advocating neoclassical geopolitics, predicted future conflicts driven by environmental scarcity, affecting subnational groups and threatening state survival. He later expanded on this in his 'revenge of geography' theory (Kaplan 2009, 2012).

One additional approach to geopolitics builds on the premise that 'distance in international relations … is an "illusion"' (Henrikson 2002: 441). Thus, *critical* geopolitics emphasises that, in addition to existing physical geography, distinctive mental perceptions of space form and re-form images and ideas of a physical reality that influence actions (Kuus 2014). This calls for efforts to 'reveal the hidden politics of geopolitical knowledge' (Ó Tuathail, Dalby, and Routledge 2003: 3). In particular, the critical approach draws attention to how the narratives of conflict, geography, and politics impact governance (e.g., Peters, Steinberg, and Stratford 2018) and how different decision makers describe and conceive the same spatial domains (Dodds, Kuus, and Sharp 2013).

Critical geopolitics aims to 'move beyond the statist framing of power and spatiality' by including human beings and their (political) institutions that are often not considered the main focus of analysis (Kuus 2013: 381–82). Notwithstanding the broad scope of many of the critical approaches to geopolitics, the central point, in essence, is to try to break deterministic, rigid, or simple causal statements concerning politics and geography – while still maintaining a focus on geographic traits.

The Return of Geopolitics

Bearing in mind this history and these different approaches, in contemporary language today, 'geopolitics' has evolved to become a lens through which analysts, policymakers, and scholars examine the interplay between geography and politics, sometimes attempting to 'claim' an objective truth concerning the relationship between state relations and geography (Topalidis et al. 2024). It is commonly used to describe the strategic considerations and power struggles between states on the world stage and seems to have become shorthand for 'great power rivalry' or 'possible conflict'. A quick check of online search trends related to 'what is geopolitics' and 'geopolitics' (as a field of study) further highlights an increased interest since 2004, with a marked uptick from around the start of 2014 (Google Trends 2024).

The increased use of, and interest in, the term can thus be explained by several interrelated trends over the last decade. The 'world is flat' argued Thomas Friedman in 2005, perpetuating the idea that new technology has made the world increasingly smaller, removing the barriers posed by traditional geographic concerns. Alas, a backlash was inevitable. Whereas – at least in Western circles – it was popular to argue for the decline in relevance of borders and geography's waning importance as a determinant of state action in the 1990s and early 2000s (Bethlehem 2014; Ikenberry 2014), global affairs over the last two decades have led to a counterreaction.

The decline of US hegemony and the interlinked relative global power shift eastwards have spurred the notion of a 'multipolar' world where state–state interactions on a less-than-global level have increased in importance. Other developments have reasserted geography's role for understanding international relations in the 2020s and beyond. For most of Europe, the re-emergence of Russia as a traditional security threat from around 2007 to 2008 is linked to territory and space. The Russian war in Georgia in 2008, the annexation of Crimea from Ukraine in 2014 with the following conflict in eastern Ukraine, and the full-scale invasion of Ukraine in 2022 highlight the importance of territory, disputed borders, and in consequence related security practices by states.

Also underscoring the role of geography and states' desire to manage the constraints that derive from it is China's increasing expansion of power through 'traditional' means such as infrastructure development, base construction, and military presence and power. The Belt and Road Initiative is perhaps the foremost example of a geopolitical, or geoeconomic if you will, development. Geoeconomics entails the attempt by states to make use of commercial and economic ventures to achieve (geo)political goals. Instead of outright application of military power, states make use of investments, ownership in companies, or direct territorial action aimed at gaining economic advantages (such as expanding territorial seas or blocking sea lanes) to ensure political goals (Luttwak 1990; Sparke 1998; Wigell and Vihma 2016). As tensions between the USA and China have increased, more traditional Cold War-style questions of territory and anti-access and area denial in the East and South China Seas have also received attention alongside concerns over Chinese intentions regarding Taiwan (C. Schofield, Sumaila, and Cheung 2016; Aspinwall 2020; Yabuki and Selden 2014).

Finally, a global backlash against 'globalisation' with the reintroduction of border and trade barriers has been apparent since the fallout of the global financial crisis in 2008 (Burgoon and Trubowitz 2023). Populist parties gaining power in Western countries leading to – amongst other things – the election of Victor Orban in Hungary (in 2010), the election of Donald Trump in the USA (in 2016 and in 2024), and the decision to leave the European Union by the United Kingdom (in 2016) are prime examples of this anti-globalisation trend. Other strongman features centred around 'protecting' national borders are increasingly prevalent, even sought after, amongst state leaders ranging from Narendra Modi in India to Recep Tayyip Erdoğan in Türkiye, who tend to portray themselves as fighting against global forces and for local and national interests (for this argument in full see Rachman 2023).

Although these trends, events, and developments are an eclectic (and cherry-picked) bundle of issues, they arguably underscore a point, namely the increased popular fascination with the role of geography in shaping global politics. One outcome of, as well as a contributing factor to, this turn away from the 'global' and towards the national and regional has been the growth of investigative non-fiction works promising the ultimate guide to understanding geography's role in shaping international affairs. The best examples are perhaps Robert Kaplan's (2012) *The Revenge of Geography*, Tim Marshall's (2015) *Prisoners of Geography*, and Ian Morris's (2023) *Geography Is Destiny*, where variants of the same argument are presented: states are destined by their respective geography, and if we ignore the underlying limitations posed by geography, we will fail in our analyses of world politics. By studying, for example, ten maps, we can 'reveal the future of the world' (Marshall 2021). State leaders, ranging from Norway to Brazil, seem to have caught on (Rachman 2023; Gahr Støre 2019).

However, as the critical approaches to geopolitics highlight, these works tend to – at times – exaggerate geography to the extent that it becomes the *primary* (or only) factor defining politics of states. The key problem is the determinism of the linkages between geography and the ability by states (and their peoples and leaders) to act (i.e., politics), which sometimes might result in political leaders being overly obsessed with geopolitics with detrimental consequences for other issues such as combatting climate change (see Toal 2024). As succinctly put by Daniel Immerwahr (2022) in a longer feature article titled 'Are We Really Prisoners of Geography?' where he critiques the determinism of the works of Marshall and Kaplan:[6]

> For Marshall, the 'tribes' of the Balkans are perpetually in the thrall of 'ancient suspicions', the Democratic Republic of the Congo 'remains a place shrouded in the darkness of war' and the Greeks and Turks have been locked in a 'mutual antagonism' since the Trojan war.
>
> Kaplan sees things similarly. Russia has always been an 'insecure and sprawling land power', he writes, its people held 'throughout history' in 'fear and awe' of the Caucasus mountains. He approvingly quotes a retired historian's theory that Russians, facing cold winters, possess an enhanced 'capacity for suffering'.

At the same time, regardless of the trendiness of geopolitics or studies of geography, or the sometimes simplified deterministic manner in which linkages are presented, it is still clear that we cannot ignore geography and its impact on international relations.[7] The fundamental assumption of geopolitics is the relative importance of geographic space and distance, that is, geographic proximity amongst states and boundaries between territories. It explores how states strategically position themselves, form alliances, and engage in conflicts based on

[6] In a similar vein:

> Though Marshall condemns Russian aggression, his tone is similar to the one Putin uses to justify it. 'They are constantly trying to sweep us into a corner', Putin said of Russia's rivals in 2014. 'If you compress the spring all the way to its limit, it will snap back hard.' One might object that Putin's ideas and attitudes, not his map, are driving Russian belligerence, yet geopolitics makes little room for such factors. 'All that can be done', writes Marshall in another context, 'is to react to the realities of nature'.

[7] For a more in-depth and academic examination of the contemporary changes to these impacts and linkages, see, for example, Bolt and Cross 2018; Tunsjø 2018; Guzzini 2012; Black 2015; Dodds 2019.

geographic considerations. Natural barriers such as mountain ranges, rivers, and deserts *can* fortify or compromise a state's security. Coastal states, with their inherent vulnerability to maritime threats, *do* formulate defence strategies distinct from those of landlocked nations. The historical use of geographical features as defensive strongholds, exemplified by the Great Wall of China or the mountainous terrain in Afghanistan, *does* underscore the impact of topography on military tactics and security doctrines.

Moreover, looking to the sea, the maritime domain with its intricate network of trade routes and sea lanes, also constitutes a geopolitical space where geography plays a role in shaping security relations – perhaps more so in the twenty-first century than ever before (e.g., Østhagen 2022b). Coastal states and those controlling strategic waterways hold considerable sway in global trade. The geopolitical significance of maritime chokepoints, such as the Strait of Hormuz or the Malacca Strait, underscores how geographical control over these passages can be a source of power and influence. At the same time, the geopolitics of the ocean is built on different sets of rights, notions of sovereignty, and states' ability to control space and uphold rights (Østhagen 2022).

In sum, the prevalence of geopolitical analyses linking geography, territory, and regions has varied over time, often in sync with fluctuations in the levels of conflict amongst states. Geopolitical diagnoses are still frequently employed to interpret international relations and the correlation between power and the organisation of space, although we should be mindful of exactly *how* we make use of theories linking space and political action, and to what extent outcomes in international relations are seen as predetermined due to geographic constraints.

This interplay, especially concerning state security interests, is what continues to inspire studies of geography, state actions, and – as a natural arena for these dynamics – regions. The question then becomes: How can we better unpack, and understand, these linkages? Can we utilise frameworks that employ approaches from international relations and lean on political geography (and geopolitical approaches) to take geography seriously as one determinant of state security and affairs, albeit without falling into the 'territorial trap'?[8]

[8] From John Agnew (1994). The 'territorial trap' refers to the limitations imposed by rigid territorial definitions that can hinder understanding and addressing complex geopolitical issues. Agnew critiques the assumption that

The polar regions are ideal examples of these challenges, where a literally melting icescape contributes to a perception of states jockeying for position to exploit resources and/or engage in regional governance efforts (e.g., Stokke, Østhagen, and Raspotnik 2022). Marshall (2019: 286), for example, looking at the Arctic, contends that the melting ice is changing geography and thus a race amongst states for territory and resources is 'inevitable'.

However, in the Arctic, as we will turn to shortly, conceptions of 'geopolitical rivalry' caused by ice melting remain too simplistic (Østhagen 2021b). The 'race' has not materialised and does not look likely to become the case in the near-to-medium term either. Symptomatic of many claims about the future of the Arctic, much work looking at this 'new' region ignores distinguishing between *different* parts of the Arctic, *different* types of resources, and *different* types of dispute.[9] These types of analysis are often under-theorised and only sparingly engage with international relations thinking on regionalism – on which they inherently lean by referring to regional dynamics, constrained by a geographic 'boundary' or limit. Therefore, to complete the picture, we need to lean on a range of theories and conceptualisations that have taken a closer look at regions – as a political process, as well as a level of analysis of security dynamics.

Regions and Regionalism

One approach to dig deeper and beyond the immediate geopolitical headlines or contemporary fascination with deterministic statements about the role of geography is to study security relations and state interactions in the specific locales (spaces) where geopolitics is taking place. On the face of it, certain geographical regions undoubtedly emerge as geopolitical 'hotspots', characterised by contested borders, historical grievances, and competing territorial claims. The Middle East, the Korean Peninsula, and the South China Sea exemplify how geography intertwines with historical, cultural, and political factors to create volatile security landscapes – and in turn a fascination with the geography of these domains. Proximity to major powers, access to

territory is a fixed, unchanging entity, and that states are the sole elements defining our relationship with territory, arguing that this perspective can lead to oversimplified analyses of state behaviour and international relations.

9 See, for example, Sale and Potapov 2010; Borgerson 2009; Østhagen 2023a.

critical resources, and the historical legacy of regional conflicts contribute to the intricate tapestry of security relations. States manoeuvre within these regional dynamics, forming alliances, engaging in diplomatic dialogues, and at times, resorting to conflict resolution mechanisms to navigate the interplay of geographic constraints.

However, we can move beyond just describing these issues and concerns in such a simple manner and study the theorising of regions in and of themselves. Whereas critical geopolitics has attempted to break the deterministic cycle between geographical attributes and state behaviour, regionalism – as a multidimensional research field not necessarily attached to one theory – has broadly speaking been concerned with the *political* processes of how regions are formed, constructed, and maintained (e.g., Börzel and Risse 2016). Strands of scholarship based on rationalist, constructivist, or other social theories have sought to grapple with the central characteristics and developments of regions (Söderbaum 2012).

Regional studies have oscillated in and out of fashion. Traditional regionalism in international relations refers to the formal process of intergovernmental collaboration between neighbouring countries to address shared challenges and goals. This process often leads to the establishment of regional organisations or institutions such as the European Union, ASEAN, or the African Union. Regionalism emerged prominently after World War II, largely driven by the need for economic recovery, political stability, and the prevention of future conflicts. It is characterised by treaties and agreements that aim to promote economic integration, enhance security cooperation, and even facilitate political dialogue and 'normative' integration amongst states (e.g., Bicchi 2006 on the EU). Traditional regionalism emphasises economic integration through the creation of customs unions, free trade areas, and common markets (Hettne 2007). It is often driven by state actors seeking to enhance economic growth and achieve political gains through collective strength.

The first wave of regionalist scholarship, emerging in the late 1960s and early 1970s, was driven by two international processes: the proliferation of new states in the wake of decolonisation and the interest generated by European integration (Väyrynen 2003). These early efforts assigned little agency to regions instead placing them as subordinates at the systemic level. This subordination was ensured by the possibility of external intervention by the primary system, which at this point was a bipolar dynamic between the USA and the USSR. The European integration process is a prime example, where member states pooled sovereignty in certain areas to achieve greater economic and political clout. This approach was plagued by problems of definition, which made it difficult to operationalise (Moon 1998). This definitional arguing and world events in the late 1970s and early 1980s led to a move away from regional scholarship (Kelly 2007).[10]

Regional studies regained their appeal with the end of the Cold War and the collapse of bipolarity. Instead of picking up where things had been left a decade earlier, this new wave of scholarship eschewed strict definitions and instead embraced an emphasis on functionalism (Alagappa 1995; Väyrynen 2003) – Oran Young (1992: 244) defined it as: 'Short of becoming a locus of regional conflict, a geographically defined area may achieve the status of a distinctive region for purposes of policy analysis and public decision making when it enters a period of political turmoil or flux as a result of the impact of realigning forces'.

'New regionalism' provides a broader understanding of regional cooperation beyond formal state-led initiatives. It reflects changes in the global political economy and is characterised by a more varied, decentralised, and multidimensional approach. New regionalism encompasses a range of actors, including non-state entities such as businesses, civil society organisations, and transnational networks, highlighting a shift from purely economic and state-centric frameworks to include cultural, environmental, and social dimensions.

From this, 'constructivist' scholars argued that regions are not naturally defined but are socially constructed and politically debated – they exist as concepts, with no universal agreement on their boundaries (Acharya 2001). As put by Iver B. Neumann (2001: 57): 'regions are spoken into existence; thus, lie where politicians want them to lie'. In essence, regions became unbundled from territory. Scholars also focused on understanding regionalism by examining ideas, identity, and the socialisation of norms within geographic areas (Amable 2022). Fredrik Söderbaum (2012: 14) argued, for example, that 'there are no "natural" regions; all regions are heterogeneous with unclear territorial margins'.

[10] There has been an inductive pushback against theories that have been downscaled from European realities and used to explain cases elsewhere. This in turn emphasises the centrality of state strength – where, for example, the real concerns are domestic insecurity and instability (Job 1992; Alagappa 1995).

This new regionalism was not one coherent bloc, however. Here the focus is on two strands: the region-building analysis that was often a counterreaction to security dynamics and concerns, and the specific 'security region' frameworks that tried to bridge this emerging chasm by arguing that there are, indeed, objective regions out there at least in terms of security interactions.

Starting from the assumption that mainstream international relations theories are normatively flawed, critical approaches believe that a theory should respond to the reality of the developing world as well as the developed. At the heart of such approaches is the idea that regions are a means of both conflict resolution and resistance to larger threats. The goal is not only, or always, to develop an analytical model but a theory of (or even to argue for) regionalisation.

From an economic point of view, Björn Hettne (2009) saw the replacement of the Westphalian state system by regions as a 'second great transformation'. Influenced heavily by developments in Europe, regions are a means of bringing order through cooperation. These regions act as shields against neoliberal globalisation and US hegemony (Hettne 1999). Thus, a normative element to this approach exists whereby regional order itself is seen as desirable. As Robert Kelly (2007; 213–14) puts it: 'Regionalism is viewed as serving progressive values like multilateralism and global governance.' Despite their break with mainstream international relations theories, these critical approaches, for the most part, adopt the basic premise of a global power hierarchy with a hegemon on top.

This means that these critical approaches implicitly accept the notion of openness, albeit in the context of something to be avoided through the process of regionalisation. Regions are often conceived of as appearing on a continuum. Kaisa Lähteenmäki and Jyrki Käkönen (1999: 214–15) posit that regionalisation begins with proximity, which in turn evolves into a shared regional identity. Hettne (2007) and Fulvio Attinà (2004) draw similar parallels, with regions culminating in a regional security community or a regional security partnership respectively. This ultimately evolves into a regional organisation and, later, a regional security community.

Proposed by Karl Deutsch in 1957, 'security communities' are neatly defined as communities where there is 'real assurance that the members of that community will not fight each other physically but will settle their disputes in some other way' (Deutsch et al. 1957: 5).

Picked up by IR scholars, in the mid 1990s, Emanuel Adler wrote that '[i]n a liberal community-region, people learn the practices and behaviour that differentiate aggressive states from peaceful states. In other words, each side develops a common knowledge of "the other's dovishness". In this sense, the democratic nature of a state becomes an indicator of its "dovishness"' (Adler 1997: 258).

Since then, a number of scholars have taken up the study of these communities around the world. The typical examples of an uncontested security community tend to be the Nordic countries ('Norden') or North American cooperation (Friis 2021; Charron and Fergusson 2022; Saxi 2019). Amitav Acharya (2001) went on to examine whether, and how, the countries in Southeast Asia (ASEAN) constitute a security community distinct from European norms and liberal values; Michael Williams and Iver Neumann (2000) discovered how NATO transformed from an alliance to a security community; Harald Müller (2006) uncovered how security communities unravel, a topic later explored by Simon Koschut (2016) more broadly; and Pernielle Rieker (2016) explored the boundaries of the European security community and the processes taking place there.

Furthermore, although a security community is not the same as a security region, the literature on security communities has noted a shift away from the geographic focus. By definition, security communities are not geographically bound. Acharya (2001: 27) argues that scholars today define regions as 'imagined communities, created by processes of interaction and socialization'. In turn, what constitutes a region might change at different times, making the 'old' idea of geographic proximity less relevant. In the early 1980s, Deutsch (1981: 54) argued along the same lines in defining regions: 'In many ways ... regions are made by culture, history, politics and economics rather than by geography alone.' Scholars of security communities not only challenge the geographically bounded focus but they also see it as less relevant to explaining states' security cooperation and behaviour. As stated by Allen, Massey, and Cochrane (1998: 2), '"regions" only exist in relation to particular criteria. They are not "out there" waiting to be discovered; they are our (and others') constructions.'

Security Regions

Alongside the increased attention given to region-building efforts and even how to transcend traditional security concerns (and geography), others argue that the regional level of analysis in terms of security dynamics and state interactions has been largely ignored in favour of the systemic, or the national/domestic, levels.

David Lake (1997: 50–51) argues that regional security is not inherently defined by geographic proximity and adjacency, as instead shared security 'externalities' bind states together, even when the states themselves are located far apart. By making it issue-specific, Lake avoided the complications that came with geography and history. In such cases, the flow of threats (or amity, as the case may be) remains paramount. For example, the decision by Australia to engage in fighting terrorism in Afghanistan in 2001 illustrates how geography is rendered less relevant with the rise of transnational threats and increased power projection through technology. In Lake's approach, Kelly (2007: 209) argues, regions become purely functional; they become simply 'anything less than global'. While this deterritorialised approach echoes some constructivist scholars who seek to break with the positivist leanings of regional studies, Lake applied rationalist neorealist principles. Security externalities translate to threat(s) projected by one state far enough to affect others. History is only important insofar as it is the history of threats; fearful behaviour overrides all else.

In another approach focusing on regional causes of war, Douglas Lemke (2002: 49) takes A. F. K. Organski's (1968) power transition theory and conceives of (security) regions as local power hierarchies, all subsumed underneath the dominant global one. In total, he found twenty-three of these local hierarchies under the larger power structure of American primacy. These regions (local hierarchies) shared the same characteristics as the international level, except openness. Lemke accepted the concept of proximity and further caveated it by saying that all states within a security region must be able to reach each other and interact. The limit of this ability to act is the boundary of a security region. If a loss of more than 50 per cent of power projection between the national capitals of two respective countries is found, a state is removed from a security region.

Finally, perhaps one of the most common approaches to regional security studies – the regional security complex theory (RSCT) developed by Buzan and Wæver (2003: 47–50) – shares some similarities with traditional security approaches, while also adding a constructivist element in terms of how security interactions emerge and are maintained. Keeping the focus on competition between states and the systemic overlay, RSCT enriched traditional international relations studies by adding another level of analysis – the regional – that was better suited to understanding the post-Cold War era. RSCT departs from international relations orthodoxy in its assertion that, when the systemic overlay recedes, regional dynamics become a key component of the international system. The emphasis is on the regional level, and an appreciation of the territorial, especially when looking at *security*: 'Security is a distinctive realm in which the logic of territoriality continues to operate strongly' (Buzan and Wæver 2003: 12).

Regional security complexes (RSCs) in turn are 'traditional, state-based military-political complexes' that are informed by a mix of neorealist principles and region-specific variables (Buzan 2000: 19). RSCs are 'mini-anarchies' that face the same problems of anarchy, polarity, and state competition as the systemic level (Buzan 2000: 4). However, RSCs are differentiated from the systemic level in several following ways. The first is the possibility of 'overlay' and 'penetration' by external powers. For example, during the Cold War, regional dynamics were superseded by the global security dilemma, as the two superpowers vied for power. This is defined as the 'openness' of a region to interference from above (Kelly 2007: 207). Penetration takes place when external powers establish security alliances with states within an RSC. Overlay occurs when the influence of major powers extends beyond basic penetration, dominating a region to the extent that the Indigenous security dynamics effectively become nonfunctional (Buzan and Wæver 2003: 46, 61).

The second is the importance of proximity. These regions, or RSCs, are formed by 'patterns of amity and enmity' between neighbouring states, that is, 'who fears or likes whom',[11] with the limits of the RSC defined by the power projection capabilities of the states in question (Buzan 1983: 190). Interactions between geographically proximate states, both positive and negative, become more intense and compound over time. This translates into regional security dilemmas informed by shared histories. Taken together, in the words of Buzan and Wæver (2003: 46), RSCT is the idea that 'anarchy plus the distance effect plus

[11] 'The pattern of amity and enmity is normally best understood by starting the analysis from the regional level, and extending it towards inclusion of the global actors on the one side and domestic factors on the other. The specific pattern of who fears or likes whom is generally not imported from the system level, but generated internally in the region by a mixture of history, politics, and material conditions' (Buzan and Wæver 2003: 46).

geographical diversity yields a pattern of regionally based clusters, where security interdependence is markedly more intense between the states inside such complexes than between states inside the complex and those outside it'.

In other words, the region is characterised by units whose securitisation processes are so intertwined that their security issues cannot be addressed separately (See Figure 2.1). Or, as put by Buzan and Wæver, 'a set of units whose major processes of securitisation, desecuritisation, or both are so interlinked that their security problems cannot reasonably be analysed or resolved apart from one another' (Buzan and Wæver 2003: 44). Echoing the scholars of geopolitics, the argument is that if we truly want to understand security interactions amongst states, focusing solely on global or national dynamics does not capture the majority of these. Simple geographic closeness leads to more security interactions among neighbouring states than those far apart. While there are occasional links across sparsely populated areas and non-territorial clusters, such as those formed around international terrorism, these are exceptions. As put by Dennis Amable (2022: 3): 'These variables – geographic proximity and meaningful interaction – have proven crucial in making the regional system the most salient level for analyzing security issues.'

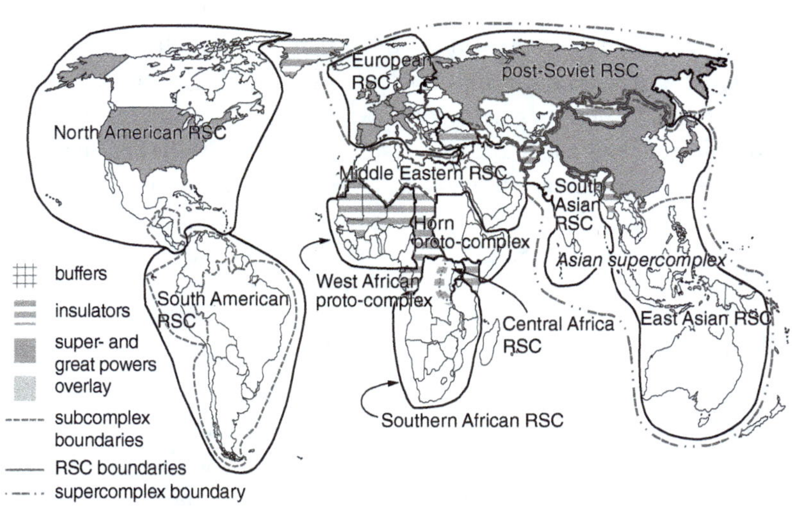

Figure 2.1 A world map of RSCs. Source: From Buzan and Wæver (2003: xxvi). Reproduced with permission of the Licensor through PLSclear.

When it comes to *how* to study (or unpack) these RSCs, Buzan and Wæver (2003; 48) highlight that they 'study the security discourses and security practices of actors, not primarily their regional(ist) discourses and practices'. The theory outlines four levels of analysis – domestic, state–state, interregional, and role of global powers – and we must consider how they interconnect (Buzan and Wæver 2003: 51). Together, these four levels form the security constellation. The key is to focus on securitisation *processes*, namely processes in which issues are framed as security threats, leading to the justification of extraordinary measures to address them. This concept emerged from Buzan and Wæver's work in the late twentieth century, emphasising the importance of discourse in shaping security perceptions and practices. Their approach moves away from traditional definitions of security, focusing instead on how security is constructed through political language and social contexts.

Some critics argue that the theory places too much emphasis on predefined regions, potentially oversimplifying the complex and dynamic nature of security relations (Sadurski 2022). Regions are not static entities, and the boundaries of RSCs can change over time. Moreover, some argue that the theory's primary focus on states and state-centric security dynamics may limit its applicability in regions where non-state actors significantly influence security matters (Ramutsindela 2015). There is also criticism that the theory neglects global factors impacting regional security dynamics, such as globalisation, transnational threats, and the influence of major global powers (Lemke 2005). Finally, there has been some systemic pushback from scholars who hold that openness and weak states counter the claims made by regionalism; regions are porous and, despite being more difficult to penetrate after the Cold War, they are still open to great power hegemony (Katzenstein 2005: 22).

An Approach to Unpack Security Dynamics in a Specific Region

Broadly conceived from this review of current literature and approaches, three broad strands link the regional and the role of geography with an emphasis on traditional security considerations as well as state practices:

1 Security studies and the emphasis on state survival.
2 Geopolitics and the emphasis on geography for international relations.
3 Regionalism and the emphasis on region-building processes amongst states.

Despite the critiques mentioned, this book draws extensively on the ideas of Buzan, Wæver, and other scholars who have explored the delineation of regions as territorial entities; as political, economic, and cultural constructs; and as arenas for intense security interactions and threat perceptions. Geographic proximity undeniably influences the security perspectives of states and, consequently, geopolitical dynamics. However, different approaches emphasise various aspects of regional security and interactions, whether concentrating on the significance of regions or the role of geography (see Figure 2.2.). This book utilises these diverse concepts of the regional by focusing on the Arctic at the start of the twenty-first century, offering more general insights into how to study areas considered security regions, where security interactions are constantly evolving.

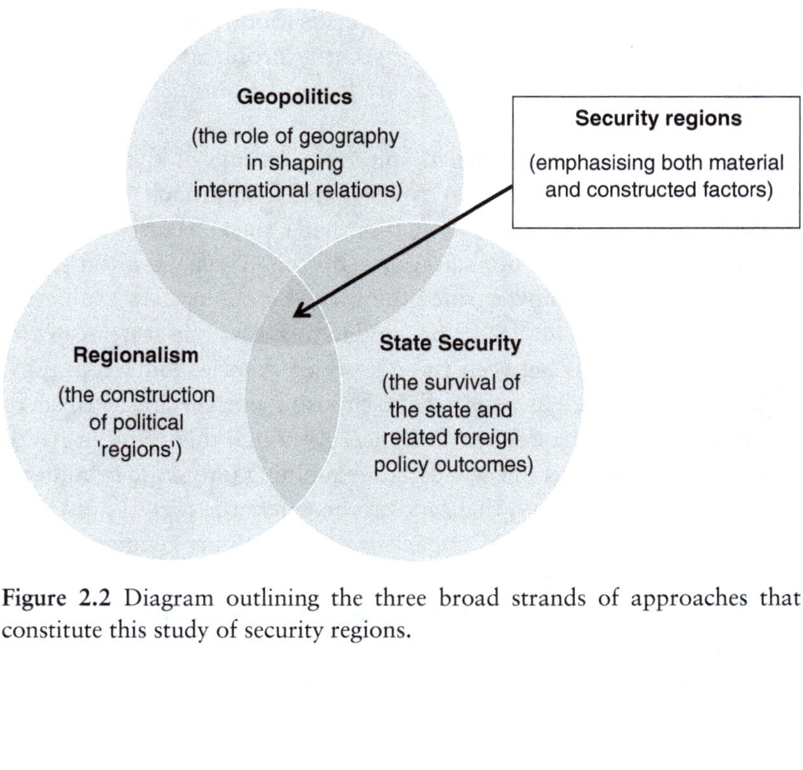

Figure 2.2 Diagram outlining the three broad strands of approaches that constitute this study of security regions.

The approach in this book builds particularly on the RSCT and the updated discussions found in *Greenland in Arctic Security* from 2024 by Marc Jacobsen, Ole Wæver. and Ulrik Pram Gad (2024b: 343) but also more broadly on what has been outlined in this chapter, namely the fundamental notions of what *security* entails for states, the role of *geography* therein (as linked to geopolitics and the study of political geography), and – finally – regions as both a political construct *and* a complex of security dynamics and practices. The following assumptions guide the onwards study of a particular 'region':

- *The centrality of geography*: regions exist, as a concept, as a territorial object, and as a venue or container for security interactions amongst states. Regions are *bounded*; they cannot contain everything and everyone and they are marked by boundaries. Inherently, these boundaries are the product of the limitations of existing states' reach (power, capability, capacity). That reach is – still – severely influenced by geography. Seas, mountains, deserts, and territorial chokepoints do matter, as does human habitation, population centres, cities, and infrastructure connecting peoples. Still, leaning on Buzan and Wæver, a starting point for analysis remains 'a set of units whose major processes of securitisation, desecuritisation, or both are so interlinked that their security problems cannot reasonably be analysed or resolved apart from one another' (Buzan and Wæver 2003: 44).
- *State-centric security focus*: despite the critique of the RSCT for being too state-centric, that is the explicit approach taken here. There are undoubtedly other actors that impact security dynamics in a regional context but states are still the primary ones in global affairs today. Moreover, since the focus here is not on evaluating the relevance of the RSCT or similar theories, but instead examining the linkages between these types of theories and geography/geopolitics, as well as regionalism, choosing one specific actor focus (state-oriented) is a deliberate choice. Although this study of Arctic security dynamics is not wedded to examining only state or military dimensions of security, that tends to be where the focus lands given its centrality in trying to unpack state behaviour, geography, and related conflict potential.

  However, when considering one threat in relation to other threats, it makes more sense with a unified concept of security rather than separating it into distinct issues or separate sectors such as 'economic security' and 'societal security' (Baldwin 1997; Gunhild Hoogensen Gjørv and Goloviznina 2014). In terms of a regional scope of analysis, 'different issues get tied together, and a world of regions is therefore less unlikely than one might at first think when listing the diversity of security issues each drawing on a particular sector' (Buzan and Wæver 2003: 49). Moreover, this approach allows the exploration of different levels of security in the Arctic, such as wider systemic rivalry and narratives of insider/outsider privileges (the global level), instilling regional cooperation to deal with issues of environmental and human security (regional level), or national concerns linked to economic or Indigenous security (domestic level).[12]

- *Merging regionalisation and security regions across levels*: region-building efforts create structures that can impact security relations amongst states in a given security region, but we need to try to separate between those efforts and related statements and where states' focus on security actually goes in terms of practices. Consequently, there is also a need to examine region-building efforts and the political constructions that derive from those efforts. Albeit constructs, these forums, institutions, and even speech and performative acts by diplomats, officials, and others are part of constituting mechanisms that arguably have tangible impact on matters of Arctic governance, cooperation, and in essence security (of states). Emphasising all three levels of analysis as outlined in Chapter 1, is central, seeing them holistically in tandem.

Moreover, this emphasis needs to be combined with a consideration of the 'naturally given' territorial dimensions that define each regional space. As Buzan and Wæver (2003: 12) emphasised more than two decades ago: 'The relevance of territorial versus non-territorial patterns of securitisation is an empirical question.' The approach put forth here positions itself somewhere in between these

[12] As put by Jacobsen, Wæver, and Pram Gad (2024c: 14):

> security is relational (Wæver 1997): it is not a quality, attribute, or possession of one unit in itself and for itself; it is always about some other(s) who are seen as threats or protectors. Barry Buzan argued (1984) that security was preferable as a central organizing concept to power or peace, exactly because the alternatives tended to become absolutist investments in the system itself: either anarchy was unchangeable (power) or had to be abolished (peace), whereas security pointed to the ongoing configuration of actors.

different ways of conceiving of geography and the role of regions, however you define and structure them. The fundamental argument is that they all do something right but can also be further improved and tailored to better explain security in a regional domain as seen from a state perspective.

- *Objects of study*: the objects of study are inherently **states** and their **security practices**. This entails studying: (1) security issues/issues that have become 'securitised'; (2) geographical traits and related restrictions imposed by states within a spatial domain; and (3) region-building efforts and attempts at constructing less-than-global, but more-than-national, structures amongst states.

  We must study a combination of foreign policy security statements and discourse and security and defence actions and investments. In states, the former tends to be dominated by, but not exclusive to, the respective foreign ministries, whereas the latter tends to be dominated by, but not exclusive to, the respective defence ministries and/or defence structures. However, these cannot be understood in isolation from each other or other non-state actors at both the national and international levels.

The concern with such an expansive scope is naturally that it becomes impossible to account for all of it, and the selection of what data and which developments to highlight will be subject to some degree of cherry-picking. Still, it is up to researchers and scholars to produce explanations and narratives that approximate a mapping of security practices within a regional context and open these up for scrutiny and evaluation. In this rather broad and eclectic approach to data, this framework deviates from the emphasis on securitisation by widening the scope of what is to be assessed, studied, and included.

What in turn this book focuses on, is trying to map and distinguish the various levels of analysis relevant to understanding the 'geopolitics' of a particular region (or subregional domain). It places an emphasis on the subregional level, or a scaled down version of the regional – so-called *subregional security complexes* (SRSCs). This emerges from the study of the Arctic in the following chapters, but it might also prove relevant when examining other parts of the world. This will be returned to in the final chapters. By exploring different security-and-state-oriented dynamics at different levels of analysis, it is easier to unpack state security, security regions, and the role of geography.

From these assumptions and considerations, when taking a closer look at a specific geographical domain – in this instance the Arctic – the most fruitful approach is to start with defining the *regional* security dynamics and then showcasing how they are separate from the *global* level, the *national* level, and the *subregional* level. This is how the book now proceeds.

3 *The Regional Level*
Exceptionalism, Cooperation, and Governance

This is the first of four chapters to distinguish between these four different levels, or sets, of political security dynamics in the Arctic. We start with the regional level, as it sets much of the frame for the other two, namely the systemic and the national. A look at the regional Arctic level, or the dynamics amongst the eight Arctic countries *in* the Arctic, is inherently what much of Arctic studies are concerned with. Thus, the goal of this chapter is to describe the specific regional political dynamics in the Arctic that relate to security and international relations.

The Arctic

What comprises the area above the Arctic Circle – the 'Arctic region' – is rather diverse in terms of a few parameters, such as population density, climatic conditions, links between the Arctic communities and the capitals located further south, and security relations; however, commonalities abound. The region's sparse population, distance to economic centres, climatic challenges – not to say the impacts of climate change – as well as security links speak of similarities across the northern part of the globe.

Before examining political security dynamics in depth, we must unpack one more essential component of Arctic fundamentals, namely the legal frameworks that govern the management of ocean space and related issues of sovereignty, sovereign rights, and jurisdiction. It is otherwise hard to distinguish between what concerns the Arctic specifically, what are global dynamics (see Chapter 4), and – crucially in an examination of the Arctic – what are unfounded speculative statements about 'lack of governance' or a 'resource race' and what the Arctic states are truly concerned with.

Figure 3.1 sets the geographical parameters for this book for clarity. The focus is on the eight Arctic states (those with territory

Figure 3.1 Map of the Arctic. Source: By CIA World Fact Book, Public Domain, www.cia.gov/the-world-factbook/static/97828fc01510ce36884bf7e07a508436/arctic_phy.pdf.

above the Arctic Circle) and the political dynamics amongst themselves and between them and states and actors located outside the region. Still, what the term 'Arctic' represents and the common and/or political usage of it are not defined quantities but the result of social construction through debate, action, and language.[1] It is also noteworthy that different Arctic countries have various traditions when it comes to referring to their northern territories.

Naturally, the Arctic states vary in size and structure. On the face of it, we can draw a rough line between the Nordic countries – all of them with territory in the Arctic – Russia and the two North American states, Canada and the USA. Starting with the Nordic states, seen from afar, they tend to look rather similar – small, highly developed democracies with a strong welfare state, market liberalism, a shared and overlapping history and, apart from Finland, a common language family. Iceland stands out as an island state with a very small population and no armed forces (apart from a coast guard force). However, when we look more closely at these countries, differences in their security concerns and approaches to the Arctic appear.

In a circumpolar context, the population patterns on the Scandinavian Peninsula, in northern Finland, and in Russia stand out as rather dense with a number of relatively large cities and towns. The most apparent explanation for this is the climatic difference between the European parts of the Arctic and the Siberian and North American parts, defined by the Gulf Stream that ensures a temperate climate even above the Arctic Circle in Europe. The Barents Sea is ice-free even in wintertime, with ice appearing to its maximum extent only around Svalbard and Frans Josef Land. These climatic conditions have led to the settlement and sustenance of large communities, especially along the coastline. Another central issue concerns the Sami, the Indigenous Peoples that inhabit the Sami land, Sápmi, that stretches across all four countries – Finland, Norway, Sweden, and Russia – in the northwestern parts of Europe.

Looking at the North American Arctic, different characteristics emerge. Greenland, the world's largest island, is part of Europe in the political sense; however, it is part of North America in a geographical sense. A population of only approximately 57,000 across a vast area entails certain challenges in terms of infrastructure and local capacity. The Canadian Arctic – or North – is equally sparse in population. Only 130,000 people reside across a vast domain that covers as much territory as Western Europe and approximately 40 per cent

[1] For more on this, see Keskitalo on the regional construction of 'the Arctic' (2004, 2007).

of Canada's land area. Most of them are not located above the Arctic Circle, but in communities in the mid-to-southern parts of Yukon, the Northwest Territories, and Nunavut as well as in northern parts of the province of Quebec. In Alaska, the situation is somewhat different – the northern state is geographically cut off from the lower forty-eight as the USA purchased the area from Russia (in 1867). The majority of the population resides in the southern part of the state, in or near Anchorage, Fairbanks, and Juneau. In North America, a range of Indigenous Peoples live close to or in the countries' Arctic regions.

Finally, the Russian Arctic territory makes up approximately half of the entire Arctic landmass. As in North America, the population in the eastern parts of Russia is sparse due to limited infrastructure, harsh climatic conditions, and lack of development. Only around 50,000 people reside in the Chukotka region bordering the Bering Sea and Strait. However, the Murmansk and Arkhangelsk regions have almost 795,000 and 1.185 million inhabitants, respectively, in the west of Russia. In total, approximately 2.5 million people live in what Russia defines as its North or Arctic. It is obvious that the sheer differences in geography, climate, and population patterns between the different 'Arctics' also have a bearing on political dynamics.

Although NATO and the EU are not necessarily competing security organisations, they tend to mark different pathways in a Nordic context. On one hand, Iceland and Norway have repeatedly rejected the prospect of EU membership while being adamant about the northern (or Arctic, if you will) role of NATO (Østhagen, Sharp, and Hilde 2018). On the other hand, Finland and Sweden have taken the lead in many of the EU's own Arctic initiatives while maintaining a close-but-not-too-close relationship with NATO (Raspotnik 2018). Only the Kingdom of Denmark bridges the two organisations, although Greenland, which is the Kingdom's Arctic foothold, left the then-European Economic Community in 1985.

However, in response to the Russian invasion of Ukraine in February 2022, Finland and Sweden took steps to initiate their NATO membership applications, with Finland joining in 2023 and Sweden in 2024. Marking a watershed in Nordic security affairs, and thus also for the Arctic, this means that seven of the eight Arctic countries now belong to a security and defence alliance that was created to counter a perceived Soviet threat. Now, the only remaining Arctic state outside the alliance is Russia.

In North America, both Canada and the USA are NATO members. The USA naturally has a global outlook on security concerns as it engages at will across the world. Canada is active in NATO missions in Europe; however, this defence alliance has often been seen as an 'out of area' dimension to Canada's defence instead of something integral to the territorial defence of the Canadian homeland (Jockel and Sokolsky 2009). Instead, for both Canada and the USA, North American Aerospace Defense Command (NORAD) provides the fundamental structure for defence of the North American homeland in the Arctic (Charron and Fergusson 2022; Lackenbauer 2021).

Russia stands outside of both the EU and NATO; however, it has its own various security arrangements and cooperation mechanisms. The Commonwealth of Independent States Free Trade Area was initiated by Russia in 2011 to promote further economic (and political) integration amongst some of the former USSR countries. After 2014, the debate over Chinese and Russian cooperation along several axes has also become more prominent and Russian leadership has rhetorically talked about increased security cooperation between the two countries' armed forces (Ebel 2022; Moe, Lamazhapov, and Anisimov 2022; Tingstad, Pezard, and Shokh 2024a). However, questions regarding the realities of this bilateral cooperation and the extent to which Russia can accept the junior role that a Sino–Russian entanglement entails remain (Hsiung 2022; Stensdal and Heggelund 2024; Lamazhapov 2024). Especially in an Arctic context, this speaks to more than just economic cooperation, but also to Russian fears of Chinese encroachment and the loss of control over key developments in the North (Røseth 2014; Almén and Hsiung 2022).

The Arctic Ocean

When it comes to maritime spatial zones, the delimitation agreements and continuing disputes are crucial to the understanding of the fundamentals of Arctic geopolitics and the background of political security dynamics. In 1982, most states agreed on a comprehensive legal regime, the United Nations Convention on the Law of the Sea (UNCLOS) (United Nations 1982). The Law of the Sea gave the legal justification for states to put in place 'new maritime zones in addition to the 12 nautical mile (NM) territorial sea, with a 200 NM "resource" or "fisheries" zone which became termed the exclusive economic zone (EEZ)' (Østhagen 2020b: 3). However, as such zones have already been implemented by coastal states since the 1950s – as some states pre-empted an international agreement – states have developed differing interpretations of how to draw maritime boundary lines (Forbes 1995: 13).

Of the eight Arctic states, seven are parties to the Law of the Sea Convention. The exception is the USA, but it generally regards the Law of the Sea as being reflective of customary international law and thus binding on all states (Roach and Smith 2012). All the Arctic coastal states – Canada, Kingdom of Denmark, Norway, Russia, and the USA – as well as Iceland have advanced broad maritime claims (see Figure 3.2) in keeping with both international law and their own national interests (Churchill 2001).[2]

Since the 1950s, the Arctic states have agreed on various maritime boundaries in Arctic waters. What stands out as a particular case of how the Arctic states have sought to insulate their region from security and governance issues, concerns their approach to the Arctic Ocean. Looking at territory and the process of establishing and demarcating borders between states in the North, most of those were settled in the nineteenth and twentieth centuries as part of the expansion of sovereignty and administrative capacity in the various Arctic countries. In contrast to many other parts of the world in which questions about borders and territories remain, in the Arctic, Canada, the Nordic states, Russia, and the USA have demarcated their northern borders and resolved sovereignty disputes over islands and archipelagos such as Hans Island, Jan Mayen, and Svalbard. The territorial questions in the Arctic – disputes over borders on land and 'who owns what' – have all been settled.

What remains are disputes and concerns over jurisdiction and rights at sea. According to a geographic definition, the Arctic is inherently

[2] These maritime claims include 12 NM territorial seas (except in respect of Greenland, where a 3 NM territorial sea is claimed). Canada, Norway, Russia, and the USA also claim contiguous zone rights out to 24 NM, although Norway's claim here does not apply to Jan Mayen Island or Svalbard. In addition, all the Arctic coastal states claim EEZs out to 200 NM (see Figure 3.1), although Norway has only provisionally implemented this around Svalbard through a non-discriminatory Fisheries Protection Zone based on the Act on Norway's Economic Zone (Norwegian Ministry of Trade Industry and Fisheries 1976).

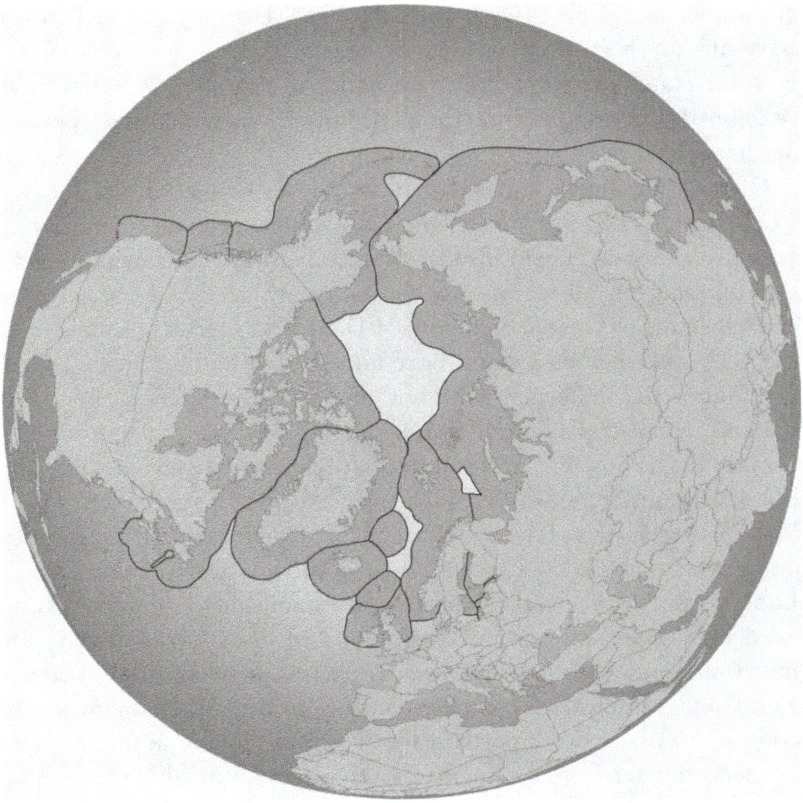

Figure 3.2 The exclusive maritime economic zones in the Arctic. Source: Created by Malte Humpert, The Arctic Institute.

an ocean surrounded by land. This entails that the 'seascape' defines many of the regional questions of resource governance, access, and subsequent security issues. Albeit the smallest of the world's oceans,[3] the Arctic Ocean (at 14 million km^2) is almost six times larger than the Mediterranean Sea (at 2.5 million km^2), five times larger than the Caribbean Sea (at 2.8 million km^2), and four times larger than the South China Sea (at 3.5 million km^2). And the Arctic Ocean represents only half of the space we call the Arctic.

[3] Historically, there are four named oceans: the Atlantic, Pacific, Indian, and Arctic. However, most countries – including the United States – now recognise the Southern (Antarctic) as the fifth ocean. The Pacific, Atlantic, and Indian are the most commonly known (NOAA 2024a).

The rapid melting of Arctic sea ice and the prediction that the Arctic Ocean might be ice-free during the summer by around 2050 bring up a range of questions relevant under the Law of the Sea, and governance more generally. In essence, the Law of the Sea sets the parameters for intra-Arctic disputes over resource management, ocean governance, and access to various parts of the Arctic. It has also been used by the Arctic states to counter claims of the Arctic as an 'ungoverned' space.

However, the Arctic Ocean remains exceptional with respect to maritime boundaries as, in contrast to the general trend, so many boundaries are *settled* (Østhagen and Schofield 2021). Looking at the Arctic 'experience', out of seven, only one boundary segment remains in dispute, between the USA and Canada in the Beaufort Sea (J. S. Baker and Byers 2012; Byers and Østhagen 2017). This highlights how the Law of the Sea provides the undisputed framework for dealing with the challenges over how to manage ocean space, including areas of overlapping maritime claims, especially in the Arctic Ocean context, where all of the coastal states have overtly indicated that they remain committed to its terms.

Fundamentally, affirming the Law of the Sea, the practical management of maritime spaces, and ultimately agreeing on maritime boundaries in the Arctic region have not only been steps taken in order to provide frameworks for ocean-based resource development but also efforts to ensure the primacy of the Arctic states (Hoel 2009, 2014), as other actors are increasingly engaged in regional affairs ranging from science to fisheries.

The End of the Cold War and a Surge in Arctic Cooperation

The Arctic states have obviously had bilateral and multilateral interactions concerning, or in, the Arctic for as long as the modern state structure has existed. As Chapter 2 explained, claiming territory and thus sovereignty and delimitating borders on land and later at sea has been an ongoing process for centuries. However, the focus here is on regional security dynamics in more recent times and the related rapid surge in cooperative mechanisms amongst the Arctic states.

Often hailed as a landmark for 'modern' Arctic relations, in 1987 in Murmansk, Secretary-General of the Communist Party of the Soviet Union, Mikhail Gorbachev, delivered a speech that became known as the Murmansk Initiative. Gorbachev called for increased cooperation between the USSR and Western countries along a range of issues, such as arms control, limiting military activity, enhancing economic prosperity in the North, inclusion of Indigenous Groups, and scientific cooperation (Gorbachev 1987). The events that followed in the period 1989–1991 profoundly changed Russia, and thus, they also changed the Arctic. Although the military and security initiatives were not met with applause from the other Arctic countries (Åtland 2008), the speech marked the start of a surge in cooperation in the North on a regional level and thus also region-building efforts (Keskitalo 2004).

As the systemic overlay of the Cold War faded, regional interaction and cooperation in the North flourished apace. While the region as a geopolitical space seemed to disappear from the radar of global power politics during the early post-Cold War period (see more in Chapter 4), interaction amongst Arctic states increased. This allowed the eight circumpolar countries to embrace creating a favourable political environment to deal with transboundary issues and stimulate economic development. In particular, Indigenous Peoples gained more political visibility and an official voice at various cooperative mechanisms.

Regional organisations such as the Arctic Council, the Barents Euro-Arctic Council, and the Northern Forum emerged in the 1990s to tackle issues such as environmental degradation, regional and local development, and cultural and economic cross-border cooperation. Military security issues were deliberately excluded; this enabled a plethora of cooperative arrangements to be agreed amongst the countries in different constellations without them getting bogged down in the high politics associated with national defence considerations and deterrence.

The Arctic Council eventually became the primary forum for pan-Arctic issues. After initiatives in the early 1990s, the Ottawa Declaration officially established the Arctic Council in 1996 as a high-level forum to 'provide means for promoting cooperation, coordination and interaction amongst the Arctic States, with the involvement of the Arctic Indigenous communities and other Arctic inhabitants on common Arctic issues' (Arctic Council 1996: 2). The note to this sentence further states: 'The Arctic Council should not deal with matters related to military security.'

The procedures and rules of the Arctic Council were adopted at the first meetings. It was decided, amongst other things, to let the chairship rotate every other year amongst the eight Arctic states and that all decisions should be taken unanimously – the role of observers and the status of Indigenous Peoples as 'permanent participants' was also specified. The permanent participants have full consultation rights in connection with the Council's negotiations and decisions; however, they do not have voting rights. The observers were and remain a highly diverse group of NGOs, international organisations, and non-Arctic states. In 2011, at a ministerial meeting in Nuuk, it was decided to establish the Arctic Council Secretariat in Tromsø, Norway. The Standing Secretariat became operational in 2013.

The day-to-day workings of the Arctic Council are handled by the six working groups. They are the producers of scientific knowledge, tasked with identifying and analysing Arctic challenges, and the venue in which the member states, permanent participants, and observers actively participate in the work of the Arctic Council (Arctic Council 2024). The working groups cover everything from the monitoring of mercury levels to formulating guidelines for Arctic shipping to undertaking a wide range of studies regarding climate change in the Arctic (Exner-Pirot 2015).

In the first period of Arctic Council, from 1996 to the mid 2000s, most of the work centred on combating pollution in the Arctic. This work has gone on to inform international conventions on pollution, including the Minamata Convention (on mercury) and the Convention on Persistent Organic Pollutants (Stockholm Convention) (Rottem 2017; Graczyk and Rottem 2020; Bloom 1999). Adding to its legitimacy, a growing number of non-Arctic state actors – initially France, Germany, the Netherlands, Poland, Spain, and the United Kingdom; and more recently, China, India, Italy, Japan, Singapore, South Korea, and Switzerland – have applied for and gained observer status on the Council (Koivurova 2010; Willis 2013; Steinberg and Dodds 2015).

Another forum that emerged in the same period, albeit with a more local/region-specific focus, is the Barents Cooperation that focuses on the Barents Region (Northern Europe). The Barents Euro-Arctic Region was launched in 1993 by the Kirkenes Declaration (Barents Euro-Arctic Council 1993). The original members of the Barents Euro-Arctic Council were the Kingdom of Denmark, Finland, Iceland, Norway, Russia, Sweden, and the European Commission, although this forum collapsed after Russia left in 2022 (returned to below). Moreover, thirteen counties or similar subnational entities form the

Barents Regional Council, highlighting the regional dimension in Arctic cooperation. In 2008, the International Barents Secretariat was established in Kirkenes, Norway.

Ergo, in the early to mid 1990s, several initiatives appeared amongst the Arctic states – the Arctic Council and the Barents Euro-Arctic Council are only two examples of this, albeit the most prominent ones, as they include a high-level component where the foreign ministers (usually) of the participating states met when concluding one chairmanship period and starting another.[4] These forums are relatively toothless if we only examine their institutional set-up and mandate, their impact is measured – as put by Svein Vigeland Rottem (2020) – by 'decision-shaping' and not 'decision-making'. All these efforts are textbook examples of the theories that emerged around the same time concerning regionalisation, and how to develop regional structures to avoid conflict and – ultimately – the pitfalls of global geopolitics and anarchy.

Moreover, by including not only the circumpolar states but also a range of state and non-state actors as observers, these forums have ensured the Arctic states' primacy when dealing with Arctic or subregional Arctic issues. This is especially the case for the Arctic Council which transformed from a relatively limited forum where foreign ministers from the Arctic states often did not attend ministerial meetings to a forum where both the Russian and the US foreign ministers would meet and portray relative harmony – despite tensions elsewhere – to signify the importance of the Council and its work. A region-building outsider–insider dynamic has been clear in terms of which states can be members of the Council; however, an underlying premise has been that it is better to have non-Arctic actors such as China, the EU, India, and the UK involved rather than exclude them.

However, as put by Jennifer Spence, the focus on the Arctic Council as the 'container' of Arctic cooperation – which is beyond what it was originally conceived as – has arguably also challenged the organisational capacity of that institution (Spence 2017). The

[4] There is a question about the conjugation of the verb 'meet' here. At the time of writing, the Barents Cooperation has been disbanded, and there are no high-level meetings in the Arctic Council. Therefore, it might be appropriate to write 'met'. Although it is not unthinkable that the Arctic Council might resume high-level meetings with the foreign ministers including the Russian in the near future, or that the Barents Cooperation is resurrected.

Arctic Council perhaps became famous because of the allure behind a regional organisation framed as 'the' Council for that region, although in reality its decision-making and enforcement abilities were, and still are, extremely limited.

Governance Vacuums and Global Interest

In the beginning of the 2000s, a set of political dynamics started to place the Arctic in the spotlight. First and foremost, global attention on the Arctic region began to show. This was initially driven by a growing awareness of climate change and the mere fact that the Arctic is one of the parts of the globe where the effects are becoming apparent at a faster rate than elsewhere.[5] Debates about Arctic sovereignty and the potential dangers associated with the 'opening' of the region remained largely academic until they intersected with the increasingly apparent effects of global climate change and popular perceptions about competition for Arctic resources in the early 2000s (Østreng 1999; Orheim et al. 2003; Stokke 2006). Arctic ice sheets were disappearing at an accelerated pace, which coincided with new prospects for offshore oil and gas exploration, as well as the opening of shipping lanes such as the Northeast and Northwest Passages (The Economist 2015).

Record lows in the extent of summer sea ice, combined with record high oil prices, uncertainty over maritime boundaries, and a much-talked-about US Geological Survey (USGS) estimate released in 2008 – suggesting that the region holds 13 per cent of the world's undiscovered oil and 30 per cent of its undiscovered natural gas (USGS 2008) – conspired to drive Arctic issues to the forefront of international politics. Scholars such as Rob Huebert (2003, 2010) and Scott Borgerson (2008, 2009) spoke of an 'Arctic arms race' emanating from *regional* resource and sovereignty issues rather than global strategic drivers. In particular, the Russian planting of the titanium

[5] A popular expression has been that the Arctic is warming 'twice as fast' as the world. In recent years (2020 onwards) the notion of 'three times as fast' started occurring. A study in 2022 showcased that in fact the Arctic has warmed four times as fast as the globe since 1979 (Rantanen et al. 2022). Many of these statements centre on Svalbard, as a particularly exposed part of the Arctic, but also a place where a relatively large amount of scientific activity is taking place (Østhagen 2024).

flag on the seabed of the North Pole in 2007 became a focal point of attention for a media narrative about an impending Arctic scramble (Dodds 2010; Nilsson and Christensen 2019).

The flag planting generated considerable media coverage, much of which was relatively alarmist in nature (BBC News 2007). This tone extended to the diplomatic arena when the Canadian foreign minister, Peter MacKay, appeared to dismiss the flag-dropping incident as a stunt, stating, 'This isn't the fifteenth century. You can't go around the world and just plant flags and say "We're claiming this territory"' (Parfitt 2007). In response, the Russian foreign minister, Sergey Lavrov, observed that 'no one is throwing flags around'; analogies were drawn between Russia's action and Edmund Hillary and Tenzing Norgay planting the Union Jack on the summit of Everest in 1953 (Parfitt 2007). Indeed, Lavrov emphasised that Russia was not acting unilaterally; its actions were 'in strict compliance with international law' (Novosti 2007).

Some of the attention that is given to the Arctic – including in terms of security and 'geopolitical' attention – sometimes seems related to a notion of Arctic 'uniqueness' or the Arctic as a 'last frontier' (e.g., Emmerson 2010). Individuals seem drawn to the unknown, mystical, and wild – especially as few such areas remain in the globe today as the impacts of human activity and settlement are increasingly evident. These notions are also part of this global fascination with the Arctic region, which, in and of itself, can lead to simplified causation and general assumptions about the Arctic (for more on this, see Sfraga 2017; Duffy 2013; Burke 2017; Medby 2014; Jacobsen 2015; Manicom 2011).

This fascination with the Arctic, although it is difficult to measure and determine its causal effects, seems to have been the instigator of a debate about Arctic governance structures that started in academic circles but gained attention in wider political circles at the end of the 2010s. Scholars were quick to examine and engage in debate about whether the existing governance structures in the North could manage all this influx of activity and political attention. Governance scholar Oran Young (2005) wrote that Arctic governance efforts were different from traditional ways of understanding regimes and international cooperation, with promising hopes for global efforts to combat climate change. Although the academic governance debates tended to, and still tend to, be rather sober with an emphasis on aspects such as resilience of current institutions or the benefits of establishing limited new forms,[6] especially those with an environmental agenda have been quick to jump to the conclusion that a completely new governance framework was, and still is, needed for the Arctic.

The argument – as explicitly stated by some government officials and environmental activists and sometimes read between the lines when non-Arctic actors discuss 'Arctic governance' – is that the Arctic is inherently the 'common heritage of mankind' and thus the responsibility of the global community and not just the Arctic states and their inhabitants.[7] Others have been eager to 'save' or 'protect' the vulnerable Arctic, perhaps without being clear about which part of the Arctic or why it needed protection in the first place (Greenpeace 2012; Østhagen 2012).[8]

One comparison that is frequently made and continues to have some leverage is that of Antarctica, which at first glance, might make sense. Both polar regions have been perceived as 'frozen landscapes' devoid of human activity, and these are the places most affected by climate change while also attracting economic interests due to their vast resource potential and global demand for those resources. While the global community managed to agree on a comprehensive governance regime to avoid conflict, resource extraction, and harm to the environment concerning Antarctica through negotiations in the post-war period, the Arctic has no such holistic governance system. However, there are marked differences between the two regions.

[6] See, for example, Rothwell 1996; Stephens 2011; Stokke 2006.

[7] For more on various debates on this, and statements, see Bartenstein 2015; Kienko 2018; Rainwater 2013; Gautam 2011. For the EU's early Arctic policy, see European Commission 2008; Østhagen 2013; Raspotnik 2018.

[8] As this author wrote – while working for North Norway in Brussels – in 2012:

> In Brussels ... the focus in last week's seminar was not on local or regional development, but on how vulnerable the Arctic environment has become as the amount of activity increases. This is by no means a change in the Brussels-discourse described in previous articles by The Arctic Institute. Instead, it stems from the geographical position of the EU-capital, located far away from the Arctic territories. It seems understandable that officials and politicians from more southern countries are not preoccupied with, or in need of, acquiring an nuanced conception of the Arctic region. Combine this with the pro-active work done by some environmental NGOs, and you get a portrait telling the story of a vulnerable, distant, and inhospitable region inhabited by different Indigenous communities that now find their traditional existence threatened (Østhagen 2012).

In fact, it would not be wrong to deem them 'polar opposites' in terms of the spaces' history, level of human activity, and their subsequent governance structures.

In essence then, while the Arctic is an ocean surrounded by sovereign countries, Antarctica is a landmass that does not officially belongs to any state, surrounded by oceans.[9] Antarctica is a continent with most of its territory falling within the Antarctic Circle (Stephens 2011: 318). Legally, the spatial extent of the Antarctic regime is specified by the Antarctic Treaty that defines its operational area as south of 60°S ('The Antarctic Treaty' 1959). Further, unlike the Arctic, the Antarctic has no history of supporting human populations and Indigenous Peoples (Rack 2018). In comparison with the Arctic, the ATS governs much of the Antarctic region and provides a fixed point in terms of governance. However, to argue that a similar system should be in place for the Arctic misunderstands the nature of the Arctic, its inhabitants, states, and the governance structures in place.

The argument might hold true if we are talking about the high seas that are the Central Arctic Ocean, around the North Pole. Although, most of the Arctic's subsea continental shelf will be subject to the sovereign rights of one of the various Arctic coastal states. The rest (and the majority) of the Arctic region is the sovereign territory of one of the eight Arctic states or comprises waters that are either territorial seas or exclusive economic zones. The 'limited governance' argument does not immediately hold true from this perspective – at least not any

[9] The establishment of a distinct Antarctic governance regime dates back to the negotiation of the Antarctic Treaty, which was adopted in December 1959 as a result of the Washington Conference on Antarctica (Wehrmann 2019). This occurred at a peak time of the Cold War – and yet, both the USA and the USSR agreed on the need to balance 'a wide range of issues including the status of sovereign claims over the continent and its demilitarization' (Stephens 2011: 344). Today, the Treaty has been acceded to by fifty-four countries. Limiting the designs of the two superpowers within the area, the Antarctic Treaty also 'froze' the territorial and – subsequently – the maritime claims of the seven Antarctic claimant states. The main purpose of the Antarctic Treaty has been to ensure 'in the interest of all mankind that Antarctica shall continue forever to be used exclusively for peaceful purposes and shall not become the scene or object of international discord' ('The Antarctic Treaty' 1959). Furthermore, the Antarctic Treaty has become the basis of a wider Antarctic Treaty System (ATS) that governs the entire area. The ATS is based on a set of binding international agreements to which all contracting parties are legally committed (Wehrmann 2019: 58).

more or less than it does in other maritime domains surrounded by states (e.g., the Mediterranean or the Caribbean).

The Arctic and the Antarctic therefore are indeed polar opposites. One difference is the presence of penguins in one domain, whereas polar bears roam the other (imagine the situation if they shared domains). Perhaps more central for the purposes of this book, the former is a landmass without human habitation surrounded by water with a Treaty that explicitly excludes military activity (Article 1), whereas the latter is an ocean surrounded by landmass with human habitation – with military activity across all eight Arctic states. Moreover, establishing sovereignty over territory, delineating borders on land, and eventually extending sovereign rights into the maritime domain are processes that do not infer a 'legal vacuum' or a lack of jurisdiction and governance in the Arctic.

In the wake of this global attention to the Arctic and Arctic governance, environmental organisations and politicians outside the region dominated debates about the ungoverned Arctic region. In Canada and Russia, nationalistic voices also demanded urgent state action to defend this 'frontier' from outside aggressors in a 'race for resources'. Such messages tended to conflate identity politics, national interests, continental shelf delimitation processes, energy security, mineral resources, and security and control over Arctic jurisdictions (Lackenbauer 2010; Sergunin and Konyshev 2014). As stated by Artur Chilingarov – a Russian scientist and explorer – after the 2007 flag planting, 'I don't give a damn what all these foreign politicians there are saying about this. If someone doesn't like this, let them go down themselves … and then try to put something there. Russia must win. Russia has what it takes to win. The Arctic has always been Russian' (NBC News 2007).

The Arctic 'Story': Exceptionalism and Cooperation

Calls for new governance structures in or for the Arctic and implications that the current system is insufficient from around 2007 provoked a response from the Arctic states. First, the Arctic states started organising events and engaging in public diplomacy (and closed-door diplomacy) vis-à-vis those actors who had been most vocal in expressing concerns about limited Arctic governance (Offerdal 2011). Amongst these were the UK (or some Members of Parliament in the UK) and the EU (or some Members of European Parliament), where subsequent efforts were made to convince critical voices of the adequacy of the current governance system, with an emphasis on the Law of the Sea (Bartenstein 2015; 'European Parliament Resolution of 9 October 2008 on Arctic Governance' 2008; Raspotnik and Østhagen 2019).

The growing importance of, and emphasis placed on, regional forums in the North and then especially the Arctic Council came about as a consequence of the increased international attention given to the region, often driven by actors external to the region (for a discussion, see Young 2010; Stephen and Knecht 2017). In response to such concerns, the five Arctic coastal states gathered in Ilulissat, Greenland in 2008 to highlight the primacy of Arctic states and their sovereign rights (Arctic Ocean Conference 2008). Senior political representatives of the five Arctic states publicly declared the Arctic to be a 'region of cooperation' (Exner-Pirot 2012). They affirmed their intention to work within established international arrangements and agreements, in particular, the Law of the Sea – highlighting a specific regional approach and coherence amongst the Arctic states (Stephen and Knecht 2017).

During this period, Jonas Gahr Støre, for example, then foreign minister of Norway, repeatedly emphasised that '[t]he Arctic – unlike Antarctica – is an ocean surrounded by national states with sovereign rights to sea areas off their coasts in accordance with international law' (Gahr Støre 2010b). Moreover Gahr Støre, together with his Russian colleague Lavrov, wrote an op-ed in the Canadian newspaper *Globe and Mail* about the importance of settling maritime boundary disputes in the Arctic (Lavrov and Gahr Støre 2010):

> the Russian-Norwegian experience proves that the United Nations Convention on the Law of the Sea provides a solid framework for addressing issues that will arise as the climate in the Arctic changes and the Arctic Ocean is transformed'. At the same time, 'we firmly believe that the Arctic can be used to demonstrate just how much peace and collective interests can be served through the implementation of the international rule of law (Lavrov and Gahr Støre 2010).

These are examples of an active Lavrov response from the Arctic states addressing concerns about governance in the North with an emphasis on the Law of the Sea and the well-functioning mechanisms in place in the Arctic. The same occurred when some actors started highlighting

Svalbard's governance structures and discussing whether a new treaty should be discussed – against the interests of Norway as the sovereign of Svalbard. These ideas were met with a direct pushback from Norway, eventually taking the issue somewhat off the agenda (Raspotnik and Østhagen 2014; Bladet Vesterålen 2011).

Moreover, the op-ed by Lavrov and Gahr Støre highlights a third strand of response from the Arctic states, namely the narrative of 'cooperative regional relations'. As stated in 2010: 'we believe that the challenges in the Arctic should inspire momentum in international relations, based on co-operation rather than rivalry and confrontation, and we believe that important steps have already been taken towards this goal' (Lavrov and Gahr Støre 2010). This idea of the Arctic as a special place for international cooperation was propagated internationally in tandem with the primacy of the Law of the Sea to underscore that the Arctic is not a legal vacuum or an arena for immediate conflict.

Another relevant dimension of Arctic region-building in this period concerns how all the Arctic states have regularly put forth so-called Arctic policies or Arctic strategies, often repeating the mantra of cooperation in relatively streamlined Arctic policy and/or strategy papers (Rottem 2010; Heininen et al. 2020). The typical phrases found in these policy documents highlight the 'importance of international cooperation' in the Arctic, the need for 'sustainable development', the importance of respect for traditional livelihoods, and the fundamentals of the Law of the Sea and its primacy in the Arctic.

Despite the Arctic countries being rather different in terms of setup, structure, and the characteristics of their Arctic regions, repeating the listed points concerning governance and cooperation became an essential part of Arctic diplomacy efforts, counteracting or downplaying the security developments that were taking place at the same time. Playing into this, at the Nuuk, Greenland ministerial meeting in 2011, the Arctic Council presented a specific set of criteria to be fulfilled for possible observers to the Council (Senior Arctic Officials 2011).[10] These criteria were meant to further reiterate the

[10] The text from 2011 states that:

> In the determination by the Council of the general suitability of an applicant for observer status the Council will, inter alia, take into account the extent to which observers:

points concerning 'sovereignty, sovereign rights and jurisdiction' in the Arctic (Guo 2012). The criteria also served the purpose of underscoring the unity amongst the Arctic countries – including Russia – while dealing with external interests to the region and emphasising the 'Arctic story'.

The deterioration in the relationship between Russia and the other Arctic states in 2014 did not fundamentally change the commonality of interests when it came to Arctic-specific issues, or the Arctic states' desire to portray the Arctic as a 'low tension' region, despite, for example, the cessation of military contact with Russia by the other Arctic states (Østhagen 2016; Byers 2017). Although security concerns became much more prominent and fears of Russian erratic behaviour were suddenly a worry, all Arctic states including Russia reconvened in Greenland in 2018 and repeated promises of cooperation and protection of the Law of the Sea (Jacobsen 2018), which, after all, gives the Arctic states sovereign rights over large parts of the Arctic Ocean.

In fact, this focus on the cooperative features of the Arctic region, despite global tensions involving the same actors in other parts of the world, has been deemed the essence of an 'Arctic exceptionalism'. In the book *Arctic Politics: Conflict and Cooperation in the Circumpolar North* Oran Young (1992: 6) noted that 'Arctic exceptionalism' had already emerged 'as a powerful force in the world'. Young argued

- Accept and support the objectives of the Arctic Council defined in the Ottawa declaration.
- Recognize Arctic States' sovereignty, sovereign rights and jurisdiction in the Arctic.
- Recognize that an extensive legal framework applies to the Arctic Ocean including, notably, the Law of the Sea, and that this framework provides a solid foundation for responsible management of this ocean.
- Respect the values, interests, culture and traditions of Arctic Indigenous peoples and other Arctic inhabitants.
- Have demonstrated a political willingness as well as financial ability to contribute to the work of the Permanent Participants and other Arctic Indigenous peoples.
- Have demonstrated their Arctic interests and expertise relevant to the work of the Arctic Council.
- Have demonstrated a concrete interest and ability to support the work of the Arctic Council, including through partnerships with member states and Permanent Participants bringing Arctic concerns to global decision making bodies.'

that the concept stemmed from outside commentators 'accentuating the exotic and unique features of the Arctic', which had 'the effect of setting the region aside from the mainstream concerns of most fields of study' (Young 1992: 6). The collective action problems associated with Arctic environmental issues, which no one state could address alone, were particularly apt to being tackled through this approach.

For most commentators, the idea of Arctic exceptionalism became inextricably linked to the assumptions that the region was a cohesive and cooperative space insulated from geopolitical tensions elsewhere, and it was 'exceptional' when compared to other regions (Gunhild Hoogensen Gjørv and Hodgson 2019: 4). Heather Exner-Pirot and Robert Murray defined the concept as 'the successful effort' both 'to maintain cooperation in the region despite internal competition for resources and territory' and 'to compartmentalize Arctic relations from external geopolitical tensions'. They argued that the Arctic regional order was exceptional insofar as Arctic states and those states with involvement in the area worked 'to negotiate an order and balance of power predicated on norms such as cooperation and multilateralism' (Exner-Pirot and Murray 2017: 47–48).

In other words, Arctic exceptionalism regarding regionalism and region-building is directly linked with norms-based multilateralism and institutionalist approaches. This concept has served two purposes: (1) it has been a focal point of Arctic studies looking at the unique situation in the North *and* (2) it has been an important aspect of the narrative created by the Arctic states' foreign ministries and Arctic policy documents when attempting to combat the alternative narrative depicting the Arctic as a region 'up for grabs' or doomed by an impending scramble for resources (Grindheim 2009; Wilson Rowe 2018).

The Arctic spirit of uniqueness must also be understood in the context of the discussion that emerges from the social construction of where to draw the line between the Arctic and the rest of the world and the idea of being 'near' the Arctic. As the region has attracted attention from countries located south of the Arctic Circle, proving one's Arctic credentials seems to be a popular endeavour (Steinveg 2017; Bekkevold and Offerdal 2014). In fact, it was the United Kingdom that first actively used an approximation of the term near-Arctic in an official document – its 2013 Arctic policy titled *Adapting to Change: UK policy towards the Arctic*. The policy states, 'The United Kingdom is not an Arctic State, but we are the Arctic's nearest neighbour' (UK

Government 2013: ii). The UK Parliament's Select Committee on the Arctic defines the 'UK as a near-Arctic state' (UK Parliament 2015).

However, the term gained notoriety when it became linked with China. In this context, the focus is to underscore that what is 'in' the Arctic or has proximity to this somewhat arbitrary boundary is likely to have political connotations and can be subject to dispute. Also the EU repeatedly highlights and justifies its Arctic interests due to its impact and influence on the Arctic region in everything from changes in climate to trade (Raspotnik 2018).

Another debate that has caused some friction amongst the Arctic states has been the emphasis on being a coastal state with a direct coastline facing the Arctic Ocean. That approach excludes Finland and Sweden and, depending on where one draws the line in the North Atlantic, also Iceland. The Arctic Five (A5) club gained traction at the beginning of the Arctic 'era' in the twenty-first century, that is, 2006–2008. This culminated in the five Arctic coastal states gathering in Ilulissat in 2008. However, the idea of maintaining the A5 as an exclusive club of only coastal states has wavered. In 2010, Canada organised another gathering; however, it was rebuked by the USA and Secretary of State Hillary Clinton for not inviting all Arctic states (Blanchfield 2010). Ensuring the primacy of *all* eight Arctic states in Arctic debates – the foremost arena being the Arctic Council – seems to have outweighed debates regarding geographical relevance and coastal features, without discounting the practical relevance of only gathering the coastal states (Kuersten 2016).

In 2018, when the Kingdom of Denmark invited the other states to reconvene for a ten-year anniversary of the Ilulissat declaration, all the eight Arctic states were included (Jacobsen 2018). This is symptomatic of traditional 'region-building' efforts, namely that it matters more to represent a common regionalised approach, even identity, than to focus on only functionality and specific issues at hand (e.g., Neumann 1994). Ergo, although Finland and Sweden are less relevant when discussing most Arctic Ocean issues, they are crucial to have around the table to present a pan-Arctic unity and the distinctness of the Arctic region.

Still, in a break with this Arctic-states-first approach, one effort that stands out as an example of both how the Arctic states have approached ocean governance and what is seen as a way to deal with potential future problems before they arrive (Wegge 2015; Young 2010) is the Central Arctic Ocean Fisheries Agreement. Fisheries management challenges are emerging further south due to the ongoing changes in the distribution of various Arctic fish stocks,[11] which showcase the difficulties of reaching an agreement when interests are entrenched. In order to avoid this situation in the Central Arctic Ocean, the agreement – officially known as the Agreement to Prevent Unregulated High Seas Fisheries in the Central Arctic Ocean – represents a multinational effort to safeguard the Arctic Ocean environment (Agreement to Prevent Unregulated High Seas Fisheries in the Central Arctic Ocean 2018). Signed in 2018, the agreement includes the six Arctic states with direct stakes in Arctic fisheries – Canada, the Kingdom of Denmark (on behalf of Greenland and the Faroe Islands), Iceland, Norway, Russia, and the USA. But it also added China, the European Union, Japan, and South Korea. Together, these states account for a substantial portion of the global fishing industry.

The primary purpose of the Central Arctic Ocean Fisheries Agreement was to pre-emptively protect the Central Arctic Ocean (the area beyond national jurisdiction) from unregulated and potentially harmful commercial fishing activities. The agreement enforces a moratorium on commercial fishing in the high seas portion of the Arctic Ocean for an initial period of sixteen years. This moratorium is intended to allow for comprehensive scientific research aimed at better understanding Arctic marine ecosystems and the potential impacts of fishing (Balton 2021, 2022). While the agreement is a proactive measure with global significance, questions remain about implementation and enforcement, particularly given the vast and remote nature of the Arctic Ocean. Monitoring compliance and ensuring that states adhere to the moratorium presents logistical and technological challenges (Balton 2021). The agreement's initial duration (until 2037, as it came into effect in 2021) could also be seen as insufficient given the ongoing and unpredictable nature of climatic and ecological transformations. The agreement focuses solely on fisheries, but the Arctic Ocean faces other environmental threats such as increased oil and gas exploration, shipping, and pollution.

Also, for all the progress in terms of developing new governance mechanisms in the Arctic or reiterating existing ones, as well as efforts in settling maritime boundary disputes within 200 NM of the coast, broad areas of overlapping assertions to outer continental shelf rights

[11] For more on this, see, for example, Berkman and Young 2009; Hoel 2009; Stokke, Østhagen, and Raspotnik 2022; Stokke 2017.

underlying the majority of the Central Arctic Ocean remain unresolved. Although the flag-planting manoeuvre by Russia in 2007 has often been used as a starting point for the 'new era' of Arctic geopolitics (Østhagen 2021b), what it concerns in essence is extended continental shelf submissions in the Arctic Ocean. However, planting a flag infers no rights or claims through the Law of the Sea. Instead, Law of the Sea Article 76 lays down complex criteria regarding continental shelf areas seawards of 200 NM whereby the outer limits of the continental shelf may be determined with assistance from a scientific and technical body established through the Law of the Sea Convention – the Commission on the Limits of the Continental Shelf (CLCS).[12]

All Arctic coastal states have been active in gathering the data required to formulate submissions. For example, some – such as the USA and Canada – have cooperated amongst themselves, to facilitate joint surveys (J. S. Baker and Byers 2012). All the Arctic coastal states, except the USA (as a non-Law of the Sea party), have made submissions to the CLCS. The USA did, however, put forth its own unilateral claim for an extended continental shelf in late 2023 (US Department of State 2023; Tingstad 2023). It appears from these submissions and the US claim that, should the Commission be in agreement, the vast majority of the seabed of the Arctic Ocean will form part of the outer or extended continental shelf of the coastal states. More likely, a dispute concerning overlapping claims will emerge. Delineating the outer limits of the continental shelf seawards of 200 NM limits is considered challenging because of numerous 'complexities and ambiguities' associated with Article 76 of the Law of the Sea (Macnab 2004b, 2004a; Cook and Carleton 2000) as well as issues concerning the way in which the Commission works (McDorman 2002).

Issues such as the legal status of the Northwest and Northeast Passages are other examples of ocean disputes that have received considerable attention in the Arctic. However, these are not related to

[12] This complexity arises because continental shelf entitlements seawards of 200 NM limits are delineated not solely by reference to a distance formula (Ø. Jensen 2014b). These areas of continental shelf seawards of 200 NM limits are often referred to as 'outer' or 'extended' continental shelves, although legally there is only one continental shelf (Ø. Jensen 2022). Then, two maximum constraint or cut-off lines are applied: a limit of 100 NM from the 2500-metre depth isobath (depth contour) or 350 NM from the coastal state's baselines at the coastal state's discretion (Article 76(5)).

the delineation of maritime space. Instead, they concern whether or not the passages can be deemed internal waters (giving Canada and Russia extensive rights to manage them) or international transit lanes (limiting the restrictions Canada and Russia can place on foreign vessels) (Lackenbauer 2010). Especially for Canada, this has been an important issue (Huebert 1995; Kraska 2007; Pharand 2007; Byers and Lalonde 2009). Canada treats the passages as internal waters subject to Canadian jurisdiction, whereas the United States (and others) maintains that they are international straits in which foreign vessels, including military vessels, have the right to transit (Lalonde and Lasserre 2013).[13]

In sum, the essence here is that in a study of the Arctic's role in international politics and relations, and especially its governance relations, we must bear in mind the way the Arctic has figured in various environmental campaigns as a symbol for climate change and the effects of human inaction. Moreover, this engagement and interest have led to calls for new governance structures in the Arctic. Although many of the structures that have emerged or been re-energised are relevant for dealing with various issues in the North, calls for pan-Arctic governance treaty structures similar to those in Antarctica are based on a false premise, namely that the two polar regions share political traits and challenges. The Arctic states 'fought' back through public diplomacy and statements, trying to convey the message that the Law of the Sea regime and the sovereignty and sovereign rights of the Arctic states are the pillars of the governance system in the Arctic in the twenty-first century, although disputes still exist concerning how to regulate and govern various maritime issues in the North.

Arctic Security Cooperation

This book is primarily about security dynamics and their existence in a regional setting. So far, security concerns and threat perceptions have been remarkably absent in the discussion of Arctic exceptionalism,

[13] The importance the Canadian government attaches to questions of sovereignty has led to some questionable policy decisions in the past, such as the forced relocation of Inuit families from northern Québec to the High Arctic Archipelago in the 1950s to 'give evidence of "occupation" as well as the presence of authority' that could be used to support Canada's sovereignty claims (Shadian 2007; Østhagen and Lackenbauer 2023).

governance, and cooperation. That is, however, exactly the point: the regionalist or region-building approaches and efforts performed by Arctic states (by their officials and politicians, as well as the academic community (Steinveg 2017)) must be interpreted as a counterreaction to security issues dominating the Arctic regional agenda during the Cold War (see Chapter 4), as well as the uptick in global interest with the region sometimes founded on the incorrect premise of an impending conflict or resource race.

Therefore, a specific dimension is absent from the previous discussions related to both the growth of regional cooperative mechanisms and the counternarrative of exceptionalism that emerged from 2007 to 2008 onwards: military or state security. As highlighted, this was explicitly excluded when the Arctic Council was formed. Many have attributed this exclusion to the Council's success in convening the Arctic states despite tensions elsewhere (Graczyk and Rottem 2020; Grønning 2016; Young 2010).

However, this does not mean that there have been no security challenges in the Arctic, either on a circumpolar level or via a regional lens. As Chapter 4 will focus on global dynamics impacting the Arctic, the increase in military activity and related tensions has more or less paralleled the focus on cooperative measures and the counternarrative, as outlined. Yet these dynamics are driven by a different logic, with a different set of (global) drivers, and they – to some extent – involve different actors, or at least different ministries, from the Arctic states.

Before turning to these dynamics, it is worth maintaining the focus on regional dynamics and how Arctic states have attempted to regionalise Arctic security issues. In many ways, the reason why this has been difficult, or has not fully served a purpose, is because military security concerns in the Arctic are not circumpolar in nature, they are rather national and subregional – which is a point we return to in the following chapters. Still, attempts have been made to engage in security dialogue, especially before 2014, where engagement with Russia was the main goal. Before 2014, a common perception amongst the Arctic countries was that a *regional* dialogue on security policy was appropriate to avoid misunderstandings and reduce tensions.

The Arctic Security Forces Roundtable (AFSR) was established in 2010–2011 through the initiative of the USA and Norway. This forum included senior officers and bureaucrats from the eight Arctic countries as well as some European countries. The Northern Chiefs of Defence Forum was established in 2012 through Canada's initiative for joint meetings between the chiefs of defence of the eight Arctic countries. Both forums were non-structured ways of engaging in informal dialogue on Arctic issues with Russia, albeit without much expected output beyond confidence-building and information sharing. However, after the Russian annexation of Crimea and the conflict in Ukraine in 2014, the CHOD was discontinued, while the ASFR has been convened without Russian participation. The Arctic Council has been raised as a possible arena for security policy dialogue; however, the Arctic countries have taken a clear stand against this, as the success of the Arctic Council lies in including all the Arctic countries in softer forms of cooperation.

The Case of the Arctic Coast Guard Forum

Perhaps the best case to study when unpacking Arctic *regional* security cooperation is the efforts made to establish a coast guard forum, because it was meant to serve a clear regional purpose with a security component and it also illustrates the difficulties in pan-Arctic or circumpolar cooperation on such issues.

As the limited capabilities of coastal Arctic states surfaced with the increasing national and international attention given to the Arctic, multilateral cooperation became the focal point to alleviate pressure. In 2011, the eight Arctic countries negotiated an agreement for search and rescue under the auspices of the Arctic Council. A similar agreement on oil spill preparedness was negotiated two years later in 2013 (Arctic Council 2013, 2011a). Although these agreements have been used to indicate how Arctic states embrace multilateral cooperation, they have been criticised for having limited impact in the Arctic, mainly outlining existing responsibilities and boundaries (Rottem 2014).

As a response to limited Arctic capacities and growing activity, the idea of establishing a dedicated forum for Arctic coast guards was proposed (Conley et al. 2012: 37). Modelled on existing forums for the North Pacific and the North Atlantic, American and Russian officials subsequently took the first steps towards the establishment of an 'Arctic Coast Guard Forum'. The Russian annexation of Crimea and start of the conflict in eastern Ukraine in 2014 halted the process as representatives from Russia were not included in two experts' meetings hosted by Canada in Nova Scotia. Still, the next experts' meeting in March 2015 was held in Washington, DC, and this time the Russian Coast Guard was present. The eight countries decided to push ahead, and the Arctic Coast Guard Forum (ACGF) was formally established on 30 October 2015, at the US Coast Guard Academy in New London, Connecticut (Melia 2015).

Some argued that the Forum could serve as an arena to tackle security dialogue with Russia at a time when other channels were unavailable (Conley and Rohloff 2015; Grønning 2016). However, others worried about attaching too much significance to the Forum, that it would hamper practical cooperation in favour of creating yet another Arctic 'talk-shop' (Sevunts 2016). Moreover, some fundamental challenges limited what tasks multilateral cooperation – via the ACGF – could embark on.

An overarching function of most coast guards is the protection of the coastal state's sovereign rights. Coast guards uphold sovereignty through naval presence and the enforcement of national jurisdiction (Till 2004: 330–47). For example, fisheries inspections are an integral part of protecting a state's sovereign rights through the management of its own marine resources (Skram 2017). This authority cannot be shared without the coastal state ceding some of its sovereignty. In addition, significant differences have been found between individual coast guards operating in the Arctic region (Østhagen 2020a). Each coast guard is tailored to national interests, geography, and institutional cultures; no single unified Arctic coast guard structure exists.

The variety of mandates and structures consequently challenges dialogue of a potentially sensitive and/or military character. As a result, the Arctic Coast Guard Forum had to resort to a 'lowest common denominator' approach (Østhagen 2015). If collaboration was to expand beyond an immediate response, the relatively wide variations in mandates and competence acted as barriers. Effective collaboration with defence actors is often dependent on a similar set of structures with similar size and priorities (Diesen 2013; Valasek 2011). The lack of commonality in organisational structures and mandates as well as the lack of harmonisation between coast guards in the Arctic offer significant obstacles for future collaboration.

Finally, the geographic vastness of the Arctic was in itself a challenge to effective on-the-ground cooperation. Although this point should be self-explanatory, it tends to be overlooked in the broader public discourse, especially when politicians make a point to praise international cooperation in the North. Arctic geography involves large maritime domains, where few emergency incidents occur and limited or no capabilities are present. Sharing responsibility or handing over tasks to other countries is often not even an option as states struggle to provide sufficient capabilities in their own areas of interest. The option of practical or onsite cooperation between, for example, Norway and Canada is non-existent, given the great distance between these two parts of the Arctic region (Østhagen 2020a).

In addition to the outlined challenges, the integration of Russia in any security dialogue or security forum is needed on the one hand and challenging on the other. In the initial pre-meetings before the Arctic Coast Guard Forum's establishment, the importance of Russian participation for the Forum's relevance was clear (Pincus 2015). This was partly because of Russia's extensive coastline and responsibilities along the Northeast Passage and partly because of Russia's shared maritime borders. In the Bering Sea, the Chukchi Sea, and through the Bering Strait, Russia and the USA share an extensive maritime boundary settled in 1990. Similarly, in the Barents Sea, Russia and Norway share an extensive and predominantly ice-free maritime boundary settled in 2010. Finland shares a maritime boundary with Russia in the Gulf of Finland, though it is smaller in size and not in the Arctic. In all these border domains, cooperation with Russia on both a practical as well as a governmental level is essential to effectively manage straddling fish stocks and provide environmental protection and search and rescue response. Moreover, all these countries have some form of direct line with Russian authorities or the Northern Fleet in order to manage any unfolding crisis scenario.

A possible expansion of coast guard collaboration in these geographic areas depends on both diplomacy *and* cooperation with Russia. In most cases, coast guard affairs constitute so-called low politics, which states tend to separate from larger diplomatic affairs (Østhagen 2016). In particular, the sharing of non-sensitive information and the identification of 'best practices' were areas of focus. Developing a community of experts that not only sees value in sharing information and cooperating but also develops trust and 'socialises' into preferring collaborative solutions had the potential to advance coast guard cooperation further, a point we examine more in the final chapter.

We have seen examples of these mechanisms at work at a lower level in the case of Norway–Russia in the Barents Sea (Hønneland and Jørgensen 2015). It is interesting to note that, while more in-depth forms of collaboration prove difficult for coast guards due to national jurisdiction and sovereignty concerns, the opposite is the case with the lower level forms of collaboration. The soft security nature of coast guard tasks works to their advantage. As a matter of fact, low-level collaboration between coast guards dealing with tasks of a non-sensitive nature, such as sharing of information, joint exercises, training, and procurement, could easily be facilitated. However, even amongst fellow NATO member states, seemingly insurmountable obstacles arise as soon as the nature of the tasks changes – more specifically, when we talk about higher level forms of military-to-military collaboration.

Simultaneously, having Russia around the table led to a bumpy start for the Forum, initially postponing its launch due to Russian actions in Ukraine in 2014. Even before 2014, collaboration with Russia had its inherent limitations. Sensitive information could not be shared on the same level as between the other seven NATO members in the Arctic (Canada, Norway, USA, Iceland, Kingdom of Denmark – and by 2024 – Finland and Sweden). After 2014, the other Arctic states also placed strict guidelines on the role and mandate of their representatives when they met and worked with their Russian counterparts. Further, with Russia's full-scale invasion of Ukraine in 2022, the Arctic Coast Guard Forum was suspended (Underwood 2022). Norway, however, overtook the operations of both the website and – following its chairship of the Arctic Council from May 2022 – the chair of the Forum (ACGF 2024; Kystvakten 2024). However, as with the Arctic Council, Russian participation has not been completely excluded and it remains a member of the Forum.

The Allure of Regional Cooperation

Zooming back out to the regional level of Arctic security dynamics and practices, a relevant question here is: Why have these various cooperative initiatives – ranging from the Arctic Coast Guard Forum to the Arctic Council and various other mechanisms – emerged? Chapter 2 outlined how rationalist approaches within international relations place emphasis on the goal-seeking rational behaviour of states driven and constrained by material factors as well as institutional structures.[14] Cooperation (between states) is inherently a product of conflicting interests and/or policies that are brought into alignment. Institutions – as tools set up by states to manage potential disputes – are arguably rational design consequences of states' interests and a result of a bargaining process between the same states (Fearon 1998; Koremenos et al. 2001). By distinguishing harmony from cooperation and discord, Keohane (1984: 53–54) argues that '[c]ooperation takes place only in situations in which actors perceive that their policies are actually or potentially in conflict, not where there is harmony'. Thus, discord provides a platform for potential cooperation. When states cooperate in the international system, they often develop an explicit or implicit international regime.[15] Regimes help 'improve the contractual environment and thus stabilize cooperation' (Levy, Young, and Zürn 1995: 288).

Furthermore, Haas (1989) – through the idea of 'epistemic communities' – argued that a small issue area of international cooperation could become dominated by technical experts who not only represent their own states but also start to share common beliefs and approaches to solve the problems at hand. 'Regimes are not simply static summaries of rules and norms; they may also serve as important vehicles for international learning that produce convergent state policies' (Haas 1989: 377). In essence, these communities take on a life on their own.

Evaluating these rather stylised conceptions of the emergence of international cooperation, it is clear that what emerged amongst Arctic states in the 1990s served several purposes. The primary motivation, as stated in, for example, the Ottawa (Arctic Council) and the Kirkenes (BEAC) declarations, was to deal with environmental and economic issues in the Arctic and Barents areas respectively. From a 'top-down' perspective, there obviously has been a realisation

[14] See, for example, Fearon 1995; Keohane 1984; Moravcsik 1997.

[15] Krasner's characterisation of a regime from 1982 has become the go-to definition of it in studies of international cooperation. He defined a regime as 'a set of implicit or explicit principles, norms, rules, and decision-making procedures around which actor expectations converge in a given area of international relations' (Krasner 1982: 2). Regimes tend to – by most definitions – be more specific than international organisations and their related policy areas. Regimes are thus issue-specific, such as 'clean water', 'Indigenous rights', or 'sustainable fisheries.'

amongst decision-makers and experts in the Arctic states regarding common or shared interests across the circumpolar region that have led to the establishment of several agreements, forums, and even institutions – or governance structures – to deal with the problem at hand. In other words, interdependence has been evolving (e.g., Keohane and Nye 1977).

Arctic states not only recognised shared interests in everything from emergency response management to best practices for local and Indigenous economic development in the North. As the states collaborated to solve these problems or deal with the plethora of Arctic issues that arose as activity increased in the North, they were also becoming increasingly connected and dependent on each other (Byers 2017). For example, as fisheries co-management mechanisms between Norway and Russia or – to a more limited extent – between Russia and the USA expand (Østhagen 2016), maintaining them becomes a priority in and of itself, not only because of the positive-sum yields they produce (in this example, sustainable total quotas).

It is in turn not sufficient to only highlight the Arctic Council and its emergent status as the primary forum for Arctic issues; we must conceive of the wider spectrum of 'governance' in the circumpolar area (and regions/areas linked to it). All the various agreements, whether of bilateral or multilateral character, that govern everything from fish stock quotas to border crossings, Indigenous issues, or environmental concerns constitute 'Arctic governance'. In this frame, the various forums with the Arctic Council as spearhead are only a minor part of a larger system, albeit the part that attracts the most attention and symbolism as a lightning rod for Arctic governance writ large. For example, a range of Arctic cities set-up the Arctic Mayors' Forum in 2019, adding further to a local level of governance in the Arctic (Németh Winther 2025).

However, there is another aspect of the emergence of these forums, namely, the intrinsic power structures embedded in them as well as their role in regional or global relative power relations. One cannot ignore that Russia is by far the most dominant Arctic actor, along parameters ranging from demography to economic investments and military capacity in the North. Yet in an Arctic cooperative setting, Russia faces seven other Arctic states that are all 'Western' liberal democracies and – after Russia's invasion of Ukraine in 2022 – part of NATO. The notion, as sometimes expressed by various Nordic countries, that these forums were meant to help curtail Russia (Skedsmo 2007; Rowe 2018; Onsager and Stuvøy 2022) is obviously related to power and the asymmetric relations between Russia and each of its Nordic neighbours.

In addition, an insider versus outsider perspective on regional dynamics is relevant in explaining the emergence of such forums from a relative power or strategic perspective. The increased international attention given to the region also led the Arctic states to become more conscious about their regional primacy and how this can be maintained and upheld (Bertelsen 2019; Pedersen and Steinveg 2024). Ensuring regional primacy, or power, vis-à-vis other countries must be seen as an underlying motivation for why both the Arctic Council and other forums have moved up on the foreign policy agendas of the various Arctic states (Stokke 2017; Steinveg 2024).

Finally, as is often the case with international relations and foreign policy dynamics, we cannot ignore the immaterial dimension of these developments. In contrast to the top-down analyses presented here, an element of 'bottom-up' exists when it comes to Arctic governance that is – albeit not unique – an important trait in the regional affairs of the North (Steinveg 2017; Young 2005). Awareness of the effects of both climate change and especially local/regional pollution became apparent in the 1980s and 1990s, and instigating low-level forms of cooperation on a practical level (i.e., working groups) became a solution to an emerging problem. Thus, the notion of a northern 'epistemic community' comes into play – and some would argue that the emergence of these forums was the result of decades (from the 1970s) of work on managing local and regional northern challenges exemplified by the Agreement on the Conservation of Polar Bears from 1973.[16]

We also cannot ignore the wider foreign policy objectives related to identity and ideas amongst the 7+1 in the Arctic. The inclusion of post-Soviet Russia in liberal or Western structures was meant to enable Russia to become more similar to the West. By including the new Russian Federation into various low-level cooperative forums in the 1990s, it was hoped that ideas of democracy and

[16] Signed in Oslo, 15 November 1973, by the five nations with the largest polar bear populations: Canada, the Kingdom of Denmark, Norway, the United States, and the Soviet Union.

liberal society would transform Russia and 'make them more like us' (Hønneland 2012a; L. C. Jensen 2017; Rowe 2015). The Barents Cooperation especially signifies this, with the relatively large amounts of funds being spent by the Nordic countries on various projects in Russia aimed at democracy, civil society, and societal transformation (Skedsmo 2007). The effects of these efforts can be debated; however, as many have pointed out, decades of Barents regional cooperation and cross-border collaboration in this specific part of the world has fostered a certain kind of identity and similarity in the North that transcends what was previously the border between East and West during the Cold War (Hønneland 1998; Onsager and Stuvøy 2022; Steinveg and Medby 2023).

In essence, it seems clear that there are apparent positive-sum effects of Arctic states cooperating – by sharing information, improving each other's capacities, and assisting possible emergency scenarios. Yet there is concern about states having the ability and capacity to coordinate and capitalise on this apparent potential. Furthermore, it seems clear that multilateral *pan-Arctic* security cooperation between Arctic states was not developed to serve an immediate demand that states could not deal with on their own or through bilateral modes of cooperation. As shown in the case of coast guards, there are marked limitations to what can actually be achieved in a multilateral setting – even before Russia invaded Ukraine in 2022.

We can conceptualise various reasons for the establishment of multilateral cooperation. First, the concept of 'signalling' is relevant to this debate. Initially introduced in international relations as it relates to game theory and games with imperfect information, signalling helps explain actions that do not initially serve a specific purpose on their own (Snidal 1985; Mock 1992). We can conceive of Arctic states' desire to signal at two different levels, where the message in both cases is clear, along the lines of 'we are taking appropriate steps to deal with a dire situation'. Given the warning calls from scholars and experts concerning Arctic maritime emergency potential leading to political and public demands for government action,[17] setting up multilateral forums is done to *signal* that efforts are being made to alleviate the situation.

On the one hand, the audience is domestic, soothing public demands for action. On the other hand, the audience is international,

[17] See, for example, Byers 2016; Funston 2014; Ivanova 2011; and Contenta 2010.

showcasing to other countries that the Arctic states are effectively managing the changing environment in the North. In particular, concerns over the Arctic described as lacking political governance structures, voiced – amongst others – by some Members of the European Parliament as well as by some Chinese officials,[18] triggered a response by Arctic governments emphasising and reiterating the robustness of the 'Arctic regime'.[19] The following chapter will discuss it in detail. By signalling to both external and internal actors, the Arctic states were seen to be dealing with an issue of growing apprehension, even if we can question to what extent a multilateral forum alleviates actual capacity concerns.

We can read the emphasis in the Arctic states' various policy documents (i.e., Arctic policies) on 'cooperation', 'Law of the Sea', and the regional governance structures, such as the Arctic Council, as a direct response to the alternative narrative of the Arctic as an ungoverned space (Heininen 2012; Heininen et al. 2020). Rather quickly, this became part of the fundamentals when officials from Arctic states discuss the Arctic. The criteria set out for observers to the Arctic Council at the Nuuk ministerial meeting in 2011 also underlined these points, namely that outsiders to the region must 'Recognize Arctic States' sovereignty, sovereign rights and jurisdiction in the Arctic' and 'Recognize that an extensive legal framework applies to the Arctic Ocean including, notably, the Law of the Sea, and that this framework provides a solid foundation for responsible management of this ocean' (Senior Arctic Officials 2011).

In addition, we can conceive of cooperation as having an effect beyond the coordination of policies and signalling to other actors. As Alastair Johnston (2001) and Jeffrey Checkel (2005) argue, actors partaking in international cooperation over time 'do not emerge the same'. The role of expert communities (or epistemic communities) – such as coast guard officials – in national policymaking should not be underestimated either. The aspiration of creating an additional layer of Arctic governance through the inclusion of another expert dimension in the

[18] See 'European Parliament Resolution of 9 October 2008 on Arctic Governance' 2008 and Byers 2013: 125.

[19] For in-depth studies of these efforts, see Grindheim 2009; Offerdal 2010; Keil and Raspotnik 2014; Raspotnik and Østhagen 2014; Stokke 2014a; Bekkevold and Offerdal 2014.

already complex Arctic system is evident from the initial conception of the Arctic Coast Guard Forum (Neffenger 2014: 42).[20]

Understanding *how* such mechanisms can serve a purpose beyond policy coordination amongst partaking states can help us better comprehend why they are set up in the first place, also with an eye on regional security dynamics and separating those from international and national ones. However, after the Russian invasion of Ukraine in 2022, the idea of instigating additional cooperative mechanisms with Russia in the Arctic seems unlikely – maintaining those that already exist is a core challenge for Arctic states.

The 'Region' Unravels?

Looking at the regional dimension of political and security dynamics in the Arctic, some of the following conclusions can be drawn. It is apparent that the Arctic territory and the Arctic Ocean are more or less segmented into spaces of state sovereignty, or spaces where states have exclusive sovereign rights, based on the Law of the Sea. There is little to argue about when it comes to boundaries, although disputes between states do exist (Lackenbauer and Lalonde 2019). Thus, despite territorial land grabs in other parts of the world, a 'race' for Arctic resources or territory is not likely to unfold in the foreseeable future. Arctic-based conflicts, where Arctic or non-Arctic states claim a limited number of out-of-bound offshore resources (many of which are likely to remain unexplored for the next few decades), are neither economically nor politically viable and thus are not an immediate cause for concern (Østhagen 2023a).

As also discussed here, it seems plausible to argue that low-level forms of regional interaction and cooperation have contributed to low tension in the North, even when they did not deal directly with security matters (Stephen and Knecht 2017). Since 1996, the emergence of the Arctic Council as the primary forum for regional affairs serves as a platform from which its member states can portray themselves as working harmoniously towards common goals (Exner-Pirot 2015). What we have seen since the start of the new millennium is that the Arctic states have actively worked to shelter regional security dynamics: from initiating regional cooperative forums in the 1990s and

[20] See Stokke 2006, 2014b; Exner-Pirot 2013; Keil 2014.

responding to international attention in the early 2000s to instilling a somewhat artificial separation between security issues and non-security governance in response to increased tension in other parts of the world from 2014 onwards. Still, the deterioration in relations between Russia and the West in 2014 as a result of Russian actions in Ukraine also changed the security dynamics in the Arctic.

However, the effects of this change depend on what issue or level we focus on (Rahbek-Clemmensen 2017; Byers 2019). Although all the other seven Arctic states were quick to condemn Russian actions in spring 2014 and put in place sanctions in line with the EU's/USA's regimes, cooperation in the frame of the Arctic Council, the Barents Euro-Arctic Council, and a range of various multilateral and bilateral mechanisms continued uninterrupted. There were some voices calling for a ban on Russian participation in some Arctic Council activities, especially from Canada (Berthiaume 2015); however, the effect was limited in the end. Only a year after the annexation of Crimea, all Arctic states *jointly* established an Arctic Coast Guard Forum. Indeed, the foreign ministries of all Arctic Council members, including Russia, kept proactively emphasising the 'peaceful' and 'cooperative' nature of regional politics after 2014 (Wilson Rowe 2020; Heininen et al. 2020). This was perhaps more of a desire or goal than necessarily a reflection of reality at the time.

From 2014, the divergence between the regional dynamics that this chapter is devoted to and the security dynamics that link to the global level as well as to the national or subregional level, as is outlined in the following two chapters, became more apparent. As such, the traditional security and defence issues in the Arctic became separate from the wider 'Arctic policies' of the various Arctic countries. As Chapters 5 and 6 explore in-depth with Norway and the USA as cases, this dual-track approach seems almost contradictory at times. On the one hand, the foreign ministries and diplomats kept emphasising the governance structures and the cooperative regional spirit of the Arctic, and on the other hand, the various Armed Forces and Ministry of Defence officials kept highlighting the increased military activity and tension in parts of the Arctic (for more on this dichotomy, see Wilson Rowe 2020).

Consequently, the idea of an Arctic exceptionalism or the Arctic being in some ways sheltered from both the geopolitical concerns and from the spillover effects from other parts of the world, started to crack. Although there are few remaining maritime disputes and no territorial issues in the Arctic – a stark contrast to many other parts of the world – the limited disagreements or disputes that remain suddenly were interpreted and analysed in a new light after the Russian regime's apparent willingness to use military force against another country to achieve its strategic and geopolitical goals. Thus, it was not issues in the Arctic per se that led to a rethink of Arctic geopolitics but a rethink of Russian intentions and Russian calculations more broadly.

Simultaneously, the spill-over from NATO–Russia tension (and to some extent USA–China tension) into *specific* parts of the Arctic became increasingly apparent from 2014 onwards both in terms of military operations in the region and rhetoric and statements from various officials – sometimes crossing, or even shattering, the imaginary separation between Arctic diplomacy issues and Arctic security issues. A clear example of this was the speech given by US Secretary of State Mike Pompeo in Finland in 2019, just in advance of an Arctic Council Ministerial Meeting, arguing that security matters should be given more attention in the Arctic (US Department of State 2019).

Then, in 2022, Russia fully invaded Ukraine on 24 February. On 3 March the same year, the other seven Arctic countries suspended cooperation with Russia in the Arctic Council. Cooperation was also suspended in the Barents Euro-Arctic Council. However, some forms of bilateral mechanisms were kept in place, especially for those Arctic countries bordering Russia. Although Russia did not immediately change its military posture in the Arctic, as with the effects in 2014, the most immediate change was a reinterpretation of Russian intent and possible scope for escalation and aggressive action.

At the same time, high-level political dialogue through the Arctic Council or writ large, initiating new forms of cooperation, or even touting existing forms, is currently not possible. What has also become clear since, is that efforts to continue pan-Arctic dialogue and cooperative structures is almost impossible, due to the lack of trust and dialogue between the seven Western Arctic states and Russia. It is therefore timely to ask whether the region-building efforts of the 1990s and 2000s, which also continued albeit dampened after 2014, are now being reversed? Is the Arctic 'region' unravelling as a political construct, or are there characteristics of the Arctic, and of its related actors, that still demand some form of particular Arctic approach?

Still, the effects of the Russian war in Ukraine have not been a complete breakdown of the various cooperative mechanisms and regimes that exist in the Arctic. Especially when looking at functional or practical cooperation at sea, contact and dialogue remain. Norway and Russia uphold their co-management regime for fish stocks in the Barents Sea (Oma Vyvial 2023). Russia continues contact with both the US and Norwegian Coast Guards in order to deal with incidents. The Arctic Council resumed working group cooperation after Norway took over the chairship from Russia in 2023 (Arctic Council Secretariat 2024). And the parties to the Central Arctic Ocean Fisheries Agreement continue to meet, including Russia. Others are musing about the Arctic potentially being one arena for future engagement with Russia through science diplomacy and other low hanging fruits that are particularly evident in the Arctic (Paul 2024; Parlato et al. 2024).

Thus, the main spill-over effect from the Russian Ukraine invasion was the increased tension in the field of security and military concerns that now occupy more space in Arctic discussions than ever, and the distinction between security affairs and 'high-level' politics and other more technocratic and 'low-level' issues is clearer than ever. However, it is apparent that these security concerns and related interests and practices do not necessarily derive from changes in the Arctic, but rather the central role *parts* of the Arctic play in systemic or global security relations – something we examine in the following chapter.

4 *The Global Level*
Climate, Submarines, and China

This is the second of four chapters that distinguishes between different levels, or sets, of security dynamics and related practices in the Arctic. Here, we look at the global, or systemic, level. These are dynamics and trends that are inherently determined beyond the borders of the Arctic region. Whether price levels for commodities, security interactions between NATO and Russia, or China and the USA, these issues and the political dynamics related to them are not to be confused with how the Arctic states interact *in* the Arctic region, as explored in Chapter 3.

Global Trends, Arctic Impact

At the start of the new millennium, certain developments and trends in international politics and global markets helped raise the Arctic on the international agenda. First and foremost, awareness of climate change and its marked effects in the Arctic spurred an environmental agenda concerning the region. The Arctic was promoted as the proverbial canary in the coal mine, showcasing the effects of human CO_2-emitting activities. In tandem, global awareness of the impact on Arctic wildlife – such as imagery of drowning polar bears due to loss of ice habitat – spurred engagement in how to 'Save the Arctic' (Greenpeace 2012). In contrast, or in addition, to a bottom-up approach to regional concerns as was showcased in Chapter 3, a renewed or increased interest in the Arctic – including from outside the region – emerged from around 2004–2005 onwards. This is reflected in the range of media attention, journalistic in-depth pieces, and scholarly work that looks at the Arctic in the early 2000s.[1]

[1] For an analysis of the media portrayals, see Nilsson and Christensen 2019; Christensen 2013; Padrtova 2019; Klimenko, Nilsson, and Christensen 2019.

In addition to climate attention, another factor spurring global interest in the Arctic was high commodity prices. In July 2008, the price of crude oil peaked at USD 147. Coinciding with the high prices for resources, the Arctic region was thought to have a resource abundance and various numbers about the resource potential were flaunted. In 2008, the US Geological Survey estimated that the whole Arctic region held 13 per cent of the world's undiscovered oil and 30 per cent of its undiscovered natural gas (USGS 2008). These reports have become the staple of any media article about the Arctic's 'riches' and resource potential, despite their rather limited value to say anything about whether this potential is actually realisable. The reports only emphasise the probability of petroleum resources being located in various offshore parts of the Arctic. Crucially, the emphasis is on 'undiscovered'.

This book is not about the oil and gas resource potential in the Arctic or the impediments to extraction;[2] hence, it will suffice to highlight that the extreme operating conditions in some of these areas, the subsequent fall in commodity prices, and the limited ability of the Arctic coastal states and companies to develop infrastructure connected to potential offshore fields are factors that have made these numbers around the resource potential less relevant. However, in the first decade of the second millennium, they helped spur attention and interest from large oil and gas companies (and their states) northward. And even today, the myth and allure concerning oil and gas in the Arctic is alive and well as internationally reputable media publish extensive features on the Arctic.[3]

We also see that at the beginning of the 2020s, the general interest in oil and gas has, to some extent, been replaced with an interest in other natural resources. However, the dynamics are the same. The region's resource abundance is coupled with global commodity prices and demand, prompting speculative statements about its future. Notions of a race to the Arctic, a scramble for resources, or an impending conflict continue to make headlines despite the situation being much more complex and nuanced in the Arctic.

This is not to say that the Arctic does not hold relevant resources, be it minerals or petroleum. Several parts of the Arctic are hubs for oil

[2] See instead Harsem, Eide, and Heen 2011; Keil 2014; Claes and Moe 2018.

[3] See, for example, Deutche Welle's documentary from late 2022 (Deutsche Welle 2022).

and gas extraction, which reaps profits for the relevant companies and regional/local governments. Fisheries, and their link to global markets for seafood, are similarly an international market-oriented product with enormous regional relevance in parts of Arctic waters. Moreover, other types of industry, for example, traditional metals industries that rely on parts of the Arctic that have cheap and stable electricity or new forms of industry-related battery plants or data centres, are relevant in some parts of the Arctic.

In addition, some Arctic countries have been particularly active in stoking interest from afar when it comes to the Arctic, working to highlight the region's resource and economic potential. Thus, an argument that attention on the Arctic is only related to international or external factors disregards the agency of Arctic actors themselves. In turn, clearly distinguishing between the global, regional, and national drivers when it comes to economic prospects and resource development, or attention given to the region as part of a larger discussion on climate change remains difficult. Ultimately, the 'Klondike'[4] that continues to lure one-liner descriptions never took place and will not come to fruition. The complexity of the region makes a pan-Arctic 'gold rush' impossible. However, global attention continues to set the frame for both the political and the economic discussions concerning the Arctic and driving attention northwards.

The Arctic's Strategic Role for Russia, and Therefore NATO

The Arctic's inherent cooperative spirit or 'exceptionalism' as explored in Chapter 3 has continued to clash against another global trend, or perhaps more accurately a certain effect of structures and developments in international politics. There are two dimensions to this: (1) the region's role in security relations between Russia and 'the West' due to its strategic importance for Russian security policies; and (2) how the Arctic increasingly plays a role in the global competition between the USA and China.

[4] The term comes from the place 'Klondike' in Yukon, Canada. It refers to the gold rush that occurred in the late nineteenth century, also related to the Arctic.

The Cold War

It is worth taking a quick look at the Arctic's importance during World War II, and since then, to understand why it has regained strategic importance for Russia (and the world). During World War II, the Barents Sea became a central part of Allied relief shipments to the USSR in the fight against Nazi Germany with convoys of ships traversing north of the occupied Norwegian Arctic. Nazi Germany occupied parts of the Kola Peninsula as part of the larger eastern offensive, whereas the Petsamo-Kirkenes offensive from the USSR became part of the Red Army's liberation from Eastern North Norway.

Furthermore, Iceland held a strategically important geographical position, leading to its pre-emptive invasion by the UK in 1940 to avoid Nazi Germany occupying the island. By 1941, the USA had taken over the occupation of Iceland – then part of the Kingdom of Denmark. World War II also marked an incident in US history in which a hostile force invaded US soil in a northern or almost-Arctic context: the Japanese Empire attacked the Aleutian Islands of Unalaska, Attu, and Kiska, which were territories of the USA.

The Arctic played a prominent role in the political and military competition between the two superpowers during the Cold War. The region was not primarily important due to conflicts of interest within the Arctic itself however, but because of its strategic role in the systemic competition between the USA/NATO and the USSR (Åtland 2008; Huebert 2013a). Key to this was the nuclear arms race and, given the geographic proximity between the two states in the North, undersea and aerospace capabilities were prioritised in the Arctic, as the shortest route of missile attack between the two superpowers went over the North Pole.

Related to this, the Northern Fleet in the USSR eventually became the largest of the Soviet Fleets. The concentration of bases on the Kola Peninsula was, and is still today, largely a result of the fact that the southern Barents Sea is ice-free in winter, enabling year-round access to the Atlantic for Northern Fleet vessels. During the Cold War, Soviet naval strategy in this part of the Arctic revolved around a 'bastion' concept. This consisted of a maritime domain (a bastion) relatively close to Soviet territory, where ballistic missile submarines could linger while being protected by various conventional Soviet forces (Breemer 1989). Two bastions existed – one in the Barents Sea centred around the Northern Fleet and the other in the Sea of Okhotsk centred around the Pacific Fleet (Wezeman 2016). After the fall of the USSR, the concept is thought to have survived, albeit in a reduced form. The focus also shifted away from the Pacific Fleet and to the Northern Fleet (J. Kjellén 2022: 40).

Norway was one of only two NATO member states (the other being Turkey, which joined NATO in 1952) that shared a land border with the Soviet Union during the Cold War. As a consequence, the North Atlantic became NATO's 'northern flank', crucial for monitoring and surveillance of Soviet military activity on the Kola Peninsula. Norway, with the help of the USA, built up its Armed Forces and force structure to be able to hold off a Soviet attack long enough for NATO allies to come to the rescue, albeit likely by ceding parts of its northern regions bordering the USSR. As is further explored in Chapter 5, Norway also imposed certain restrictions on itself regarding military activity and presence on its territory to 'reassure' the USSR that Norway posed no direct military threat (Skogrand and Tamnes 2001).

One prominent geopolitical dimension was the possible role that the Norwegian archipelago of Svalbard might play in a large-scale conflict involving the Arctic (or parts of it). Since World War II, sensitivities concerning the use of the archipelago for military purposes have fluctuated in conjunction with East–West tensions. This possible military use of the islands was even discussed during World War II, albeit on a limited scale (Holtsmark 1993: 21–22; Dean and Lackenbauer 2017). During the Cold War, the USSR was to some degree concerned about the possible military use of the archipelago, demanding adherence to the Svalbard Treaty's ban on the use of the islands for warlike purposes or the construction of fortifications or naval bases (Svalbard Treaty 1920). When the archipelago was placed under NATO's Atlantic Command in 1951 (Tamnes 1987: 4:25; Baudu 2023), the USSR criticised this action as preparation for further Allied activity in the High North (Holtsmark 1995: 466–75; Skogrand and Tamnes 2001: 81). Soviet leaders were concerned with Svalbard being utilised by NATO to hinder Russian access to the wider Atlantic via the Greenland-Iceland-UK (GIUK) and Norway gaps.

Norwegian analysts and policymakers thus thought, particularly in the late 1970s and onwards, that if war were to break out, control, and the possible use of Svalbard, would be the primary motivation for the Soviets – both to limit possible NATO command and to potentially use Svalbard as a forward operating area in order to protect the maritime domain of their strategic submarines with nuclear ballistic missiles (Tamnes 2020: 155).[5] The Soviet interest in Svalbard expressed through, for example, the 1944 proposal to scrap the 1920 Svalbard Treaty (see more in Chapter 5) and also a debacle concerning the retrieval of the black box from a crashed Soviet reconnaissance aircraft in Hopen in 1978 probably contributed to the notion that Svalbard held an important position in strategic thinking in Moscow.[6]

In the wider North Atlantic, the GIUK gap became an operational area where the emphasis was on the ability to monitor and deny the movements of Russian submarines with strategic nuclear capabilities (J. Smith, Hendrix, and Kaplan 2017). Thus, the US presence increased markedly throughout the Cold War in both Iceland and Greenland, with US bases being established at Keflavik (Iceland) and Thule (Greenland). The USA never truly left Iceland after the occupation in 1941, and by 1949, Iceland became a founding member of NATO on the condition that they did not have to establish an army. Instead, the USA established the Icelandic Defence Force in 1951, which remained in place until 2006 (Ingimundarson 2011). The USA also took over the defence of Greenland in 1941, as Denmark was occupied by Nazi Germany in 1940. Military usage of the Thule airbase to the northwest of the island was expanded, which continued after the end of World War II. By the early 1950s, Thule Air Base was constructed, and it became a crucial component of the USA's (and NATO's) strategy to monitor and possibly intercept Soviet nuclear threats.

In Canada, the North – or the Arctic – was seen as a potential liability due to its extreme vastness and limited human population. The fear was that Russian bombers in particular could quickly reach North American cities by taking the shortcut across the North Pole. Given this vast, sparsely populated, and often inaccessible region, it perhaps comes as no surprise that Canada's primary Arctic security and foreign

[5] However, Rolf Tamnes (2020: 149) argues that, during the Cold War, Norway had no intention of actively defending Svalbard in the case of outright conflict, but would rather seek to evacuate the population.

[6] Sven Holtsmark, Rolf Tamnes, and Jørgen Holten Jørgensen make a distinction between Soviet political leadership (and their top bureaucrats) and military strategic thinkers, where the latter were less concerned with Svalbard and its use. For more on the Hopen incident, see Arlov 2003: 409–10; Børresen, Gjeseth, and Tamnes 2004: 42; J. H. Jørgensen 2010: 33–34.

policy concerns came to revolve around issues of sovereignty (Huebert 1995). In the words of Rob Huebert (2011: 19–20): 'The core issue of Canadian Arctic sovereignty is control; the core issue of Canadian Arctic security is about responding to threats.'

The USA and the USSR also started developing integrated, radar-based early warning systems, with the USA establishing the Strategic Air Command in 1946, installing the Ballistic Missile Early Warning and Distant Early Warning Systems and initiating bomber missions along the North American Arctic coast to deter the Soviet Union. Alaska was in close proximity to the north-east of the USSR – albeit separated by the Bering Strait. Thus, Alaska became the location for a range of strategic US forces – not only to control possible Russian incursions by air and sea but also in terms of ballistic missiles as part of nuclear deterrent structures during the Cold War.

Based on their decision to cooperate on mutual defence – the Ogdensburg Agreement from 1940 – in response to these new challenges, the USA and Canada bilaterally established the North American Aerospace Defence Command (NORAD) in 1957 to provide joint surveillance of potential air space threats in North America. Over time, NORAD became one of the foundations of defence collaboration between the USA and Canada, with the surveillance of the maritime domain added in 2006 (Jockel and Sokolsky 2012; Charron and Fergusson 2022). However, these arrangements are what Jockel and Sokolsky deem 'remarkably informal' when framed in the broader context of North Atlantic security – namely in contrast to the formal structures of NATO (Jockel and Sokolsky 2009: 316–18). An example of this collaboration is the joint surveillance infrastructure put in place during the Cold War in the Arctic, first and foremost the Distant Early Warning line: a 63-point line of radar stations across the northern parts of Canada and Alaska, which was replaced by the North Warning System when it became obsolete (Tukker 2022).

In all of these security considerations across the Arctic, from Norway to the GIUK gap to Canada and Alaska, the USSR was the primary concern. Thus, the Arctic played a significant role as a buffer space, a front line of defence in the case of outright conflict, and perhaps most importantly, as a domain for surveillance and monitoring of Soviet activity and possible air and seaborne threats.

When the Cold War ended, the Arctic moved from a region of geopolitical rivalry and investments in military infrastructure to one where

Russia could be included in various cooperative arrangements with its former opponents. As discussed in Chapter 3, several *regional* organisations (such as the Arctic Council, the Barents Euro-Arctic Council, and others) appeared in the 1990s to deal with issues such as environmental matters, regional and local development, and cross-border cooperation (Young 2009). Although the interaction between Arctic states and Arctic peoples increased during this period, the region nevertheless disappeared from the geopolitical radar and lost its systemic or global significance. Eventually, the narrative of cooperation and stable Arctic relations came to dominate at least the foreign affairs statements from the Arctic states.

Russia's Arctic Resurgence

The Arctic holds a particularly central position in Russian security policies for several reasons. First, from an economic perspective, the region is crucial in terms of resource extraction and exports. The Russian Arctic generates approximately 10 per cent of Russia's GDP and 20 per cent of its exports (Almén and Hsiung 2022: 30). Russia is boxed along all its western and central borders, with no direct access to the world seas in those domains. Russia is less restricted geographically in the Far East, although this is not where most of the population, economic activity, or political leadership are located. Writ large, the Arctic serves as a buffer zone for the world's largest country, and it also provides unrestricted access to the North Atlantic and, to some extent, the North Pacific via the Northern Sea Route.

Since around 2005–2006 there was a marked resurgence of the emphasis placed on the strategic importance of the Arctic region for Russia. As in the Cold War, this was because Russia under the Putin regime has been committed to revamping its global militaristic and political security position, in part with an Arctic dimension (e.g., Hønneland 2016). That in turn has led to increased attention northwards. Although this book is not about the underlying motivations of Russia in the Arctic or the various challenges it faces,[7] it is relevant to highlight the new security dynamics that have emerged consequently, of which there are at least two dimensions.

[7] See instead Hønneland 2016; Wilson Rowe 2009; Konyshev and Sergunin 2014; Baev 2018; Wilson Rowe and Blakkisrud 2014; E. Buchanan 2023a; Laruelle 2014.

The *first dimension* is that in contrast to the situation during the 1990s, Russia has wanted to develop capacities in the Arctic itself to reap the benefits of resource extraction and increased Arctic shipping. The melting of the sea ice and the increased accessibility of the Russian Arctic undoubtedly pose a security problem, and Russian authorities have responded with a desire for more presence and control in their Arctic areas (Baev 2018). In some ways, as highlighted by Lackenbauer (2010), this mirrors the debate in Canada, where Canadian forces and politicians have stressed the need to increase presence and capacity in the North due to the increased accessibility of the region.

In this duality between opportunity and challenges, Russia has responded by investing heavily in various aspects of its vast Arctic area. These investments have come in everything from research stations to search and rescue and military bases along the Northern Sea Route (i.e., with the small uptake in sailings along the Northern Sea Route) (Moe 2014; Humpert 2014; Gunnarsson and Moe 2022) to the vast hydrocarbon projects on the Yamal Peninsula (Sergunin 2014). While other Arctic countries, in particular Canada, the USA, and the Kingdom of Denmark, have been slow or reluctant to push for large-scale Arctic projects, Russia has pursued these with a vigour that bears the mark of both symbolic and strategic rationales, in addition to economic (Hønneland 2016). In other words, the Arctic has become an important area of investment in terms of military capacities.

As Russia placed the Arctic higher on the domestic or national agenda, followed by investments with a defence element, the other Arctic states as well as international media focused on the potential threats emanating from these investments. Despite the Arctic 'cooperative spirit', calls for other Arctic countries to respond to Russian investments were made from the 2010s onwards – often used in various domestic contexts to focus attention on Arctic issues in that specific setting. In particular, the 'they-have-more-icebreakers-than-us' argument referring to Russia (Byers 2012; Humpert 2020) – without necessarily taking into account geography or icebreaker needs and where these are being used and what for – showcases the role of Russia in the other Arctic states' domestic Arctic capacity and investment debates (Herrmann and Hussong 2021; Østhagen 2020a, 2023a).

Similarly, notions of Russia 'militarising' the Arctic appeared around 2007 in conjunction with the flag-planting stunt on the bottom of the North Pole seabed (Parfitt 2007). However, the problem with some claims about Russian aggressive militarisation of the Arctic was that they seldom made a distinction between the different types of military capacities in which Russia was investing and the purpose and intent behind these investments (Borgerson 2009; E. Buchanan 2023b; Huebert 2013a). There was no doubt that Russia was increasing its military capability in the Arctic between 2006 and 2007 and 2022. Also, politically, the Russian leadership singled out the Arctic (Devyatkin 2023). First, in December 2014, a Joint Strategic Command was formed at the Northern Fleet headquarters. By presidential order in 2021, the Northern Fleet had the status of being one of the main military-administrative entities of the Armed Forces (Boulègue 2022).

However, a few questions arose: Was this a capability that could be used to, for example, stage attacks on the Canadian or Norwegian Arctic areas? Or were the capabilities – often consisting of small units in reopened or newly constructed bases in remote locations along the Northern Sea Route – rather intended for Russia to project presence and have the ability to monitor activity and respond to emergencies along the Northern Sea Route? Naturally, both dimensions held true to varying degrees. Still, confusion or even obfuscation was arguably done to serve a purpose, namely raising awareness of Arctic issues and drawing attention to Arctic security issues in the domestic debate in other Arctic countries.[8]

The point here is to try to further nuance the image of Russia doing only *one* thing in the Arctic. Pavel Baev, for example, showcases how Russia was actually doing different things along three rationales: realist/militaristic, institutional/cooperative, and diplomatic management (Baev 2018). Others have explored similar multifaceted approaches to unpacking Russia's Arctic actions (Konyshev and Sergunin 2014; Sergunin and Konyshev 2014; Hønneland 2016; Zysk 2013; Boulègue 2019; Sergunin and Konyshev 2017). Looking at military activity and expenditure solely, it is also relevant to note a simple divide between assets in the Barents Sea, the central Arctic region of Russia, and

[8] An example of this is the map produced by a US Senator from Alaska, Dan Sullivan, highlighting every single Russian military base in the North, but toning down all the NATO/Western military bases on the same map (Gramer 2017): https://foreignpolicy.com/2017/01/25/heres-what-russias-military-build-up-in-the-arctic-looks-like-trump-oil-military-high-north-infographic-map/.

those located close to the Bering Strait/North Pacific (J. Kjellén 2022; Boulègue 2022: 8).

In other words, if we unpack the idea of the 'militarisation' of the Arctic, at least three different strands of Russian Arctic military investments and capacity development seem to exist simultaneously:

- To establish or re-establish capacities that primarily serve the purpose of sovereignty enforcement, emergency preparedness, and domain awareness. For example, including the bases in Russia's Far East/Eastern Military District (see Figure 4.1).
- Military defence of Arctic-located important military locations (i.e., Severomorsk, the location of the Northern Fleet) and Russian northern territory where it borders NATO countries (Norway and the USA, and from 2023, Finland). Here, the 'bastion concept' still holds relevance. For example, this includes the base structures established on Franz Josef Land (J. Kjellén 2022: 41; Boulègue 2019: 36).
- Strategic competition and enhancing the Russian nuclear deterrent vis-à-vis NATO/the USA. For example, this includes the rearmament of the Northern Fleet and investments in submarine capabilities such as the new Borei-A class (Nilsen 2021c).

As such, the investments in Franz Josef Land stand out as perhaps the most obvious example of Russian militarisation of the Arctic, with a clear defence or security purpose *in* the Arctic. But the 'aggressive' nature of this can be questioned as its goal is primarily defensive – to assert control in that specific part of the Arctic. Still, these capabilities may undoubtedly be used aggressively, and they are part of a wider trend, namely increased military activity and installations in the Arctic, especially the Russian Arctic.

It is vital to distinguish between these Arctic-specific and Arctic-intended investments by Russia and the *second dimension* to the importance of the Arctic for Russian security policies, namely its strategic role. This is also what links this section on Russia to the overarching theme in this chapter, namely, the international level and the global security dynamics that have an Arctic impact or effect. As mentioned, the Arctic is not only, or even primarily, important for Russian security, but also – in the eyes of a Russian security strategist – critical to Russia's nuclear deterrence strategy vis-à-vis NATO/the USA. As outlined in the Cold War section, this is a result of Russian geography and its reliance on the Russian Northern Fleet that houses the state's strategic submarines.

Figure 4.1 Russia's main base structures in the Arctic. Source: Created by Jonas Kjellén, published by Arctic Review on Law and Politics: https://arcticreview.no/index.php/arctic/article/view/3338/6318 (CC-BY 4.0).

The Russian Navy includes four fleets: the Northern Fleet, the Pacific Fleet, the Baltic Fleet, and the Black Sea Fleet. Its nuclear-powered strategic submarines, equipped with ballistic missiles, are deployed with the Northern Fleet and the Pacific Fleet, with the Northern Fleet being the most crucial for its strategic deterrence capabilities (J. Kjellén 2022). The Pacific Fleet's base in Vilyuchinsk on the eastern seaboard of the Kamchatka Peninsula is eastern Russia's closest permanent naval base to Arctic waters. It is primarily home to nuclear-powered submarines, in addition to some surface ships.

Increasingly, naval operations in the Bering Strait and the Chukchi Sea by the Pacific Fleet are also becoming common, adding to Russia's Arctic presence, although it is still the Northern Fleet and the Barents Sea that constitute the primary strategic role for Russia in the North. Russia's strategic posture is weaker in the North Pacific than in the North Atlantic: neither Pacific Fleet assets nor Kamchatka installations are in particularly good condition (Boulègue 2022: 23), while climate change is also impacting infrastructure there (Conley *et al.* 2021). However, as we will return to in Chapter 7, there are developing security dynamics involving China that also make the Pacific Fleet a central component of Russia's great power rivalry with the West.

Moreover, in addition to submarines, some of Russia's long-distance bombers that can carry nuclear warheads are increasingly forward deployed at various bases in the Arctic as airbases and landing strips have been modernised. These play a role in showcasing Russian capabilities, reaching targets across the North Pole in North America, or in Western Europe – although some argue they pose only limited threat to North America (Lasserre and Têtu 2016).

In other words, this book distinguishes between intra-Arctic militarisation and Arctic-based militarisation, which is not intended for the Arctic. As argued by Jonas Kjellén, 'although the Northern Fleet can be considered a latent Arctic force, in reality most of its vessels are rarely, or never, employed on Arctic missions' (J. Kjellén 2022: 43). Yet this is perhaps an artificial divide, as investments in one capability (the submarines) impact the other (conventional forces in the Arctic to protect the submarines). As Kjellén puts it, regarding Russian military investments in the Arctic before the 2022 war in Ukraine, 'All these measures combined have certainly strengthened the defence of the Kola Peninsula and the provision for sustaining a Submersible Ship Ballistic Missile Nuclear "bastion" strategy and ensure that surface combatants can safely exit into the North Atlantic' (J. Kjellén 2022: 42).

Key here is again the bastion concept, as it was central during the Cold War (Olsen 2017). However, Ina Holst Pedersen Kvam argues that the current Russian force posture does not allow for a full-fledged 'bastion defence'; instead, a rather limited version where new technology has enabled Russia to operate further behind its defensive lines (Kvam 2021). The bastion itself is probably a much smaller area than the wider bastion defence (see Figure 4.2), something that has become more apparent since Russia's war in Ukraine from 2022. Kristian Åtland, Thomas Nilsen, and Torbjørn Pedersen, for example, find that after February 2022, Russian naval exercises in the North Atlantic/Barents Sea have shifted to concentrate on areas closer to the Russian Northern Fleet (Åtland, Nilsen, and Pedersen 2024). Nevertheless, the fundamental point here is that military investments and activity have increased. This is due to various rationales that we must unpack to understand so-called Arctic security practices.

Moreover, these military investments have varying degrees of relevance in different parts of the Arctic. Chapter 5 discusses the effects of the bastion when looking at one country – Norway. Based on these security practices by Russia, it seems apparent that if we truly want to understand the complexities of Arctic security dynamics and the linkages to geography, it is by including a subregional (i.e., less-than-circumpolar) lens that we get a full picture of the geopolitical situation. In this instance, that concerns the Barents Sea/North Atlantic region and the Bering Sea/Strait region where we see the effects of this increased military attention given to the Arctic, whether for strategic (offensive) or defensive purposes.

NATO's Response

It is clear that Russia's increased military emphasis on the Arctic stems both from the changing Arctic geography, which leads to increased shipping and activity, and the importance of the Arctic to the Putin regime's overall strategic plans and ambitions (Sergunin and Konyshev 2017; Todorov 2020a; Hønneland 2016). Retaining focus on the international level, or the effects of systemic competition and its Arctic-specific impact, in response to Russia, other states have looked to the North to counter the Russian presence there by increasing their

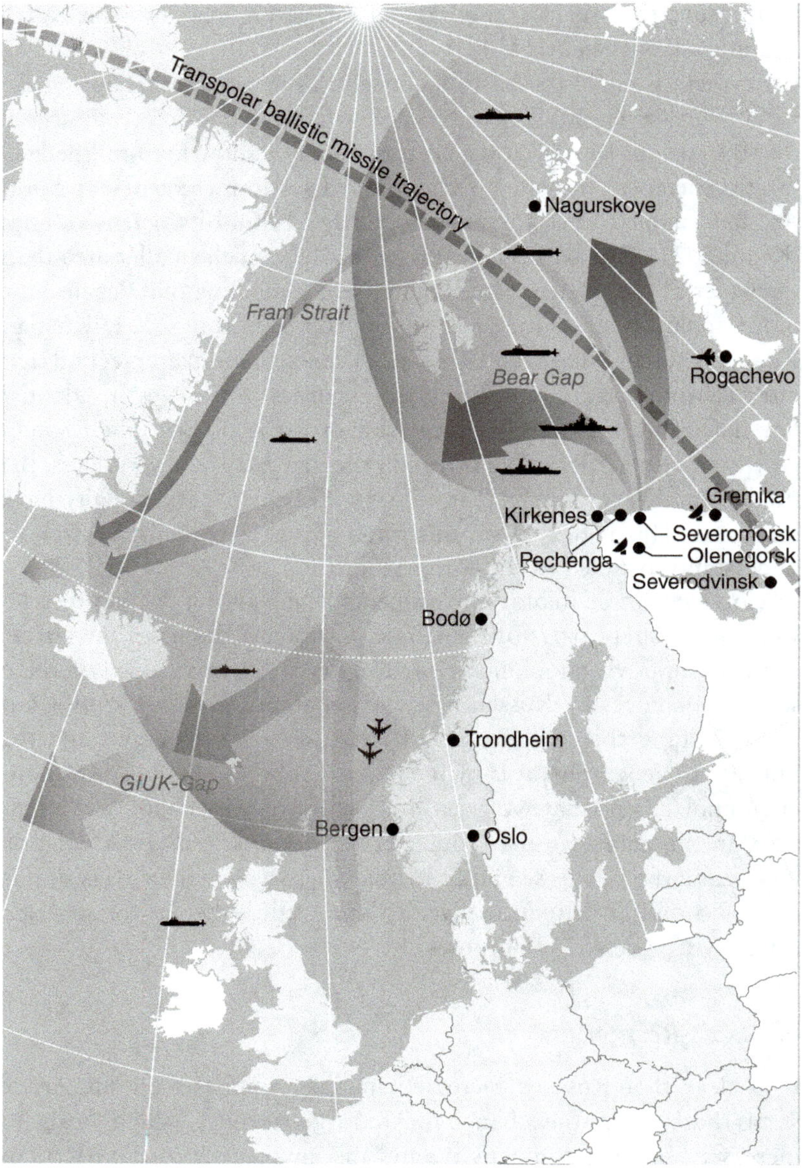

Figure 4.2 Map of naval chokepoints in the North Atlantic and European Arctic, and typical Russian military activity in this space. Source: Norwegian Defence Commission 2021 (Norwegian Ministry of Defence 2023: 8).

military presence through exercises or maritime security operations in the Barents Sea.

Initially, the much-hyped 2007 flag planting led to concerns over Russian expansionist behaviour in the North. (Although, as already discussed, the planting of a seabed flag does not achieve anything per international law.) After the speech given by Russian President Vladmir Putin at the Munich Security Conference some months earlier that same year, where he attacked the West and US policies (Government of the Russian Federation 2007), coupled with the Russian invasion of Georgia in 2008, the fear of a revisionist Russia in the Arctic started to emerge. As examined further Chapter 5, Norway took a leading role in attempting to bring NATO attention to the Arctic, or more accurately, what it deems the 'High North'.

However, at the same time, the notions of exceptionalism, cooperation, and the Arctic as a relatively peaceful domain were actively used by all Arctic countries (and academics (Käpylä and Mikkola 2019; Hoogensen Gjørv and Hodgson 2019)) to counter ideas of an Arctic resource race or scramble. On a practical level, Norway, for example, continued to conduct an annual military naval exercise with Russia – named Pomor – up until 2014 (NATO-Russia Council 2013). This duality – on the one hand countering alarmist rhetoric concerning the conflict potential in the Arctic, while on the other hand increasingly worrying about Russian increased military investments and activity in the North – characterises much of the period from 2007 to 2014 (Łuszczuk 2016).

Although some NATO countries engaged in the Norwegian-led exercise Cold Response organised every second year since 2006, Allied engagement and interest in the Arctic – or more accurately the European parts of the Arctic/the High North – had been relatively limited. An exception to this became the NATO-led air policing mission in Iceland. In 2006, the United States Armed Forces announced that it would leave Iceland and dismantle the US Iceland Defence Force operating out of Keflavik Naval Air Station. The Defence Force, established after a joint NATO decision, had been present since 1951. The Icelandic government was vocal in highlighting its disappointment with the Americans' decision (White 2006). The USA still had a stated bilateral security responsibility for Iceland (Icelandic Ministry of Foreign Affairs 2006); nonetheless, Iceland was left with a security predicament as the country does not have armed forces of its own.[9]

From 2007, Russia started showing more assertive military behaviour in the North Atlantic, particularly with regard to Cold War-style missions with Russian long-range bombers aimed at reasserting Russian long-range strike capabilities (Bases 2008). This highlighted the strategic role of Iceland in the so-called GIUK gap and as gatekeeper (geographically) to the North Atlantic and North American East Coast. For this reason, it is also used as a 'practice target' by Russian strategic long-range bombers intent on showcasing Russian military capabilities. Since 2008, NATO has thus taken over the responsibility of Iceland's defence and air policing. NATO's efforts evolve around a rotary schedule of two to three joint air policing exercises annually, each for a period of three to four weeks. A range of NATO allies have contributed to this effort, such as France, Portugal, the USA, and Germany. The Nordic allies (Norway and Denmark) have also taken a considerable share of the burden.[10]

With Russian resumption of Cold War-style military activity in the North Atlantic, combined with reinvestment in military bases and capabilities in the Arctic, the need to address NATO's relative inattention took on a greater urgency (Depledge 2020: 296). The reduced ability of NATO to conduct anti-submarine operations in the North Atlantic and High North was central to these concerns, following years of declining training and exercise activity (Gade and Hilde 2015). During the Cold War, this capability was seen as critical to protecting the Trans-Atlantic Sea Lines of Communication between Europe and North America (Olsen 2017). As explained by Duncan Depledge (2020: 296):

> Starting in 2012, NATO responded by reinvigorating its maritime exercise programme in the North Atlantic and High North. It did so by amalgamating what remained of national and small group training and

[9] Except the Icelandic Coast Guard (semi-military) and some special forces/national security teams (Icelandic Coast Guard 2016).

[10] From 2013, NATO renamed the Icelandic mission as Airborne Surveillance and Interception Capabilities to meet Iceland's Peacetime Preparedness Needs (NATO SHAPE 2023). In 2013, non-NATO members Finland and Sweden also joined the air policing mission, albeit outside of the NATO framework. Instead, they participated under the umbrella of the Nordic Defence Cooperation, together with Denmark, Norway, and Iceland (Kinnunen 2012).

> exercises into a single NATO exercise involving undersea, surface and airborne assets: DYMON [Dynamic Mongoose]. The new NATO-led exercise allowed the Alliance to begin regenerating its joint anti-submarine warfare capabilities in the North Atlantic and High North ... As such, DYMON marked the return of NATO to the North Atlantic and High North, although any link to changes unfolding in the Arctic could only be inferred by its proximity to the Greenland-Iceland-UK Gap and the focus on anti-submarine warfare.

Depledge's points further underscore the importance of *parts of* the Arctic in wider security concerns, not necessarily linked to the 'changes' in the Arctic (i.e., resource extraction, climatic changes, etc.), but rather to the wider relations between NATO countries and Russia under Vladimir Putin's leadership. Norwegian and Icelandic concerns over Russian resurgence are part of the explanation; the other part relates to Allied concern more generally over the lack of focus on the North Atlantic/High North part of the Arctic since the end of the Cold War. During Barack Obama's presidency, for example, in 2011 the US 2nd Fleet out of Norfolk with responsibilities for the North Atlantic (i.e., Russia) was *deactivated.*

However, a point central to this book, these developments and NATO member states' (slowly emerging) interests in High North security affairs were not in unison across the Arctic region. In Canada, for example, as an Arctic security interest came to the fore under the Conservative government from 2006, the focus was less on countering the Russian nuclear and naval threat, and more on domestic politics and developing adequate northern capacities. Prime Minister Harper made Arctic security an important component of his campaign and emphasised upholding Canada's sovereignty (MacAskill 2007). Throughout his years in office, the prime minister made Arctic security and Canada's sovereignty central elements of his party's brand: he made yearly visits during Operation Nanook, a 'sovereignty operation exercise' initiated in 2007 (National Defence 2024); he funded and heavily publicised efforts to find Sir John Franklin's ill-fated Arctic expedition; and he ordered Canadian scientists to revise their submission to the Commission on the Limits of the Continental Shelf to include the North Pole (Byers 2014b).

During Harper's time in office, potential threats to Canada's Arctic were highlighted. In particular, Russian bombers and fighter jets flying close to Canadian Arctic airspace were used as an example of an imminent threat to Canadian sovereignty in the North (CBC News 2014; Lackenbauer 2010; Wallin and Dallaire 2011). These flights had become infrequent following the Cold War but became more common after 2007. In response, the Conservative government increased the defence budget and initiated procurement projects, including stealth snowmobiles, Arctic drones, and Arctic offshore patrol vessels (Huebert 2013a, 2011). The subsequent drop in oil prices compounded the financial strain on the government, however, and several of the Arctic projects were put on hold or scaled back (Cox and Speer 2016: 4). This back-pedalling on Arctic investment led some to label Harper's approach as 'all talk, no action' (Byers 2014a).

Others, however, debated the credibility of the threat posed by Russian aircraft flying close to Canadian Arctic airspace. A comparative study of Russian flight behaviour by Frédéric Lasserre and Pierre-Louis Têtu found that those aircraft flying near the North American Arctic did not violate Canadian airspace; that the number of interceptions in the North American Arctic was quite small relative to areas such as the Baltic Sea, the Sea of Japan, or the North Atlantic; and that the behaviour displayed by Russian planes was less aggressive in the Arctic (Lasserre and Têtu 2016, 320–21).

Many argued instead that a more pressing concern in the Canadian North was the persistent social and economic problems – ranging from food security to vulnerability to a changing climate – facing Canada's northern communities (Perreault 2011; Furgal and Seguin 2006; Power 2008). If anything, the soft security tasks, or safety-related concerns, were more apparent in the Canadian and wider North American Arctic despite Russia's military resurgence from 2006. For example, investing in the Canadian Rangers or the Canadian Coast Guard was argued by some to be a more accurate response to Arctic security concerns (Lackenbauer 2013a; Mitchell 2013).

Even the concerns posed by the outstanding boundary dispute with the USA and the status of the Northwest Passage were relatively minor when considered in the context of the wider relationship between Canada and its neighbours (F. Griffiths 2011a: 10–11, 2011b). From 2006 to 2015, the Conservative government's strong rhetoric thus stood at odds with the lack of credible military security threats in the Canadian Arctic. Had the threats been more pressing, it is unlikely that the proposed investments in Arctic security would have been scaled back as significantly as they were, or that the government in Ottawa would have resisted a more pronounced role for NATO in the Arctic (Østhagen, Sharp, and Hilde 2018). As Whitney Lackenbauer (2011a: 73) surmised it: 'Anxiety about "using or losing" our Arctic inheritance is more revealing of the Canadian psyche ... than of objective realities.'

These developments in the North American and Canadian Arctic stood, at the time, in sharp contrast with the European Arctic. The Russian annexation of Crimea in March 2014 and ensuing conflict in Eastern Ukraine showcased the willingness of the Russian regime to use force and its disregard for international law. These events had an Arctic impact. As Depledge highlights in his article 'Train Where You Expect to Fight', the number of military exercises in the Arctic and the number of NATO allies that partake in these exercises increased:

> the intensity of exercises in and around the High North has increased since 2006: from one exercise every one-two years to two or three exercises each year, peaking with four in 2019. While this may not look dramatic, it implies an increased military presence and a higher tempo of activities involving non-Arctic states, which not only raises the costs for those involved, but also ties up capabilities that might be needed for military operations elsewhere in the world (Depledge 2020: 294).

A feature in this phase (2014–2022) of military response to Russian actions and activity was the NATO-led so-called high visibility exercise *Trident Juncture* in 2018. It became the pinnacle of Allied attention to the High North, although Norway worked actively to dampen antagonistic elements of the exercise, and most of it took part in the southern parts of Norway, namely below the Arctic Circle (Østhagen 2023b). Russia in turn responded with an increasing amount of military exercises – both scheduled and so-called snap exercises intended to throw Norway/NATO members in the High North off balance (Nilsen 2021b, 2022; Danilov 2020).

In fact, in an analysis of where and when Russia issued notice to airmen (NOTAM) warnings in international waters and airspace in the High North in the period between 2015 and 2021, Åtland, Nilsen, and Pedersen (2022) found that the number of NOTAMs steadily increased to a peak in 2019 and subsequently dropped in 2020 and 2021. This drop could be explained by the onset of the COVID-19 pandemic and its effects on military activity throughout the world. They also found a correlation between Russian NOTAMs and military exercises involving a NATO component in 2018, 2019, and 2021, suggesting that Russia most likely intended to disrupt these exercises.[11]

With Russia's full-scale invasion of Ukraine in February 2022, the security environment in the Arctic became even more tense, but these fundamental dynamics and practices did not change. In an updated study, Åtland, Nilsen, and Pedersen (2024) found that 'Russian NOTAM events in the High North now take place in the Barents Sea, rather than in the Norwegian Sea. The primary purpose of the activity now seems to be to bolster the Northern Fleet's bastion defense, rather than to send political signals.' Hopes of restarting security dialogue in the North to reduce the tension that emerged around 2019–2020 were dashed, and further sanctions on Russia were put in place.

Finland's and Sweden's subsequent decisions to join NATO in 2022 further solidified the divisions and spillover of tensions to the North. Moreover, not only has NATO's role in the Arctic changed with seven member states in the North (see Figure 4.3); NORAD modernisation has also received fresh attention from Canadian and US decision-makers (Tukker 2022), with pledges of considerable spending over the next decades (Coates *et al.* 2024: 21). These dynamics are further discussed in Chapter 7, where we examine the ongoing developments in the Arctic security subregions.

[11] Åtland, Nilsen, and Pedersen (2022) showcase an interesting discussion related to these Russian NOTAMs and the exercise Trident Juncture in 2018. US Navy Admiral James Foggo – who commanded the NATO exercise – later talked about these NOTAMs:

> This all came to a head during Trident Juncture, when the Russians … designed what we call a notice to mariners or a closure area just outside territorial waters from Norway. A closure area is an intent to conduct an exercise, so they overlaid an exercise on top of our big exercise. Very ungentlemanly thing to do. And so, the Norwegians were a little upset about this. They hadn't ever seen this before, that close to their waters, with Admiral Bruun Hansen and his joint staff, they said, Admiral Foggo, what do you want to do? And I said, well, those are international waters. We do the same thing. And quite frankly, unless somebody is launching a missile or doing some kind of an event that might cause an accident or collision, I'm gonna drive right through it. We're going to continue to do this exercise and we are going to project power and presence and keep those lines of communication open. So that's exactly what we did (The General and the Ambassador 2021; Åtland, Nilsen, and Pedersen 2022: 74).

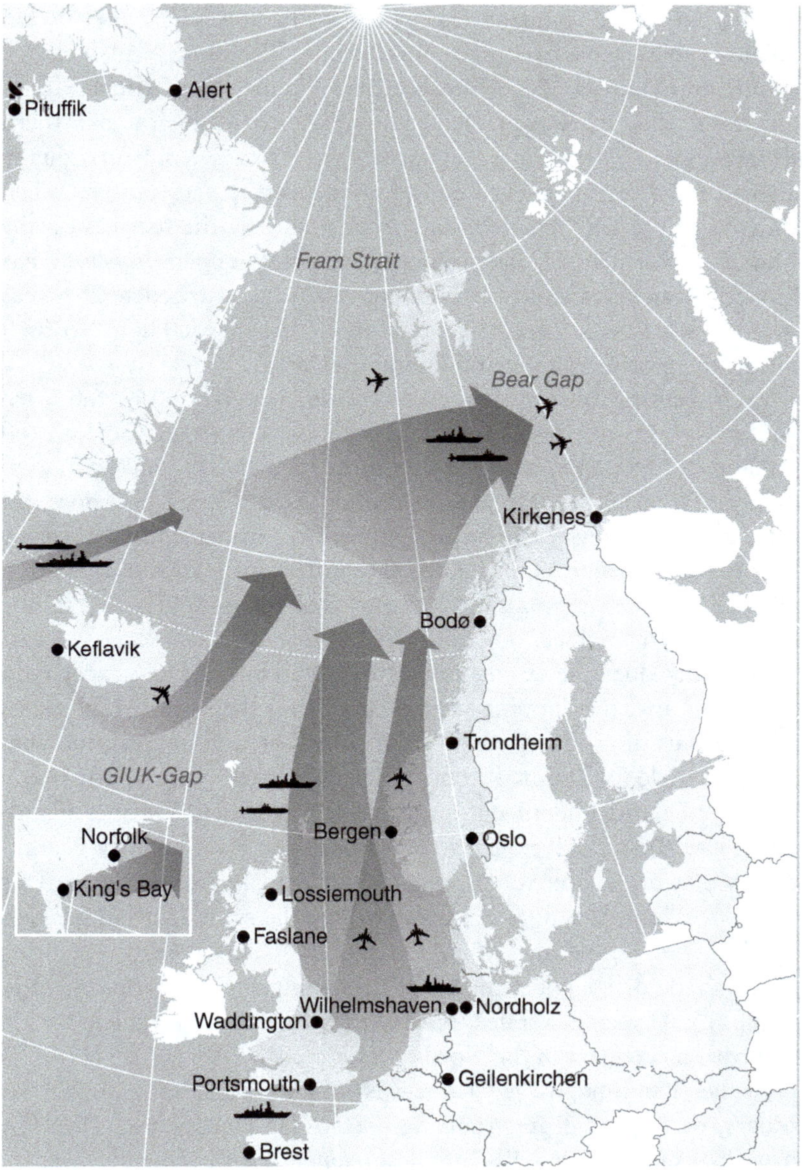

Figure 4.3 Map of naval chokepoints in the North Atlantic and European Arctic, and allied military activity in this space aimed at countering Russian activity. Source: Norwegian Defence Commission 2021 (Norwegian Ministry of Defence 2023: 9).

With all Arctic countries except Russia having joined the NATO alliance, there have also been increasing calls for a NATO Arctic command (Mottola 2023; Jonsson 2023; Palmer and Gosnell 2024). However, this leads to a challenge since, in reality, the Northern European and North American parts of the Arctic are only marginally aligned in terms of defence concerns and threat perceptions. When Canada or the USA talks about 'Arctic security', the focus is on the Canadian North and Alaska respectively. What happens in Alaska has limited, if any, relevance to Finland, Norway, or Sweden. Whereas when Norway, for example, talks about the Arctic/High North it means the North Atlantic and Barents Sea areas.

Even security concerns related to Russia, as perceived by the seven Arctic NATO members, do not completely converge. Canada and Finland, for example, have varying threat perceptions and operate in different security geographies, even after 2022. Russia also does not treat the Arctic seven as one coherent block. Instead, Russia sees its vast Arctic coastline as segmented into different theatres of operation (J. Kjellén 2022), the most prominent ones being the North Atlantic and the North Pacific.

Not only does the Arctic consist of different security spaces with limited, if any, overlapping security practices; the seven Arctic states that are part of NATO have diverging views of, and interests in, what NATO can do for national security. The Nordic countries, particularly the three Fennoscandian states that share a Northern European landmass with Russia, have viewed NATO as their main security guarantor, as they are unable to ensure adequate security on their own. This guarantee, however, does not come from policies set forth in Brussels by NATO HQ or because of NATO's Allied Reaction Force – it comes in the shape of US commitment to upholding the territorial integrity of all NATO member states. NATO thus becomes the vehicle for US security engagement in the Nordics (Hilde 2019).

At the same time, we see that this Nordic engagement with the USA occurs, to a great degree, outside the NATO framework. The USA has, for example, signed bilateral agreements with Finland, Sweden, and Norway as a basis for enhanced defence cooperation. The USA actively conducts joint exercises with the Nordic states in the European Arctic, alongside other states such as the UK and France. This is in stark contrast to what happens in the North American Arctic – there, the USA and Canada manage their own security concerns. There are undoubtedly concerns over outdated technology and the need to modernise Cold-War infrastructure and policies (Charron and Fergusson 2022). However, Canada and the USA do not need or want German, Swedish, or Turkish support to manage these issues. They have their own bilateral defence arrangement through NORAD, and NATO is not deemed necessary in their Arctic spaces.

In fact, the term High North has *deliberately* been used by NATO to underscore this divergence when it comes to the Arctic. Canada has been reluctant to use the 'Arctic' in NATO summit statements, whereas Norway has wanted NATO attention to go further north since 2008. The solution was NATO emphasis on deterring Russia in the High North, namely what Norway terms *its* Arctic areas (Østhagen, Sharp, and Hilde 2018). Some argue that the Canadian position has shifted, partly showcased by former NATO Secretary General Jens Stoltenberg's travel to the Canadian Arctic in 2022 (Brewster 2022). That does not discount the fact that within NATO the High North means something different to the Arctic. Whereas the former has been prioritised as an area that needs attention, the latter is often avoided in official statements or linked to wider concerns over climate change and general Russian military build-up.

Finally, there has also been a tendency to ascribe too much agency to NATO as an organisation in Arctic security debates. NATO HQ in Brussels has limited authority and cannot force its members to enact policy. NATO is the sum of its members, but its members' actions cannot be equated with the whole defence alliance. However, in descriptions of Arctic security and geopolitics, NATO is often referred to as conducting independent activity in the North, or we hear statements about NATO militarising the Arctic (Little 2024). This is especially prevalent in the Russian narrative (Wilhelmsen and Hjermann 2022), as a sign of how NATO is aggressive and expanding, whereas Russia is the victim – also in the Arctic.

Currently, the sole continuous NATO operation in the Arctic is the Air Policing mission conducted over Iceland, as that country lacks its own air force. The only NATO-led exercise regularly held in the Arctic is Dynamic Mongoose, an annual anti-submarine exercise in the North Atlantic. Other exercises are one-offs, such as the 2018 Trident Juncture – a high visibility exercise in response to requests from member states such as Norway for increased NATO attention and presence in the North. Numerous Western-state-led military exercises occur in the region.[12] However, these are not NATO-led initiatives. While they may involve a NATO component, such as support systems or specific NATO reaction forces, they are organised and conducted by individual countries in response to what they perceive is needed for their own security policies. Despite Russian claims to the contrary, these activities do not constitute NATO exercising in, operating in, or 'militarising' the Arctic. Rather, they reflect the preferences and activities of individual states that are also NATO members. These preferences may sometimes even diverge from those of other NATO members or involve only relevant regional coalitions.

However, there might be room for more sharing of knowledge, competence, and threat perceptions across the North Pole within the NATO alliance. This is also fundamentally the point made in many of the pieces calling for an Arctic Command, namely better situational awareness of what happens in the Arctic and better coordination amongst Arctic states. The solution to this might already partly be in place: in 2022, the Nordic Chiefs of Defence recommended that all Nordic states join the new Joint Force Command Norfolk to enhance their coordination (Gaasland 2024), especially with the USA. Meanwhile, the USA and Canada maintain their joint NORAD command in Colorado Springs.

The point here is not to argue that the Arctic is not being militarised. It has been militarised during the Cold War and it started anew around 2007. The initial driver for this, both then and more recently, has always been Russia's focus on developing its capacities on the Kola Peninsula due to its strategic importance – not because of ice melting or resources, but because it gives Russia a sheltered platform to operate its nuclear triad from, with relatively open access to Western Europe via the North Atlantic, and the North American continent writ large. Thus, for security analysts trying to understand Arctic developments, knowing where and what 'Severomorsk' is arguably matters more than knowing where Arctic shipping lanes might traverse or what resources the Arctic holds.

[12] Such as Cold/Nordic Response (in North Norway/Fennoscandia), Operation Nanook (in Canada), Northern Viking (in Iceland), Arctic Challenge (in Fennoscandia), Formidable Shield (in the North Atlantic), Joint Warrior (around the UK), Flotex (off the coast of Norway), and Arctic Edge (in northern Canada and Alaska).

Still, it is not that intra-Arctic issues are irrelevant for Arctic security practices. However, global, or at least continental (i.e., European), security dynamics are the primary cause for the Russian military build-up in the European Arctic, which in turn has led to a response from initially Norway and eventually other NATO member states such as the USA, the UK, France, and also Finland and Sweden. This in turn has caused a security dilemma in the European Arctic (e.g., Åtland 2014; Sørensen 2019; Knutsen and Pedersen 2024), where one actor's security practices to increase security create a spiral of subsequent actions raising security concerns for all the actors involved. The Arctic, or, more accurately, the European parts of the Arctic, has thus undoubtedly been increasingly militarised over the last decades. This did not, however, start in 2014, or in 2022, but before that, in the early years of the Putin regime. Moreover, it is not NATO that has led the counter-reaction and the defence response since then, it is individual NATO member states in the Arctic.

The Arctic's Role in Global Power Relations

When examining the international or systemic factors related to international affairs and security that have an impact on Arctic security practices, the role of global power rivalry is paramount. As outlined, this was central during the Cold War, due to the role of the Arctic as a buffer zone between the two superpowers. In contrast to during the Cold War, China has also emerged as a security actor in the Arctic – due to both concrete functional interests and the wider systemic rivalry between China and the USA (or the 'West').

China Emerges

From a systemic perspective, it is generally accepted that Russia is not the primary competitor vis-à-vis the West on a global level, due to Russia's limited economy, declining population, and relative lack of innovation in society (Mearsheimer 2010; Tunsjø 2018). Instead, China's global rise in recent decades has prompted what is deemed a bipolar world order or, perhaps more accurately, if we also factor in India, Russia, the EU, and other emerging powers – a multipolar one. Of these non-Arctic actors, China is obviously the most central one due to its economic size and subsequent rise as a military power.

When China asserts its influence on the world stage, the Arctic is one of many regions where its presence and interactions are components in its expansion of power through scientific research or investments in Russia's fossil fuel industries (e.g., L. Guo and Wilson 2020; Edstrøm, Stensdal, and Heggelund 2020; Koivurova and Kopra 2020; Brady 2017).

Initially, as the Arctic was attracting attention due to economic opportunities and the effects of climate change, China's interests were described primarily through an economic lens (Lim 2018), with an emphasis on shipping along the Northern Sea Route (Jakobson 2010). In the early stages of China's Arctic interest, its link to polar research and the increasingly apparent effects of climate change along the Chinese coastal areas were furthermore highlighted as additional drivers for it (Stensdal 2014). China also applied for an observer role in the Arctic Council, being accepted alongside Italy, Japan, India, South Korea, and Singapore at the Kiruna Ministerial Meeting in 2013. The deterioration in relations between the West and Russia after March 2014 further contributed to concerns over increased bilateral cooperation between Russia and China, and as a result, Russia seemed to allow China to play a larger part in the Arctic through those relations, from economic investments to science cooperation (Bertelsen and Gallucci 2016). Still, as many have highlighted, Russia has also been wary of giving China too much influence in this strategically and economically sensitive area (Røseth 2014; Moe 2021; Stensdal and Heggelund 2024).

China released its first and to date only Arctic policy in January 2018 (State Council of the People's Republic of China 2018). In it, China describes itself as a 'near-Arctic state', which some have interpreted as having not only the right to get involved but also a duty to do so (Brady 2017). The near-Arctic terminology was used for the first time in official Arctic documents not by China, but by the UK in 2013–2014.[13] The Chinese policy paper does not reveal anything noteworthy. As with most Arctic policy or strategy documents, it highlights the importance of the Arctic, the need for research, environmental and marine protection, and the legal frameworks governing the region:

[13] The 2013 UK policy states: 'The United Kingdom is not an Arctic State, but we are the Arctic's nearest neighbour' (UK Government 2013: ii). The UK Parliament's Select Committee on the Arctic defines the 'UK as a near-Arctic state' (UK Parliament 2015).

> China maintains that all activities to explore and utilize the Arctic should abide by treaties such as the UNCLOS and the Spitsbergen Treaty as well as general international law, respect the laws of the Arctic States, and proceed in a sustainable way on the condition of properly protecting the eco-environment of the Arctic and respecting the interests and concerns of the indigenous peoples in the region. (State Council of the People's Republic of China 2018)

Perhaps one interesting thing – apart from China stating all the 'correct' things about Arctic climate, resource development, and governance – is how frequently the Spitsbergen Treaty (now usually known as the Svalbard Treaty) is mentioned. This book offers a deeper discussion of Svalbard and geopolitics in Chapter 5; however, it is noteworthy that the Treaty is mentioned with reference to the rights China has in the Arctic.[14] This must be, in part, read as a way to legitimise China's Arctic interest and engagement, as an outsider to the region, and used as a starting point for Arctic engagement: 'China has long been involved in Arctic affairs. In 1925, China joined the Spitsbergen Treaty and started to participate in addressing Arctic affairs' (State Council of the People's Republic of China 2018). In fact, China was amongst the first countries to ratify the Treaty – albeit it was not part of the negotiations in Versailles in 1919–1920.

An issue that has drawn particular attention to China's interests in Svalbard has been Norwegian efforts through the Norwegian Polar Institute to better coordinate research in Ny-Ålesund (Norwegian Research Council 2019), which is a small research settlement on the island of Spitsbergen. In Ny-Ålesund, ten different national polar institutes have established their research presence, starting with the gradual establishment of the Norwegian presence in the 1960s as coal

[14] As stated in the policy document:

> States from outside the Arctic region do not have territorial sovereignty in the Arctic, but they do have rights in respect of scientific research, navigation, overflight, fishing, laying of submarine cables and pipelines in the high seas and other relevant sea areas in the Arctic Ocean, and rights to resource exploration and exploitation in the Area, pursuant to treaties such as UNCLOS and general international law. In addition, Contracting Parties to the Spitsbergen Treaty enjoy the liberty of access and entry to certain areas of the Arctic, the right under conditions of equality and, in accordance with law, to the exercise and practice of scientific research, production and commercial activities such as hunting, fishing, and mining in these areas (State Council of the People's Republic of China 2018).

mining there was declining, and the most recent one being the Indian presence through the Indian National Centre for Antarctic & Ocean Research (Moe 2020: 118).[15] Research access and activity is not explicitly tackled in the Svalbard Treaty from 1920,[16] but Norway's policy up until the 1990s was one of treating this type of activity as on par with other types of activity covered by the Treaty, given that any national of the signatory states would have free access to the archipelago (Moe 2020: 120).

Over the two past decades, Norwegian authorities have wanted better control over research activity and to counter the impression of third-party states that they have their own independent research stations based in Svalbard (as they have, for example, in Antarctica) (Hanger 2018; Moe 2020: 124–25). At the same time, a goal has been to increase international collaboration in Ny-Ålesund to improve science quality, economies of scale, and environmental impact (Norwegian Research Council 2019). This came to the fore with a government White Paper on Svalbard in 2016 (Norwegian Ministry of Justice and Public Security 2016) and a new strategy for research and education on Svalbard in 2018 (Norwegian Government 2018). In the words of the director of the Norwegian Research Council, Norway wanted to 'underscore that Svalbard is Norwegian' (Hanger 2018).[17] Tiril Vold Hansen showcases how these efforts by the Norwegian government were indeed undertaken to support Norwegian sovereignty on Svalbard, but there were also idealistic and practical motivations and interests behind the policy changes (Vold Hansen 2024).

The response from some research institutions, such as the British Natural Environment Research Council, was negative, indicating disappointment and fears that this coordination effort could undermine the British symbolic presence in the Arctic (Hanger 2018; Moe 2020: 128). However, it was the Chinese response that attracted the most attention. Norwegian efforts to improve research coordination in 2018 led

[15] The others in addition to those of Norway (1960s) and India (2008) are research institutes from Germany (1990), Japan (1991), United Kingdom (1991), the Netherlands (1995), Italy (1997), France (1999), South Korea (2002), and China (2004). It is important to note that it is not a state that establishes a research 'station' or presence, but rather its various research institutes.

[16] For a debate about whether research activity is covered by the Treaty or not, see (Pedersen 2021; Ulfstein 2021; Molenaar 2022).

[17] Author's translation.

to Chinese research actors expressing concerns over whether Norway was extending its reach too far (N. Liu 2021: 4; Hanger 2019). In an official note from the Chinese Arctic and Antarctic Administration, it was argued that these matters should have been decided through an international process and that they violate the principle of equal access and right to certain activities – despite acknowledging that research does not constitute any of those (Moe 2020: 128–29).

As Arild Moe highlights (Moe 2020: 129), Norwegian leniency since the start of the international research presence in Svalbard might have been to blame for these counter-reactions, as the idea of Svalbard research as an international 'common good' had taken root in public discourse. This mirrors Rolf Tamnes's argument (Tamnes 2020: 140, 147) about Norway's passive role when it came to governance of Svalbard in the first decades after the Treaty came into force, which in turn might have left an impression amongst other states' citizens that they have special privileges in Svalbard.[18]

We return to Svalbard and security practices there in Chapter 5. Here it is used as an example of a concrete, or functional, Chinese interest in the Arctic, which the Chinese government apparently also places a certain degree of weight on in order to legitimise its Arctic engagement. Other examples range from Chinese investments in the Yamal LNG-project in Russia to research interests in Iceland or mining interests in Greenland. However, that general engagement and interest into the Arctic policy realm has elicited counter-reactions, especially in the USA.

Reactions to China in the Arctic

China's Arctic interests have led, as per a systemic or global outlook, to the Arctic's becoming relevant in the increasing global power competition between China and the USA. In 2019, US Secretary of State

[18] A related and interesting point is how to explain *why* Svalbard was left alone by the authorities in Oslo and deemed important only by a small group of security analysts concerned with Soviet military plans. Perhaps, as explored by, for example, Ingrid Medby (2014) and Medby and Beate Steinveg (Steinveg and Medby 2020) in a wider Arctic context, the answer lies in Norway's relationship to its Arctic 'identity' (or lack thereof), and Svalbard's distance from the rest of Norway – geographically, but also politically, economically, and psychologically.

Mike Pompeo gave a speech in Rovaniemi, Finland, just before an Arctic Council Ministerial Meeting, where he lambasted both Russia and China for their 'aggressive' actions in the Arctic. China was obviously the main target in this speech: 'Beijing claims to be a "Near-Arctic State," yet the shortest distance between China and the Arctic is 900 miles. There are only Arctic states and non-Arctic states. No third category exists, and claiming otherwise entitles China to exactly nothing' (US Department of State 2019).

The fact that the speech was held just in advance of the Arctic Council Ministerial meeting, where the main keywords were 'cooperation', 'peaceful', and 'sustainable', caused concern in the Arctic community over where the Arctic Council, and Arctic cooperation at large, was going (Koivurova 2019). China also had, and still has, a seat around the table, being granted observer status to the Arctic Council in 2013. However, this speech was perhaps the final blow to the efforts made by all Arctic states – including the USA and Russia – to focus primarily on the region's cooperative traits and downplay, or ignore, the parallel increase in military activity and build-up.[19]

This was also quite in line with the Trump administration's 'America First' foreign policy (Council on Foreign Relations 2024). The 2019 warning about Beijing's Arctic interests highlighted how the USA saw the Arctic as yet another arena in which the new systemic competition between the two countries was sharpening (US Department of State 2019).[20] Arguably, this was linked to a lesser extent to Chinese actions *in* the Arctic; it was more about the United States wanting to stop China's global growth in as many areas as possible (Østhagen 2021b). However, questions about China–Russia cooperation in the Arctic and the effects this could have on regional tension are increasingly on the agenda after the sanctions placed on Russia in 2022.

[19] In the EU-focused Arctic Futures Symposium in Brussels in November 2019, the author was the chair of a session discussing security and safety in the Arctic. The statements made by Arctic Security Officer Kathleen Larkin, from the US State Department, made clear that there had been a shift in the official line from the USA concerning Arctic security. The tone concerning China and Russia was much harsher, calling out aggressive behaviour and the need to combat these. See Arctic Futures Symposium 2024.

[20] Other political leaders followed suit – the most extreme might perhaps be the French defence minister in 2019, Florence Parly, quoting Michel Rocard saying that 'The Arctic? It's a second Middle East!' (French Ministry of Armed Forces 2019: 3).

With its harsh tone and the rather inaccurate comparisons to the South China Sea,[21] the 2019 speech also seems to have led to, or been part of, a resurgence of the 'new-cold-war-conflict-is-imminent'-style articles from large media outlets. While the first wave of such stories in 2007–2010 focused on Russia, climate change, and resource conflicts (for an analysis, see Nilsson and Christensen 2019; Padrtova 2019), this time, the main antagonist was not only Russia, but increasingly China.[22]

We can question what China has done in the Arctic to warrant this focus. Some of the Chinese economic interest had already disappeared with the fall in commodity prices in 2014 (Koivurova and Kopra 2020: 26). By 2024, Yamal LNG represented the majority of Chinese maritime activity, in that there are Chinese ownership interests in fourteen of the fifteen ice-breaking LNG transport vessels delivered from South Korea (Moe and Stokke 2019). Nevertheless, as Christopher Weidacher Hsiung (2016) as well as Oscar Almén and Hsiung (2022) point out, Chinese interest in economic projects in the Arctic is modest, and the engagement in both the Nordic countries and – to some extent – Russia is limited. Moreover, '[m]ajor obstacles to further development relate not only to the harsh physical environment that the Arctic represents, but also to the response and willingness of Arctic states to welcome a greater Chinese economic presence' (Almén and Hsiung 2022: 92). As Anders Edstrøm, Iselin Stensdal, and Gørild Heggelund also argue (2023):

> In particular, transport and energy are areas where China has expressed interest. Nevertheless, activity in both areas is relatively low. At present, we see few concrete Chinese Arctic projects, and we have also seen the sidelining of projects due to cost levels. Whether the number of projects is due to low interest, a poor commercial basis under current conditions, or other factors is difficult to conclude with certainty.

[21] Pompeo stated: 'Do We Want the Arctic Ocean to Transform into a New South China Sea, Fraught with Militarization and Competing Territorial Claims?' (US Department of State 2019).

[22] See, for example, 'Arctic Rivalry May Shape a New Cold War over Resources' (Anstey 2022), 'The Return of Great Power Competition to the Arctic' (Saxena 2020), 'Freezing Cold War: Militaries Move in as Arctic Ice Retreats' (Luhn 2020), and 'A Battle for the Arctic Is Underway. And the US Is Already Behind' (K. R. Rosen 2022).

It seems clear that the role of China in the Arctic is far more complex than Pompeo's 2019 speech made it out to be (Fravel, Lavelle, and Odgaard 2022). The effects of the Russian invasion of Ukraine in 2022 have also impacted Chinese–Russian cooperation in the Arctic (and in general) (Hsiung 2022). This has a circumpolar/Arctic and a subregional/North Pacific effect, returned to in Chapters 6 and 7.

China's interests do not necessarily align with those of Western Arctic states, although when it comes to specific economic projects in the Arctic, Chinese investments and capital might even be sought after (Gåsemyr and Sverdrup-Thygeson 2017). At the same time, China's interest in Arctic-specific investments and activity have spurred Arctic states into blocking possible activity, or purchasing land, rights, or companies, in advance (Edstrøm et al. 2025). Examples of this includes fears of Chinese mining activity in Greenland (Andersson, Zeuthen, and Kalvig 2018; Jacobsen and Lyngholm Lindbjerg 2024) and fears of Chinese acquisition of land on Svalbard (Moe 2020; Østhagen 2024).

However, we must see China's interests and expansion in the Arctic as part of the global expansion of the 'new superpower' (Edstrøm, Stensdal, and Heggelund 2023), with all the challenges that entails for not only the 'old superpower' – the USA – but also the various European countries. For example, NATO has been increasingly warning that China is a competitor and rival to the West, with an Arctic component. In early 2023, NATO Secretary General Jens Stoltenberg warned about becoming too dependent on China for economic purposes and of increasing Russia–China military cooperation (NATO 2023a). Chinese 'spy balloons' notwithstanding (Chotiner 2023), concrete military expressions in the Arctic of China's interests are few and far between (Lajeunesse and Choi 2022).

The points made here regarding China and the USA showcase the core difference between a *regional* look at security relations, especially as showcased at the end of Chapter 3 when discussing security cooperation and dialogue with Russia, and the security impact of *global* affairs on Arctic security relations. In the Arctic, Russia is undoubtedly the dominant actor along a range of parameters (population, size of territory, military investments, or – as used by some – number of icebreakers).[23] However, globally, China is the

[23] As explained by the author in a previous piece (Østhagen 2023a) published by *The Arctic Institute*:

competitor – or adversary – that the USA (and the West, although that varies) perceives as the more severe threat in the long term.

Reactions to China in the Arctic also showcase another thing, namely the difference between the strategic or systemic dimensions as already outlined when it comes to Russia's military posturing and strategic positioning in the Arctic and China's Arctic interests. Russia's dependence on parts of the Arctic for its security strategies and nuclear deterrent triad results in military investments *in* the Arctic. However, China has an outside-in approach to the Arctic, increasingly displaying Arctic capabilities – such as its three polar class icebreakers (Humpert 2024b; Staalesen 2024) – and highlighting Arctic investments but nevertheless remaining firmly on the outside of the region in terms of sovereign territory and sovereign rights (Lamazhapov 2024).

Nevertheless, there is one part of the Arctic that is seeing increased Chinese activity beyond symbolic statements, namely the Bering Sea/North Pacific region. While examining Russian military activity in the Arctic and unpacking Chinese security interests and military activity, the circumpolar or 'pan-Arctic' labels are too loose to accurately fit.

Arctic-related stories love to count how many icebreakers the Arctic countries have. This is often done to showcase how Russia is dominating in the Arctic and the USA is a laggard. The main problem here is that icebreakers are contextual. A range of different factors explain why Russia has the number of icebreakers it has: the size of its Arctic coastline, economic activity in that area, and ice conditions (which is why you have a breaker-of-ice in the first place). This is beside the intra-institutional set-ups of each Arctic coastal state concerning coast guard/navy tasks and whether icebreakers are even useful for the latter. Do not get me wrong. There is no doubt that the USA is in need of more icebreakers. It currently has only two operational ones: the USCGC Healy from 1999 and the USCGC Polar Star from 1976. Moreover, icebreakers are important for a number of reasons, including their ability to support ship traffic, keep harbours open in wintertime, and conduct research at high latitudes. However, icebreakers are not the ultimate sign of strength or military power in the Arctic. Other parameters, such as military expenditure, presence and capabilities (beyond breaking ice), are better suited. In most instances, icebreakers are not even operated by the various Arctic countries' armed forces. Why does Finland have four times as many as the USA when it is not even an Arctic coastal state? Because the country needs to break ice in the Gulf of Bothnia and is attempting to reap commercial benefits from icebreaking services. As such, the numbers of icebreakers do not demonstrate the Russian Arctic threat – there are other things that are more relevant for that, ranging from military capacities (such as subsea or surface naval vessels, or airborne capabilities) to patterns of behaviour (such as military exercises, troop posture, statements).

Instead, we must also examine the subregional level, focusing on – in the case of China – what is happening in the North Pacific. In comparison to the dynamics that take place in Barents Sea/North Atlantic, this will be discussed in Chapters 5 and 7.

Other Non-Arctic States: The Rise of India

There are more non-Arctic actors engaged in, or at least showing an interest in, Arctic affairs other than China. These include the EU, France, Germany, India, Italy, Japan, Poland, Singapore, South Korea, Turkey, the UAE, the UK, and others. Yet keeping an eye on *security* dynamics, few of them tend to have much impact on security dynamics (even loosely defined to include national interests, governance issues, and cooperative elements). Often, a different set of dynamics explain their engagement: (1) either with functional interests in the Arctic related to research presence (such as being part of the Norwegian-led research station in Ny-Ålesund in Svalbard) or economic investments (such as Chinese investments in Yamal-LNG); or (2) an interest in stating an Arctic interest for partly symbolic or immaterial interests, as engaging in Arctic discussions and Arctic governance is increasingly seen as a staple of global politics and a state's global aspirations (Ayaydin 2022; Steinveg 2017; Lamazhapov 2024).

Although these strands have the potential to impact Arctic developments and related politics, they arguably have limited impact on specific security dynamics, unless the efforts involved are considerable. Increased intangible interests by non-Arctic actors might also spur Arctic states to intensify their rhetoric concerning current governance mechanisms, or initiate new efforts, as described in Chapter 3. For security-specific actions, however, is has been European non-Arctic states such as the UK and France that have engaged militarily the most *in* the region, although this tends to take place in a subregional context. Keeping a focus on the international or systemic dynamics emanating from outside of the Arctic, that might have long-term Arctic security implications, one other actor is increasingly relevant to consider, namely India.

India has been active in Arctic affairs with, for example, its research presence in Svalbard through the Himadri station since 2008, and in 2022 India released an Arctic Policy (Government of India 2022). However, only in recent years with the deterioration of security relations between Russia and the West, alongside China's increased engagement in the Arctic through its cooperation with Russia, has India begun to show an interest in the region's security and strategic dynamics (Sharma 2024).

Starting with the operational perspective, it is obviously difficult for India to mount much of a military or security presence in the different parts of the Arctic. India does not belong to the NATO alliance, nor does it figure in an outright security or defence alliance with Russia, meaning that it so far has not partaken in military exercises held by either actor above the Arctic Circle. Still, Uttam Sinha, for example, muses that increased military cooperation between Russia and India might have the 'potential to grant India access to Russian naval port facilities in the Arctic and thereby add to its overall strategic reach' (Rishikesh 2021). Italy, albeit located far away from the Arctic and having limited operational security interests there, still participates on a limited scale in northern NATO exercises from time to time (Pes 2020). It is thus not unthinkable that countries located closer to the equator than the Arctic Circle might want to engage in some form of military exercise or operation, primarily led by an Arctic state, in order to both train in an extreme environment and to build diplomatic and military relations. An example of this was the US–Indian Yudh Abhyas exercise in Alaska in 2023 (Indian Ministry of Defence 2023).

Similarly, showcasing capacity, capability, and – to some extent – symbolic interest by sending a state vessel into Arctic waters is not uncommon for non-Arctic states such Italy, France, or China. Given India's location, the North Pacific Arctic or subarctic domain is probably what is within closest reach, although still located at a considerable distance. Still, the North Atlantic is the easiest part of the Arctic to access due to milder temperatures and limited sea ice, as well as the most developed infrastructure. India–Norway relations in particular have grown exponentially in recent years, with a strong Arctic focus also linked to Indian aspirations in the maritime sector, as well as India's foothold on the Norwegian Arctic archipelago of Svalbard (Singh and Østhagen 2024).

India's Arctic ambitions are also influenced by concerns over China's growing presence in the region which India fears could divert the USA's focus away from the Indo-Pacific (Suhag 2022). Indian strategists are equally worried that new Arctic sea routes could provide China with an alternative to the Malacca Strait, thus easing what is commonly known as the Malacca dilemma (Suhag 2022). Although this could arguably be questioned as any Arctic route would also be contingent on the sailing through the Bering Strait – another potential chokepoint.

This links to the second dimension of Indian security engagement in the Arctic, namely the symbolic and/or political aspirations that tie India to Arctic geopolitics. China's acquisition of polar icebreakers primarily designed for research purposes arguably fulfils several goals at once: they perform research tasks, they enable Chinese learning and capacity-building for operating in polar environments, and they highlight Chinese global reach and interest spheres stretching *even* to the Poles (Lamazhapov 2024; Tatlow 2024a). An Indian polar class research vessel is thus rather likely as India's global aspirations, as well as Arctic and Antarctic interests, become more concrete (Sharma 2021).

Although, as described, the Arctic is the polar opposite to Antarctica concerning aspects related to population, climate, land/ocean ratio, and state sovereignty, the importance of presence and an ability to operate in both domains undoubtedly links to perceptions of great power status and global influence. India today not only has a research presence through the Himadri research presence in the Ny-Ålesund research station in Svalbard in Norway, but also has two research stations in Antarctica, the Bharati and Maitri. Thus, Bipandeep Sharma (2021) concludes that 'icebreaking capabilities will be an important requirement for India not only in terms of scientific research but equally for economic and strategic interests in the Arctic'.

Coupling apparent interests in research, economic activity, and governance questions with a clear sense of what India wants to achieve politically can lead to the Arctic being utilised for both domestic and international purposes. Notwithstanding the benefits of the scientific data and research itself, up-playing, for example, India's Svalbard presence with the first year-round occupancy at Himadri (Srivastava 2023), showcases to a domestic audience the state's competence and global aspirations. Similar actions (and showcasing of capacity) from Indian companies or other government entities operating in the Arctic are conceivable in the near future.

For a country increasingly throwing its weight around in global politics, expressions of its global engagements and interests – as with the rise of China over the last two decades – are also likely to have an Arctic component. However, how it utilises these interests and what issues India decides to address (and with what tools), remains to be seen. Generally, non-Arctic states have been wary of highlighting any strong divergence from the interests of the Arctic states. The perhaps most apparent example of outside interests wanting to challenge existing governance mechanisms and/or ongoing activity has been the EU's environmental, economic, and governance statements on the Arctic especially at the start of its Arctic endeavour (Østhagen 2013; Raspotnik 2018). Fears about China's interests and role in Arctic governance have also been raised, although China seems to have taken a relatively muffled approach in the North to sensitive issues that might irk Arctic states (Ye 2014; Edstrøm, Stensdal, and Heggelund 2023).

Finally, India's ambition to be a global power broker – positioned somewhere in between the USA, Europe, and Russia, and with an ambivalent relationship with China – comes into play in the Arctic. As highlighted above, beyond Arctic-specific activity or actions that are either directly security-oriented (e.g., sailing a frigate through the Bering Strait) or linked to wider political interests and posturing (e.g., increasing scientific missions or launching an Arctic policy), the Arctic's role in global politics as an arena for contestation and cooperation became increasingly apparent from 2018 onwards.

For the smaller Western Arctic states, there is a wait-and-see approach to what India's geopolitical ambitions entail for the Arctic. Especially given the growing friendship of convenience emerging between Russia and China in an Arctic and North Pacific context, hopes are that India can play a balancing role vis-à-vis China, and possibly also Russia, as the latter state might eventually implode due to external and internal pressures. Terming itself 'non-West but not anti-West' (Mohan 2024, quoting India's Foreign Minister Subrahmanyam Jaishankar), India could also have a central role to play in the Brazil, Russia, India, China and South Africa grouping and its possible Arctic endeavours – which Russia is increasingly seeking to utilise as a wider alternative to Western Arctic collaboration (Edvardsen 2023a).

In conclusion, the Arctic continues to be on the global political agenda both because of its importance for Russia's strategic dependence on the region and increasing Chinese interest in the region, which in turn impact Chinese rivalry with the USA. Security tensions arising from issues in other parts of the world (i.e., Ukraine in 2014/2022) or global power struggles (i.e., China–USA, or – perhaps – China–India) have a spillover effect for the Arctic: on the rhetorical level in the form of bellicose statements and on the operational level in the form of increased military presence and exercises by some NATO member states and Russia, to some extent China, and – in the future – perhaps also other actors. In Chapter 7, how these dynamics might play out on a subregional level is further discussed. Yet there is one more level we must examine to get the full picture of what is occurring in the Arctic when it comes to security and geopolitics, namely the bottom-up perspective of specific Arctic states.

5 *The National Level*
Norway

So far, this book has been about the various political dynamics that take place in or over the Arctic. However, as has been alluded to throughout this book, if we really want to unpack Arctic 'geopolitics', it does not suffice to only look at intra-regional dynamics or the Arctic's role in global power struggles. Keeping an eye on one or both of these dimensions might prove to be analytically neat and provide explanations with some level of generalisability. However, there is explicit analytical value in further unpacking empirical complexity. By looking at the domestic level, and two Arctic states in particular, it is easier to analyse security dynamics and practices in more depth. Thus, this chapter will make use of Norway – a small Nordic NATO country which is also a neighbour to Russia in the Arctic – and the USA – a global superpower also neighbouring Russia in the Arctic – to showcase the relevance of looking at national policies, interests, and practices when we talk about Arctic security and geopolitics.

Norway's Arctic Context

In a Norwegian context, the Arctic is not a far-removed or remote part of the country. In fact, the Norwegian Arctic makes up one-third of Norway's land area, is home to almost 9 per cent of the population (albeit this figure is declining), and comprises as much as 80 per cent of Norway's maritime domain. The Norwegian definition of the Arctic includes everything north of the Arctic Circle (66°34N). In Norway – a unitary state structure – this includes Nordland, Troms, and Finnmark counties, the Svalbard archipelago, and the island of Jan Mayen (see Figure 5.1), with approximately 480,000 inhabitants. This includes the Sami – the Indigenous Peoples of Norway – who primarily reside in the North. The Sami have their own Parliament, located in Karasjok, which has some political and administrative responsibilities.

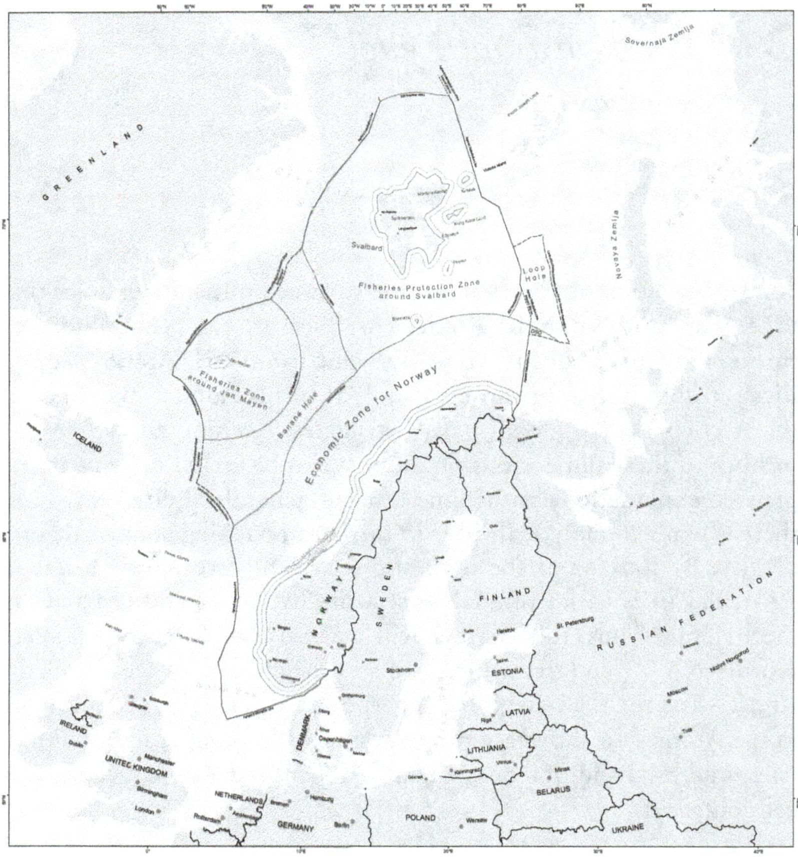

Figure 5.1 Map of Norway including Svalbard and Jan Mayen, outlining Norway's maritime zones. Source: Forsvarets militærgeografiske tjeneste (FMGT), Public Domain, www.regjeringen.no/globalassets/upload/ud/vedlegg/folkerett/20121002_fmgt_overview_norge_nordomraadene_norwegian_maritime_boundaries_7.5m_h61xw52_p_jb_ed04_ugradert_300dpi.pdf.

The population of almost half a million in the Norwegian Arctic alone is relatively high compared to the North American Arctic, although it is sparsely populated by European standards.

Note that a distinction is made between the 'High North' and the 'Arctic' here. The High North (*nordområdene* in Norwegian) has, in many contexts, been used by Norway over the last decades to denote the immediate areas in the North that are part of, or are adjacent to,

Norway. This encompasses the Barents Sea region, including Svalbard, but not the entire Polar region. The Arctic, on the other hand, is used as a term for the entire circumpolar region, or the entire area north of the Arctic Circle.

Fundamentally, the High North is central to Norway's security considerations, the primary one being its shared land and sea border with Russia. Since the end of World War II, Norwegian security policy has concentrated on managing its relationship with its eastern neighbour. In what is generally termed an asymmetric relationship, Norway endeavoured to balance its military inferiority to the USSR/Russia through its membership in NATO (founding member in 1949) and a bilateral relationship with the USA (Tamnes 2011, 1997).

The Svalbard Archipelago

A special note on the Svalbard archipelago is needed due to its unique status and the fact that currently this is probably one of the primary geopolitical dimensions in the Norwegian Arctic. Uninhabited and remote (see Figure 5.1), without any permanent occupancy, after periods in the seventeenth and eighteenth centuries with economic activity focused mainly on fisheries, whaling, and trapping, the nineteenth century saw an increase in both scientific and touristic expeditions to Svalbard to document and map this area, still referred to as 'terra nullius' (no man's land) (The National Archives of Norway 2020). In the late nineteenth and early twentieth centuries, promising discoveries of coal were made and mines were opened. This spurred further interest in establishing an administration for the archipelago.

Norway – being the closest state to the islands and having a considerable economic presence there – was keen to have the sovereignty issue settled and initiated several international discussions. From around 1890, the question of Svalbard and Norway's economic and strategic presence there gained attention in Norwegian politics (Arlov 2003: 239). Moreover, as Norway obtained full independence from Sweden in 1905, it was keen to consolidate its territory and near-abroad positioning.[1] Prior to World War I, various governance models were discussed, culminating in three international conferences in

[1] For more on this, see Berg 1995.

1910, 1912, and 1914 (Arlov 2003: 285). Then, during the peace conference in Paris 1919–1920, the matter of Svalbard was revisited and finally settled with the Treaty of February 1920 (here called the Svalbard Treaty, to avoid confusion, although the Treaty does not have any official name). The Treaty affirmed Norway's increasing demands for sovereignty over the archipelago (Svalbard Treaty 1920).

A key objective of the Treaty was, after assigning Norway 'full and absolute sovereignty' (Art. 1) (Svalbard Treaty 1920), to secure the economic interests of nationals from other states (Ulfstein 1995). This was done by including provisions that nationals of all parties to the Treaty 'shall enjoy equally the rights of fishing and hunting' (Art. 2) and 'shall have equal liberty of access and entry' (Art. 3) in certain defined areas of activity. Norway may not treat other nationals less favourably than its own citizens, and taxes levied in the Treaty area can be used solely for local purposes (Art. 8). Moreover, the islands cannot be used for 'warlike purposes', and Norway committed 'not to create nor to allow the establishment of any naval base in the territories specified in Article 1 and not to construct any fortification in the said territories' (Art. 9).[2]

Due to the ongoing Russian civil war, the Bolshevik leadership was not recognised by most other states and was not present during the Treaty negotiations. However, the Treaty expressly granted Russian nationals and companies the same rights as nationals of the High Contracting Parties 'Until the recognition by the High Contracting Parties of a Russian Government shall permit Russia to adhere to the present Treaty' (Art. 10). One question at the time was whether Soviet Russia would challenge the Treaty, given their geographic proximity to the area and claims of historic use (Arlov 2003: 385–87; Tamnes 1991; Todorov 2020a). Following the signing of the Treaty in 1920, the Russian government protested to ensure Russian interests were safeguarded. In 1924, however, the Soviet government recognised Norwegian sovereignty over the archipelago unconditionally and it acceded to the Treaty in 1935. This was in return for Norwegian de jure recognition of the Soviet government the same year (Holtsmark 1993: 14). The Treaty came into force in 1925, after Norway had implemented relevant domestic legislation and Japan – as the last of

2 For an in-depth examination of the Svalbard Treaty, see Ulfstein 1995; Ø. Jensen 2020; Arlov 2003, Chapter 11.

the original nine signatories – had ratified the Treaty.[3] Norway, as the sovereign, then changed the name of the archipelago to Svalbard. The Treaty has been acceded to by more than forty states (Norwegian Ministry of Justice and Public Security 2024: 14).

Svalbard saw some military activity during World War II, its inhabitants were evacuated and thereafter the Germans bombed and occupied the islands, before it was retaken by allied forces (Arlov 2003; Skogrand 2004; Dean and Lackenbauer 2017). The islands' strategic value was primarily linked to the ability to gather weather data crucial for operations in the North Atlantic. At the end of World War II, in November 1944, Soviet officials attempted to gain special status on Svalbard, with a proposal made by the USSR's foreign minister, Vyacheslav Molotov, to Norway's foreign minister, Trygve Lie, that the Svalbard Treaty should be scrapped in favour of a bilateral arrangement, and Bjørnøya – a small island halfway between Svalbard and the Norwegian mainland – should be Soviet territory (Holtsmark 1993: 50–76).[4] This suggestion was eventually rejected by Norway in 1947, after several years of discussions during the post-World War II turmoil (Arlov 2003: 394–96).

However, the Soviet, later Russian, sentiment that Svalbard issues should be settled bilaterally between Norway and the USSR/Russia has remained, with statements and overtures by the Russian foreign ministry at various times. Norway has continued to reject these. As stated by both Norway's foreign and justice ministers in response to Russian suggestions of a bilateral dialogue of Svalbard issues in 2020: 'Svalbard is part of Norway. It is not natural that we consult with other countries about how we exercise authority in our own areas' (Søreide and Mæland 2020).[5]

International economic interest in Svalbard had faded before World War II, and after the war, only Norwegian and Soviet mining companies conducted economic activities there. Consecutive Norwegian governments have sought to maintain the Norwegian population on the

[3] In addition to Norway, these were Denmark, France, Italy, Japan, Netherlands, Sweden, the United Kingdom, and the USA.

[4] Only a month earlier, in October 1944, Soviet forces had started their campaign to liberate Northern Norway, crossing the border into Nazi-occupied Norway. See further descriptions of this event, as well as the larger relations between the USSR and Norway in Holtsmark 1995, 2015.

[5] Author's translation.

islands, predominantly by subsidising coal mining with the state-owned company Store Norske Spitsbergen Kulkompani and supporting the largest community, Longyearbyen (Pedersen 2017). Similarly, governments in Moscow have worked to maintain a small Soviet population through the state-owned mining company Trust Arktikugol, in the settlements of Barentsburg, Pyramiden, and Grumant – of which only the coal mine in Barentsburg is active today. The Cold War saw different periods of both strategic concern and economic interest in the archipelago, as one of the few places where Soviets and Westerners had the opportunity to interact (J. H. Jørgensen 2013; Bykova 2023). By 1990, the 2,407 Soviets on Svalbard outnumbered the 1,125 Norwegians nearly 2.5 to 1 (Bykova 2019).

By January 2024, the total population of Svalbard, spread across seven locations, was 2,993 (Norwegian Ministry of Justice and Public Security 2024: 11).[6] The largest settlement Longyearbyen had a population of 2,600; 36 per cent are not Norwegian citizens and come from more than fifty countries (Norwegian Ministry of Justice and Public Security 2022b: 7, 41). Ny-Ålesund had a population of forty-four scientists, whereas the population in Barentsburg has been steadily declining and was 354 as of 2024. Pyramiden – abandoned in 1998 – is an open-air museum with a population of twelve (Norwegian Ministry of Justice and Public Security 2024: 10), although the Russian state-owned company Trust Arktikugol launched ambitious plans to revive the former settlement with tourism and research activities in 2023 (Edvardsen 2023b, 2023a).

National Priorities and Policy Entrepreneurs

Looking at the Arctic – or High North – in general, it has been Norway's foremost priority in its foreign and security policies since the early 2000s. We can divide Norwegian High North policy – with particular regard to foreign and security affairs – into three pillars (Østhagen 2021a). The *first pillar* deals with resource extraction and economic potential. The Arctic moved to the forefront of Norwegian policymaking through a series of studies and parliamentary reports

[6] It is often said that there are more polar bears on Svalbard than humans, but the number of polar bears is probably much lower (Polar Bear Science 2018; Staalesen 2021).

from 2003 to 2005 that highlighted the development potential of the region (Norwegian Ministry of Foreign Affairs 2005; Orheim et al. 2003; Brunstad et al. 2004; ECON 2005). This interest was particularly spurred by economic pursuits in the Barents Sea from the petroleum sector, as fields further south in the North Sea were depleting.

The first High North policy document – *Opportunities and Challenges in the North* – was released in 2005 by the conservative coalition government (Norwegian Ministry of Foreign Affairs 2005) and bears the stamp of optimism around business development and oil and gas extraction. Still, it was the 'red–green' coalition government led by Jens Stoltenberg that took office a few months later that elevated the northern area's portfolio to Norway's 'most important strategic focus area'.[7] Foreign Minister Gahr Støre in particular led the government's High North drive (L. C. Jensen and Hønneland 2011).

The *second pillar* deals with cooperation in the North, with an emphasis on both the bilateral relationship with Russia and multilateral forums, especially the Arctic Council.[8] During the Stoltenberg government, the elevation of the High North was part of the Norwegian Foreign Ministry's deliberate focus on circumpolar cooperation, which was designed to counterbalance the bellicose statements concerning the potential for conflict in the North (L. C. Jensen and Hønneland 2011; Grindheim 2009). In both the foreign and domestic media, Foreign Minister Gahr Støre and Prime Minister Stoltenberg frequently stressed the region's uniqueness as an area for cooperation (Gahr Støre 2011; Moe, Fjærtoft, and Øverland 2011; Gahr Støre 2012). Notably, when the Russian scientist and parliamentarian Chilingarov planted a flag on the North Pole seabed in 2007 and helped draw worldwide attention to the region, Gahr Støre used the opportunity to emphasise multilateral cooperation (Grindheim 2009: 6–10).

Moreover, Norway was pursuing diplomatic and multilateral efforts to help ensure so-called low tension in the High North (Stephen and Knecht 2017). To this end, Norway promoted the inclusion of other non-Arctic actors, such as the EU and China, in Arctic discussions (Offerdal 2010; Wegge 2012), while also emphasising the primacy of

7 Consisting of the Labour party (red), the Socialist Left party (red/green), and the Centre party (agrarian green).

8 See Ø. Jensen and Rottem 2010; L. C. Jensen, Jensen, and Rottem 2011; Rowe and Hønneland 2010; L. C. Jensen and Hønneland 2011; Hønneland 2012b.

Northern countries when dealing with Arctic issues. The emergence of the Arctic Council in the wake of the Cold War as the primary forum for regional affairs in the Arctic plays into this setting (Rottem 2017), as Norway worked to get the secretariat permanently located in the northern city of Tromsø (Graczyk and Rottem 2020).

The political focus on the High North from around 2005 also emphasised the need to build a pragmatic bilateral relationship with Russia to deal with cross-border issues, from trade to fish stocks, and to improve people-to-people cooperation at the local and regional levels, especially within the framework of the Barents Cooperation (Hønneland and Rowe 2010; Leland et al. 2008; Onsager and Stuvøy 2022; Haugseth 2017).

A high point in this work came in 2010, when Norway and Russia agreed on a maritime border in the Barents Sea after almost four decades of disagreement and uncertainty (Gahr Støre 2010a). In 1957, Norway and the then-USSR delimited the first Arctic maritime boundary from the terminus of the two countries' land boundary on the coast for 24.35 NM through the Varangerfjord (Norway–Soviet Union 1957). However, a maritime boundary dispute arose in the 1960s when both Norway and the Soviet Union drew on the 1958 Geneva Convention to claim continental shelf rights beyond that point (Henriksen and Ulfstein 2011).

Then, the dispute escalated in 1977, when both countries asserted 200-NM EEZs encompassing both water column and seabed resources. In 1978, an agreement concerning the disputed areas, the Grey Zone Agreement, was signed and it was renewed annually until 2010 (Riste 2005: 200). It recognised the rights of Norway and the USSR/Russia to fisheries in the area without tackling the boundary dispute as such (Stabrun 2009). A breakthrough occurred at the beginning of the 2000s. First, in 2007, the two countries reached an agreement essentially replacing the 1957 Varangerfjord treaty, extending the delimitation line to 39.41 NM (Russia–Norway 2007). However, further north, in the Barents Sea and Arctic Ocean, overlapping claims to continental shelf encompassing an area of approximately 175,000 km^2 remained (Moe, Fjærtoft, and Øverland 2011). At the core of the dispute was Norway's preference for a median-line solution versus Russia's preference for a sector line.

The breakthrough came in 2010 when the two countries committed to an all-purpose boundary being drawn (Norway–Russian Federation

2010; Norwegian Government 2010). The agreement contained provisions aimed at continued cooperation over fisheries (Henriksen and Ulfstein 2011: 1); provisions on the co-management of any hydrocarbons that straddle the boundary were also made (Byers 2013: 43–44; Fjærtoft et al. 2018). Moe, Fjærtoft, and Øverland (2011), as well as Holsbø (2011) and Solstad (2012), held that Russia's desire to affirm the primacy of the Law of the Sea regime and also 'tidy up its spatial fringes' help to explain the 2010 agreement beyond economic interests.

Indeed, Russia has benefitted greatly from the right of every state to an EEZ because of its extremely long coastline that predominantly exists in the Arctic. As already outlined in Chapter 3, Russia's foreign minister, Lavrov, and his Norwegian counterpart, Gahr Støre, subsequently co-authored an op-ed in the Canadian newspaper *Globe and Mail*, in which they asked Canada to take note of the following: 'if there is one lesson that the biting cold and the dark winters of the Arctic should teach us, it is that no one survives alone out there for long' (Lavrov and Gahr Støre 2010).

This message could be considered equally applicable to Norway's efforts to keep its allies engaged in Northern affairs. The *third pillar* in Norwegian Nordic policy also deals with Russia but from a security and defence perspective. Concerns related to Russia never completely disappeared after the Cold War but were seen as less pressing in the early 2000s. From the 1990s to 2005 (when the first High North report was released), and to a large extent from 2005 to 2007, traditional security aspects were almost absent in High North politics (Norwegian Ministry of Foreign Affairs 2006, 2009).

While cooperation was still highlighted in Norwegian foreign policy in general and High North policy in particular, there was a clear shift in Norwegian security and defence policy in 2007 and 2008. This shift primarily occurred because Russia, from the mid 2000s, under President Putin, began to strengthen its military activity in the North and especially on the Kola Peninsula (Hilde 2014; L. C. Jensen 2014).

With the renewal of the Russian Northern Fleet and especially the resumption of bomber flights resembling patterns during the Cold War, Norway was (again) faced with a more challenging security policy situation in the North. At the same time as the second Stoltenberg government continued to emphasise the need for good neighbourly relations with Russia, it also took the decision to modernise the

Norwegian military (e.g., Håkenstad and Bogen 2015).[9] In this context, it is important to point out that a close bilateral relationship with the USA has been one of the main components of Norwegian foreign and security policy and also features in this new security policy situation since the turn of the millennium (Hilde 2019).

In the same period, Norway launched a new multinational military exercise every second year, from 2006, named Cold Response. Initially, the focus was on engaging allies in cold weather operations and shifting some of the alliance's focus northwards – albeit without antagonising Russia and leading to further High North tension (Depledge 2020: 295). Here, the Norwegian work to secure NATO's and allies' attention regarding Norwegian concerns in the North is central, through amongst other things the 'Core Area initiative' launched by Norway in NATO in 2008 (Haraldstad 2014). The proposal gained support from other allies in NATO as well as the USA and the UK – which had initially been sceptical because of the ongoing operations in Iraq and Afghanistan (Hilde and Widerberg 2010; Depledge 2020: 296). Norwegian efforts to increase the relevance of Nordic Defence Cooperation emerged around the same time, although these efforts varied and at times floundered, as the Nordic countries have had diverging security approaches and organisational memberships (Saxi 2019).

Russia was the obvious reason for Norwegian concerns, although not all the other Arctic NATO countries perceived this so clearly (e.g., Østhagen, Sharp, and Hilde 2018). At the same time, the Norwegian side continued to emphasise that the Norwegian policy was related to 'deterrence and reassurance' with self-imposed restrictions to not allow nuclear weapons or other countries' military bases on Norwegian soil and not conduct military exercises too close to Russia to avoid unnecessary provocation (e.g., Holst 1966; Søreide 2017; Oma 2021).

How to Manage Russia

A recalibration of Arctic expectations occurred in Norway in 2014. The drop in the price of oil and natural gas, combined with the dramatic

[9] Over the past decades, the Norwegian Parliament (Stortinget) approved the acquisition of new frigates (2000), new missile torpedo boats (2003), new fighter aircrafts (2008), a new logistics vessel (2008), new coast guard vessels (2013), new maritime surveillance aircrafts (2016), new submarines (2017), and new tanks (2021), as well as a new battalion in Brigade Nord.

events in Ukraine in spring 2014, were key reasons for this shift. Starting in 2014, as NATO gradually returned to emphasising collective defence at home, Norwegian security and defence policy became more detached from its High North policy as it shifted towards more traditional Cold War issues and geography (Norwegian Ministry of Defence 2015). The Norwegian authorities began to openly refer to Russia as a possible threat that should be deterred – a shift that was in many ways a return to a 'normality' in Norway–Russia relations (Rowe 2015).

In terms of security, the traditional – for Norway – defence issues related to geography and Russia in the North became more explicitly pronounced, while joint exercises with Russia and forums to discuss Arctic security policy challenges were cancelled. In 2015, the Norwegian Chief of Defence, Haakon Bruun-Hansen, went out forcefully in the media to point out the 'new' threat from Russia and the necessity of increasing the defence budget (Higraff et al. 2015; NTB 2015). At the same time, a review of the Norwegian National Intelligence Service's threat assessments shows how the descriptions of Russia gradually change after 2014, albeit not with immediate effect. In February 2015, the assessment points out that 'Despite the significant military and security policy changes in 2014, the Norwegian Intelligence Service maintains its assessment that Russia in the current situation does not pose any military threat to Norway' (Etteretningstjenesten 2021: 16). In 2016 and 2017, the threat assessment regarding the northern regions was summarised by saying that 'Russia is particularly concerned with the Arctic and the northern regions in light of the area's untapped resources and military strategic importance. This has led to increased military investment in the North, primarily of a defensive nature against threats from the north' (Etteretningstjenesten 2016: 25, 2017: 28).[10]

In this context, the Norwegian minister of defence, Ine Eriksen Søreide, was clear that Norway was in a new security situation. In February 2015, less than a year after the Russian annexation of Crimea, she stated in an interview with CNN (Krever 2015): 'We are faced with a different Russia. I want to warn against the fact that some people see this as something that is going to pass. The situation has changed. And it has changed profoundly. There is no going back

[10] Note that these texts are originally in Norwegian and translated by the author.

to some sort of normality or some sort of back to normal business. Because that normality does not exist.'

At the same time, instead of promoting NATO's involvement in the 'Arctic', Norway highlighted maritime security issues in the North Atlantic/Barents Sea (Østhagen, Sharp, and Hilde 2018). In a longer article published by the US National Defense University, Søreide emphasised the threat posed by Russian nuclear submarines with ballistic missiles to the USA and its allies, and it explained Norway's desire for a maritime presence and ability to use force in the northern regions (Søreide 2016: 56). Although both cooperation with Russia and the desire to preserve peaceful conditions in the Arctic are mentioned, the message is clear: 'We would like to see a more frequent presence of allied forces in the North, training and taking part in exercises' (Søreide 2016: 54). In autumn 2016, Ash Carter became the first US Secretary of Defense to visit Norway in eleven years (Skeie and Pettersen 2016), which included a trip to the Arctic city of Bodø to discuss the security policy significance of the High North (US Department of Defense 2016).

As a response to the requests from Norway, as well as concern over Russian and, to some extent, Chinese interests in the Arctic, the USA and the UK, amongst others, showed an increasing involvement in the North through exercises and military operations in the North Atlantic. In 2016, following a request from the USA, a rotating presence of the Marine Corps in Norway was established (Hilde 2019: 68). In the aftermath of that decision, Søreide was criticised in the Arctic newspaper *Nordlys* for having deviated from Norway's balancing policy in its search for allied commitment in the northern regions, to which she herself gave a response emphasising the need for cooperation with allies in the North (Søreide 2017). Norwegian Prime Minister Erna Solberg's trip to the USA in 2016 and subsequent meeting with President Barack Obama was also clear about the Norwegian request: 'From the podium, Solberg asked for continued attention to the High North, as well as the Nordic relationship with Russia. Norwegian authorities want, amongst other things, an increased maritime presence from NATO in the north' (NTB 2016).[11]

[11] Another narrative was that the USA and the UK in particular have an increasingly vested interest in getting involved in the 'Arctic' for both security and symbolic reasons (Depledge and Østhagen 2021).

As a result of the security situation with Russia after 2014, Norwegian security and defence policy gradually became more detached from the Norwegian High North policy. This mirrors the development in general across Arctic region-building efforts, as described in Chapter 3. The High North initiative consisted, to a greater extent, of foreign policy attempts to preserve forms of cooperation in the North in areas such as environmental cooperation and fisheries management, not least within the framework of the Barents Cooperation and the Arctic Council (Rottem 2017; Onsager and Stuvøy 2022). In addition, a greater domestic political focus existed on regional instruments for economic development in the three northernmost counties – Nordland, Troms, and Finnmark – after oil and gas projects did not materialise as expected (Pedersen 2018).

In 2018, the new security picture in the North is also clearer, as stated by Norway's Intelligence Service: 'Russia is increasingly making its mark, and these areas have been central to Russia's military modernisation programme. Norway notices this through increased Russian activity, and it may move towards a new military normalcy in the north' (Etteretningstjenesten 2018: 24). In 2020 and 2021, the Intelligence Service's description of possible aggressive Russian behaviour and risks for Norway became even clearer in what is deemed a tense relationship: 'A direct consequence of the military modernisation is that Russia's ability and willingness to use military power as a foreign policy tool has increased' (Etteretningstjenesten 2020: 43). And: 'The Arctic and the northern regions have increasingly become an area of focus and investment for certain parts of the Russian Special Forces. In Norway, military infrastructure and allied military presence, as well as civil infrastructure that supports allied reinforcement, will be relevant targets in a conflict situation' (Etteretningstjenesten 2020: 66).[12]

[12] In 2021, it is also mentioned that: 'the attention Norway receives in the public Russian conversation has increased since 2017. Moscow claims that the Norwegian authorities justify increased military activity in the north and closer integration with the United States by constructing an exaggerated Russian threat' (Etteretningstjenesten 2021: 51). In 2021, the danger of accidents is highlighted, especially when it comes to Russian nuclear weapons in the North, and in addition, Chinese capabilities with regard to the Arctic are mentioned for the first time in the open threat assessments (Etteretningstjenesten 2021: 77).

In the same period, the effects of Norway's request for attention to the North become clear along two axes. First, *operational* interest in the Arctic and the North Atlantic from allies becomes more obvious in terms of presence. In addition to the rotating forces, the USA demonstrates its commitment to the defence of NATO's 'northern flank' through exercise activity and military operations (Depledge 2020; Depledge and Østhagen 2021). As mentioned in Chapter 4, the largest military exercise in Norway after the end of the Cold War – the NATO-led exercise Trident Juncture – was held in autumn 2018. Also, in 2020, American interest in the High North culminated with the US Navy carrying out so-called maritime security operations in the Barents Sea together with the British Navy (in May) and with the British, Norwegian, and Danish Navies (in September), which led to claims of a 'dangerous game' in the Arctic (Sevunts 2020). In 2020 and 2021, American Seawolf-class nuclear-powered submarines marked a presence outside Tromsø, and in 2021, American B-1 bombers operated in the Nordic region from Ørland air station. This eventually led to a debate about local interests in the use of Tønsnes harbour outside Tromsø for submarine landings (Mortensen and Tønset 2021; Sveen 2021) and Norway's role in a possible conflict in the North (Påsche 2021).

In the summer of 2018, US Defense Minister James Mattis visited Norway, and then Norwegian Minister of Foreign Affairs Søreide stated that (Grønningsæter 2018): 'The Americans are clear that our investment in and organization of the Armed Forces is the right way to go. They see the same strategic picture as us. There has been too little attention to the North Atlantic in NATO, but both we and the US are happy that more is going in that direction.'

At the same time, in Norway, work was underway on a new additional agreement with the USA to regulate the American military presence in Norway. The agreement was presented in April 2021 and joined the debate about the government's balancing act between the desire for an American presence to deter Russia and at the same time to maintain stability (Bye 2021), calm, and – so as far as possible – Norwegian control in the northern areas (see especially Heier 2021). In the wake of these debates about the USA and the High North, the Norwegian government also stated more clearly that the goal was to 'coordinate allied activity in the North' to avoid misunderstandings and unintentional escalation (A. B. Johnsen and Bringedal 2021).

Svalbard Geopolitics

One particular dimension here is (again) the Norwegian archipelago of Svalbard. In tandem with increased interest in Arctic affairs, this northern archipelago, closer to the North Pole than to its capital, Oslo, has received considerable scholarly and journalistic attention for more than a decade. Increased Russian military activity in the European Arctic since around 2006 again made discussions of Svalbard's geostrategic location relevant, mirroring those during the Cold War (see Chapter 4). The argument put forth was that the archipelago's proximity to the Russian Northern Fleet located at Severomorsk on the Kola Peninsula and Svalbard's strategic position as a potential location for anti-access and area denial operations in the Barents Sea and the North Atlantic remain the primary rationales for Russian security interests (Åtland and Pedersen 2009; Wither 2018; Olsen 2017). The concern (again) was that Svalbard could become subject to Russian attempts at control if a full-scale conflict between NATO and Russia were to erupt (Nilsen 2017; Stormark 2017; Baudu 2023; Boulègue 2022).

However, scholars have questioned whether Svalbard truly holds such a strategic position – both during the Cold War and today (Holtsmark 1993: 3). As Rolf Tamnes (2020) points out, holding Svalbard territory would not provide any significant advantage; moreover, even in the case of an outright NATO–Russia war, such an operation would be costly and relatively disadvantageous. Rather, the objective would be to deny the counterpart the use of the islands for strategic or operational purposes – again referring to the position of Svalbard as the location from which Russia and/or NATO can control and deny access in the North Atlantic. This was the central component of the USSR's 'bastion' defence concept (Breemer 1989; Gjelsten 2013: 1, 16); it is still thought to be a pillar of Russia's defence strategy today (Olsen 2017; Forsvarskommisjonen av 2021: 146).

Another thorn in the side concerning Svalbard concerns the maritime zones. The Svalbard Treaty itself refers only to 'territorial waters' (Svalbard Treaty 1920), initially four and today twelve nautical miles from the baseline.[13] There is, however, a dispute concerning whether the 200 nautical mile zone and the continental shelf around the islands

[13] For an interesting discussion on the difference between territorial waters and extended zones, and what the intentions behind the Treaty from 1920 were, see Rolf Einar Fife (2021).

are covered by the provisions in the 1920 Treaty.[14] Although claiming to have a right to establish an exclusive economic zone around Svalbard, Norway has yet to do so (D. H. Anderson 2009; Tiller and Nyman 2017). In 1976, Norway established its EEZ in the Barents Sea, a year later it decided to establish a Fisheries Protection Zone (FPZ) around Svalbard for the conservation and management of living marine resources (Churchill and Ulfstein 1992; Norwegian Ministry of Justice and Public Security 2016). This avoided potential challenges from other interested parties to the Norwegian claim, leaving Norway to protect and manage what is the main nursery of Northeast Arctic cod stock – the largest cod stock in the world (Ø. Jensen 2014a).

The other treaty signatories have so far acquiesced, although Iceland and Russia, and to some extent Spain and the UK, have been particularly outspoken in their criticism of what they perceive to be discrimination by Norway against their fishing vessels in the area.[15] For Russia, which is cooperating with Norway in the co-management of five shared fish stocks in the Barents Sea, the issue has primarily been that the Norwegian Coast Guard has inspected and arrested Russian fishing vessels. The Soviet Union and later Russia have argued that Norway had no right to take such a measure unilaterally and that Russia should have been consulted about any such regime. Although for all practical purposes, Russia has acquiesced in the Norwegian regulatory and enforcement regime in the FPZ as it has been in Russia's own interest to manage fish stocks sustainably and get a share of the quota (Jørgensen and Østhagen 2020; Østhagen 2018).

Subsequently, the position – albeit rather ambiguously specified by Russian officials – seems to be that the area is international waters,

[14] The concept of extended maritime zones was not in place at the time the Treaty was signed, and therefore they are not mentioned in the Treaty. Norway states that the maritime space is not covered by the Treaty's provisions and that the intention of the Treaty was never to extend beyond the territorial sea. Since Norway is the sovereign of Svalbard, per the Law of the Sea the maritime space around Svalbard constitutes an area where Norway has exclusive sovereign rights not limited by the provisions in the Svalbard Treaty (Norwegian Ministry of Justice and Public Security 2022b: 16; see also Churchill and Ulfstein 2010; Anderson 2009). Others have argued that the Treaty must be interpreted dynamically, and that although Norway has jurisdiction in this maritime space, jurisdiction must be based on the same principles as activity on land per the Treaty (Molenaar 2012).

[15] See Pedersen 2008b; Molenaar 2012; Østhagen 2024.

where only flag states have jurisdiction over vessels (A.-K. Jørgensen and Østhagen 2020; J. H. Jørgensen 2010). The continental shelf, on the other hand, Russia has argued, is covered by Treaty provisions (Østhagen, Jørgensen, and Moe 2020; Todorov 2020a), albeit restricted by the so-called Svalbard box (i.e., the coordinates set out in the Treaty that creates a box that includes both the land area of Svalbard and parts of the maritime domain surrounding it).[16]

With the establishment of the co-management regime between Norway and Russia concerning transboundary fish stocks in the Barents Sea, the issue of quota distribution in the FPZ became less acute (Hønneland 2014), and it has historically been fishers from Iceland and Spain that have mounted the largest challenge to the FPZ itself (Østhagen and Raspotnik 2019; Skram 2017). For all practical purposes, the fisheries co-management regime between Norway and Russia that was established in the 1970s makes the question of the Treaty's applicability to the maritime zone less relevant for fisheries relations between the two states, as the whole Barents Sea (including the Norwegian EEZ, the Russian EEZ, and the FPZ) is seen as one zonal area for fisheries (Hønneland 2012a; Hønneland and Jørgensen 2015).

Shelf matters, however, are a different story, and Russia has protested against the possible opening of the shelf for oil and gas activity by the Norwegian government (Bolongaro 2017; Tamnes 2020; Churchill and Ulfstein 2020). At the same time, Russia has shown an interest in keeping this dispute from escalating, due to what is generally thought to be viewed in Moscow as the beneficial current setup (Todorov 2020a; Østhagen, Jørgensen, and Moe 2020; Tiller and Nyman 2015; Tiller 2009).

Great Power Interests and Regional Balancing

A second axis of development in the High North's security environment from around 2018, which does not necessarily coincide with Norwegian interests, is that the symbolic value of the Arctic and the

[16] However, by entering the 2010 delimitation agreement with Norway (Norwegian Government 2010), which sets a multi-purpose (zone *and* shelf) maritime boundary all the way from the Norwegian–Russian border on land and up towards the North Pole, it seems that Russia recognises that Svalbard indeed generates a maritime zone (Henriksen and Ulfstein 2011; Moe, Fjærtoft, and Øverland 2011) and is not to be considered as international waters.

High North is put into a great power perspective. Where the first wave of international Arctic enthusiasm in 2005–2008 was characterised by misconceptions about an imminent 'resource race' in this 'new' region (see Exner-Pirot 2018), from 2018 there were other drivers of international attention, namely Russia's military threat and rivalry with China (as outlined in Chapter 4).

The Arctic was then portrayed as the new arena for rivalry and possible conflict in a multipolar world order (Østhagen and Rottem 2023), where the Norwegian High North played a prominent role as a front line between NATO and Russia. Although great power politics has undoubtedly become more prevalent in the northern regions in recent years both in the form of rhetoric and in the form of military activity, such conclusions are too simple and do not capture the nuances of security policy dynamics in the North (Moe, Østhagen, and Rottem 2022).

To contend with such simplifications, Norwegian foreign ministers have had to carry out the 'art of balancing' that has been firmly in the job description since the end of World War II (see, e.g., Skogrand and Tamnes 2001), with an emphasis on the desire to preserve the High North as a low-tension area. This 'balancing act' has largely been used to describe Norway's strategic approach to defence policy in a trade-off between deterrence and reassurance towards Russia (for the first time, see Holst 1966; see also Bjur, Hilde, and Eggen 2020). However, in an Arctic and Norwegian context with regard to foreign policy, this term can also be used in a broader sense to denote a balance not only between defence policy considerations but also between defence policy and foreign and High North policy.

Indeed, Norway's foreign policy is more complex than an either/or description of balancing and the NATO–Russia relationship does not set the whole frame for Norwegian foreign and security policy. Norway's bilateral relationships, especially linked to the Barents Cooperation and practical forms of functional relations, are examples of activities that are not necessarily characterised by, or that are attempted to be shielded from, systemic power balance and global tensions (Østhagen and Rottem 2023). Local and national interests are also relevant here, such as the interests in trade across the border between Eastern Finnmark and Murmansk. It is also obvious that historical conditions, identity, and the effect of Norway–Russia cooperation for three decades after the fall of the Soviet Union are relevant dimensions (L. C. Jensen 2017; Onsager and Stuvøy 2022). As Steinveg and Medby (2020) point out, the *narrative* of cooperation and low tension in the North has been important for the Ministry of Foreign Affairs' development of a High North policy.

In addition to the deterrence that Norway has had as its main focus in the North in line with increased Russian military activity, Norwegian foreign ministers since Jonas Gahr Støre have emphasised two things to keep the northern areas as a region defined as 'High North, low tension': counteract notions of a race or new cold war in the North and further develop local and so-called soft cooperation with Russia (Østhagen 2021a). Therefore, a large part of the work with the High North for Norway has been dedicated to communicating the political and legal situation in the Norwegian High North and explaining that Norway cooperates with Russia in a number of areas despite tension in other areas (see The Arctic 'Story' section in Chapter 3).[17]

We also see signs in the period 2019–2022 that the relationship with Russia was entering a new phase or a new 'normal state' (Rowe 2018), characterised by both political and military tension at the same time as cooperation and dialogue resumed in some areas. In 2019, Prime Minister Solberg met Russian President Putin for the first time since 2014, together with Søreide at the Arctic Forum conference in St Petersburg (Lohne, Myrvang, and Oddstad 2019). Six months later, Russian Foreign Minister Lavrov was in the town of Kirkenes together with Søreide and Solberg as well as Norwegian King Harald for the 75th anniversary of the liberation of Eastern Finnmark by the USSR. During the same period, Norway and Russia negotiated an addendum to the Incidents at Sea Agreement from 1990 to prevent dangerous incidents at sea (Bjur, Hilde, and Eggen 2020).

In the period 2019–2021, Norway also held the chair of the Barents Council. As highlighted by then Foreign Minister Søreide regarding the value of cooperation with Russia in this format: 'Cooperation across borders in the north, where regional actors and Indigenous Peoples

[17] An example of this was Foreign Minister Søreide's response to Bobo Lo, a researcher from Australia who thought the time had come for an Arctic treaty, during the Arctic Frontiers conference in January 2020 (Holm 2020). As a response, Søreide said 'I don't think there is a need for it (a treaty, ed.). And I think that what we have to continue to do is to actually put a lot of emphasis into the cooperation that we have, as we did through the period of the Cold War' (McGwin 2020).

contribute actively, has, now as before, great importance for trust-building and is an important peace-building function' (Søreide 2020: 458). And '[in] a time of increased geopolitical tension and closed borders as a result of the pandemic, we must protect the forms of cooperation we have in order to maintain trust and cooperation with our neighbouring countries' (Solvik 2021). Onsager and Stuvøy (2022) show how local actors in the Barents region in particular have in many contexts had a vested interest in highlighting the Barents Cooperation as an additional dimension to Norwegian foreign policy in the North.

At the same time, the relationship with Russia has several dimensions to it. Norway had to deal with the particularly difficult Frode Berg case after a retired Norwegian border inspector was arrested in Moscow in December 2017 on charges of espionage. It took almost two years before Berg was returned to Norway. Moreover, in connection with the centenary of the Svalbard Treaty in 2020, Russia breathed life into a dispute about who has the right to what in the maritime zone around the Svalbard archipelago. Simultaneously, Russia invited Norway to bilateral consultations over Svalbard (Ministry of Foreign Affairs of the Russian Federation 2020) – something they know is in direct conflict with Norwegian Svalbard policy (Moe and Jensen 2020: 516).

In November 2020, the government launched yet another High North policy. This document built on the earlier mix of regional and economic development priorities as well as more general foreign policy ambitions. Norwegian High North policy – as a specific portfolio under the Ministry of Foreign Affairs – gradually turned more towards 'soft' security issues and regional development. Other involved ministries, such as the Ministry of Local Government and Modernization and the Ministry of Trade, Industry and Fisheries, were gradually given a greater role in policy development (Pedersen and Skagestad 2020). This High North policy, even more explicitly than in previous iterations, focuses on how value creation and regional growth in the North is a goal in itself, which in turn will support not only the local and national economy but also the foreign and security policy goals of Norway (Utenriksdepartementet 2020).

A slight shift occurred less than a year later, when a Labour-party-led coalition took office in the autumn of 2021 – after eight years of a conservative-led government. The new government pledged that they would work to ensure 'low tension' in the Arctic by considering establishing new structures for security policy dialogue in the North (Norwegian Government 2021: 79), as well as engaging in bilateral cooperation with Russia on 'low-level' issues such as forestry and research to improve bilateral relations. We must also read this as hopes for improving the rather strained security environment in the High North region in 2020–2021. However, these hopes were dashed only a few months later when Russia invaded Ukraine in February 2022.

'High North, Low Tension' No More

What this previous section has outlined is how Norway – perhaps aptly exemplified by the differences in rhetoric and tone from Ine Eriksen Søreide as first defence minister, and then foreign minister – has continually had to balance between maintaining cooperative and functional dialogue with Russia, while at the same time increasingly engaging allies to deal with fears of Russian military expansionism in the same area. Russia's full-scale invasion of Ukraine in 2022, however, changed some things in the Norwegian High North.

Ideas concerning an expansion of collaboration or cooperation with Russia, mused by the Norwegian government in office from October 2021, were shelved. The belief that regional cooperation and local dialogue can create peace, or at least stability, was – and still is – being questioned. Instead, emphasis has been placed on continued allied engagement in the North through military exercises and statements about the importance of the Arctic in order to deter Russian general aggression. As already outlined in Chapters 3 and 4, cooperative structures such as the Barents Cooperation were placed on hold and eventually Russia left it, while increased military activity in the North Atlantic and Barents Sea through exercises and signalling expanded after 2022.

At the same time, Norway maintained bilateral mechanisms to deal with co-management of the shared fish stocks in the Barents Sea, emergency preparedness and response in the border region at sea, and dialogue in Svalbard, where 300–400 Russians citizens live as they can according to the 1920 Svalbard Treaty. The ongoing cooperation in managing shared fish stocks has indirectly led to Russian trawlers being exempted from the Norwegian port ban on Russian vessels, allowing them to deliver catch to three Norwegian ports, highlighting the complexity of sharing borders – and resources – with Russia, as well as the sometimes-perceived necessity of having to engage and coordinate with Russia on certain issues. This policy has faced criticism from some Western allies and domestic actors (Klausen, Wilhelms, and Fjørtoft 2022).

From the Norwegian government's perspective, keeping ports open to Russian vessels was based on the need to be a predictable neighbour (to maintain low tension), as well as the importance of sustainably managing some of the world's largest fish stocks (Hønneland 2024). However, this arrangement has also had security implications for Norway, as there were several reports of Russian trawlers engaging in surveillance and intelligence activities while docked, further complicating the situation (B. M. Pettersen et al. 2023). By 2025, as Norway decided to evict two Russian companies operating vessels in the Norwegian EEZ due to fears over espionage and hybrid activity, Russian officials were threatening to cancel fisheries cooperation with Norway in the Barents Sea (Hønneland and Jørgensen 2025) – highlighting the increasing sensitivity of fisheries relations in the High North.

In light of events after 2022, a few questions have arisen for Norway: How much influence can Norway have in one way or the other on maintaining the North as a low-tension area? Has the extended balancing vis-à-vis Russia there had any effect? Russia's invasion of Ukraine and the subsequent tension in the northern regions have been linked to everything from cable cutting outside Svalbard to imminent Russian exercise activity (Fredriksen et al. 2022; Nilsen 2022), perhaps showing the limitations in Norway's ability to reduce tension in the High North. At the same time, the need for tension-reducing measures has never been greater in recent times for Norway. Moreover, unity in NATO, deterrence of Russia in Europe, and China's global role continue to be on the agenda as they have been since 2014, but now with more acuteness. Therefore, security policy in the High North has become even more important for Norway and also NATO.

Moreover, by 2025, it seemed to become clear that great power rivalry increasingly defines international politics, with the role of traditional power considerations dictating not only policy responses in Moscow and Beijing, but also Washington, DC under the second Trump administration. The fear has been that Trump's statements questioning everything from NATO to free trade are leading to the demise of the so-called liberal world order. Spheres of interest and geopolitics are also back on the international agenda to an extent not seen since before the end of the Cold War. So this is what this book in essence is about, namely 'regionalisation'; that economic and security policy developments are defined within the geographical framework of specific regions.

As already outlined in this chapter, Norway stands in the middle of this in an Arctic context, given its centrality in the part of the Arctic that is subject to most interest from and activity by great powers. There is thus a marked difference between Northern and Southern Norway, both politically and symbolically. In the High North, Norway is a regional power. This is supported by the Norwegian capacities in the North related to everything from infrastructure and business activity to universities and defence. Svalbard is the crown jewel of Norway in the Arctic. At the same time, Norwegian interests are being challenged, as the Arctic and the High North have become, and will increasingly be, an arena for power projection, geoeconomics, and symbolism.

In response, in August 2025, the Norwegian government presented a new High North strategy, entitled *Norway in the North: High North Policy in a New Reality* (Regjeringen 2025a). Everything from health to foreign policy must be stitched together by the two ministries – the Ministry of Local Government and Regional Development and the Ministry of Foreign Affairs – that coordinate Norway's High North initiative. It is, however, in the description of the Armed Forces in Northern Norway and the upcoming investments there that the strategy is most concrete, which also reflects the security policy situation in the North (and in Europe in general).

Several elements further underscore the focus on security issues and concerns, such as decisions about a new fibre cable to Svalbard and Jan Mayen, the location of NATO's new centre for air operations outside Bodø, and the comprehensive plans for the Armed Forces including a 'Finnmark Brigade' and a task force in the Home Guard in Troms. The language of the strategy is also more security-oriented than in previous years, as it underscores that:[18]

- 'Norway should be a leader in the alliance in terms of knowledge of Russian and Chinese activity in our immediate areas, and should contribute actively to NATO processes with our expertise' (Regjeringen 2025b);

[18] Author's translation.

- Norway 'should reduce the risk of conflict and misunderstandings that could lead to escalation and increased tension, and we should contribute to allied understanding of conditions that are particularly important for Norway' (Regjeringen 2025b);
- 'In order to safeguard central Norwegian interests and contribute to stability and security in the High North, we still need contact with Russian civil and military authorities in areas such as border administration, search and rescue cooperation and fisheries management cooperation in the Barents Sea' (Regjeringen 2025b);
- 'Norway shall have a restrictive attitude towards Chinese activity in the High North' (Regjeringen 2025b).

Whether these measures are sufficient to reverse a trend where the population (both total and relative to the rest of Norway) is declining, is difficult to say (Østhagen 2025a). In the part of the strategy to address demographics in the form of education, business, and health, great ambitions and goals are mentioned. However, it is unclear what the result of these goals will be or whether they will be followed up. As Arne O. Holm points out (2025), many of the promises lack a price tag. The most obvious change is the formulation that Norway should have a restrictive attitude towards Chinese activity in the North. This is not surprising in itself, given concerns related to everything from Chinese research activity in Svalbard to possible investments in critical infrastructure in Northern Norway. What is new is that it is formulated so bluntly.

It is natural that it is defence and various instruments for regional development and societal structures that are emphasised in Norway's High North strategy, given the security policy upheavals that are taking place in the North in relation to Russia. At the same time, there is a larger geopolitical backdrop – as the strategy also points out with regard to 'a sharp increase in global interest' – which will increasingly come to define Norway's room for manoeuvre in the High North. In the strategy, the focus on the research project Arctic Ocean 2050 is one example of a measure to deal with this: both because it is a core interest for Norway and because Norwegian research communities must be world leaders in what is happening in its immediate geographic vicinity. In essence then, the division between traditional security and defence interests and foreign policy ones, which already started to appear around 2007 and became pronounced after 2014, is now less apparent. Defence and security issues and interests now also dominate the traditionally 'softer' High North policy of Norway.

In sum, it is clear that for Norway, the High North (or *its* part of the Arctic) is not just a resource base, or a defence region where Russia must be contained and population decline managed. The High North has also become a geopolitical arena and resource for Norway in the face of a shift in international politics. The examination here of Norway's approach to Arctic geopolitics underscores a multifaceted strategy embedded in security, economic priorities, and diplomatic balancing. As this chapter of the book has elaborated, Norway navigates the complex interplay of regional (Arctic), global, and national developments in the Arctic by continuously adapting its policies in response to shifting geopolitical dynamics.

The strategic emphasis on the High North aligns with national interests and reflects Norway's pursuit of maintaining stability in an increasingly volatile security region. This is further reinforced by Norway's emphasis on regional cooperation, especially marked by initiatives within the Arctic Council and bilateral agreements with Russia. However, the reigniting of tensions with Russia from 2014 signals a realignment wherein security concerns about military investments and regional stability take precedence. Simultaneously, Norway's focus on maintaining cooperative conduits, such as fisheries management with Russia, exemplifies an approach to preventing escalations.

This highlights how Norway's security practices are oriented not towards the whole Arctic 'region' or, for that matter, all of 'Europe'. There is little that links Norway's security considerations with those of Greece, or Canada, or even the security milieu in the Pacific. Instead, the focus as showcased in this chapter is on a much smaller immediate geographic vicinity, linked to the Russian border and maritime domain in the High North and the Barents Sea. Still, as Norway continues to advocate for low tension in the High North, the overarching narrative of cooperation and geopolitical balancing remains a central theme, linking the immediate northern space of this one Arctic country with wider regional and global geopolitical developments.

6 *The National Level*
The United States of America

All states are not equal. There is obviously a difference between a global superpower of almost 350 million people and Norway, with a population of 5.5 million people. Still, both *are* states, both are in the Arctic, and both face geographically determined security issues in their immediate Arctic domain. It is thus both highly relevant and instructive to compare how these two states have approached their respective Arctic security dynamics. Whereas Norway and Norwegian politicians and bureaucrats in many ways responded to international attention and shifts by emphasising the High North and taking lead on a range of security initiatives (or governance initiatives) in the Arctic, US interest in Arctic security affairs over the last two decades has been a conundrum. The Arctic has not played the same vital security role in North America as it historically has for the Nordic states or even Russia.

The US Arctic Context

In a US Arctic context, the Arctic refers to Alaska and the Beringia regional space. Alaska has, however, only to a limited extent attracted the attention of federal decision makers in Washington, DC, or in the so-called lower forty-eight (see Figure 6.1). Alaska has 'only' a population of around 730,000 and most do not reside in the northern parts of the state. The strategies of US presidential administrations since the mid 2010s seem to reflect this as they have focused more on climatic and industrial considerations than on security and defence. Commentators thus declared that the USA's ostensible lack of interest in security changes in the Arctic until the mid 2010s has made it a 'reluctant Arctic actor' in the post-Cold War era (Huebert 2009; Herrmann 2019; Conley et al. 2020; Nilsson 2018).

However, over the last decade, the USA has paid increasing attention to great power politics in the various subarctic and Arctic regions.

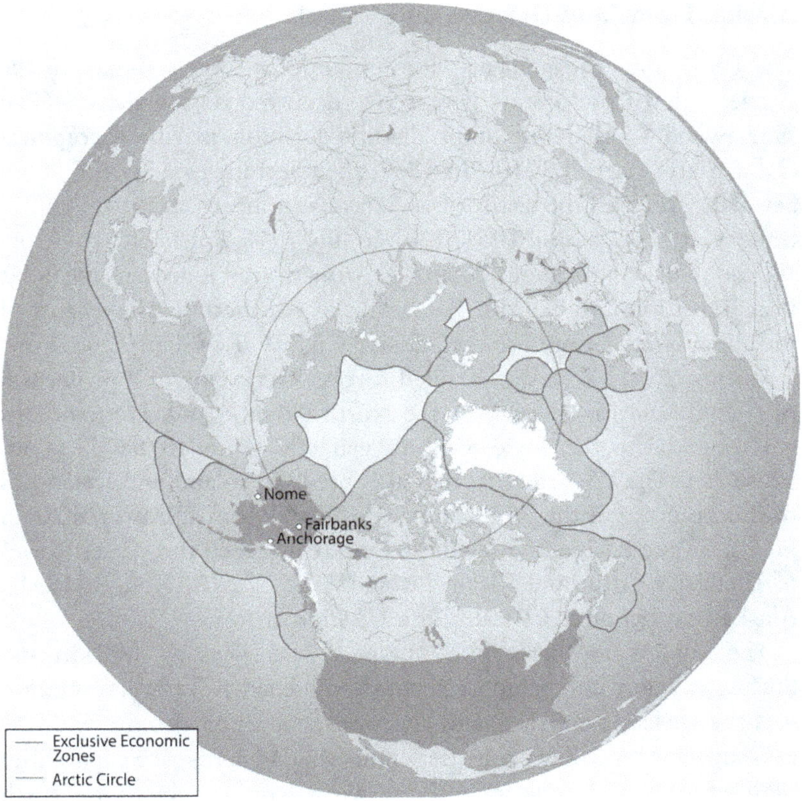

Figure 6.1 Map of the Arctic with the USA and Alaska highlighted. Created by Malte Humpert, The Arctic Institute.

This has materialised on many levels: US troops are exercising in various parts of the Arctic to showcase readiness and Arctic capabilities, each branch of the US military has launched new or updated Arctic strategic documents, and US politicians have been increasingly vocal about Arctic security concerns and the USA's need to engage with the circumpolar region more strategically. We must study the various approaches, or levels, to the USA's Arctic security engagement. Given the global role of the USA in international relations, the Arctic serves as a particularly interesting policy domain where the domestic (i.e., Alaska) meets the immediate-near abroad (i.e., Canada/the North American continent), as well as the wider systemic (i.e., the role of European and Asian actors in defining US security priorities).

Alaska, Homeland Defence, and Canada

The USA's first engagement with circumpolar security issues in the Arctic region from a state perspective occurred during the US Civil War, when Confederate ships disrupted Union maritime commercial activity (Carroll 2010). In 1867, then-Secretary of State William Seward purchased the territory of Alaska as a means to strengthen US trade relations (Caswell 1951: 220). In July 1918, President Woodrow Wilson sent a contingent of 5,000 US Army troops, known as the Polar Bear Expedition, to fight the Bolsheviks in Northern Russia as part of the allied intervention into the Russian Civil War (Willett 2003: xix).

During World War II, the allied forces' Arctic convoys' protection of friendly maritime traffic in the North Atlantic and Arctic oceans was essential to the Soviet Union's war efforts under the US Lend-Lease Act. The US Armed Forces also used Greenland and Iceland as transit points after the USA's entry into World War II (Edwards 2002: xi–xii). In the Pacific, the Japanese Empire attacked the US territories of Unalaska, Attu, and Kiska during World War II (Roucek 1951). Alaska only acquired its status as a US state in 1959.

The Cold War ushered in a bipolar system whereby the USA and the Soviet Union competed for regional influence. Key to this ambition was the nuclear arms race, and given the geographic proximity of the two superpowers, they prioritised undersea and aerospace capabilities as the shortest missile attack route between them was across the North Pole (Weir 2002; Grossman 1946). As outlined, parts of the Arctic have long been a base for the Soviet Union's (and later Russia's) sea-based nuclear-deterrent in the Northern Fleet (J. Kjellén 2022). After World War II, the USA and Soviet Union began conducting bomber missions along the North American Arctic coast to deter each other.

Official Arctic strategy began under President Richard Nixon. His 1971 National Security Decision Memorandum 144 stated that the US would 'promote mutually beneficial international cooperation in the Arctic; and will at the same time provide for the protection of essential security interests' (National Security Council 1971). Congress's Arctic Research and Policy Act of 1984 expanded on this policy and declared, 'as the Nation's only common border with the Soviet Union, the Arctic is critical to national defense' but also stressed the need for cooperation with Arctic-rim countries in research and controlling pollution (National Science Foundation 1984).

The first comprehensive US Arctic security policy was issued in 2009, shortly before President George W. Bush left office. The principal interests listed in National Security Presidential Directive-66 / Homeland Security Presidential Directive-25 (NSPD-66/HSPD-25) are 'missile defense and early warning, deployment of sea and air systems for strategic sealift, strategic deterrence, maritime presence, and maritime operations, and freedom of navigation and overflight' (The White House 2009). Former President Barack Obama's visit to Alaska in 2015 marked the first time a sitting president travelled north of the Arctic Circle in the USA (Eilperin 2015).

While the US government began publishing a series of strategic documents to address the issues raised in the Presidential Directive during the Obama administration, the security rhetoric concerning the Arctic only began to noticeably escalate during Donald Trump's first presidency (2017–2021). President Trump's musings in purchasing Greenland (Neuman 2019), and then Secretary of State Michael Pompeo's refusal to sign the 2019 Arctic Council declaration – officially due to its mention of climate change (The Guardian 2019) but unofficially due to tensions with Russia and China – exemplified a new, assertive US approach towards Arctic security. From 2017 to 2018, the Trump administration applied its view of great power competition to the Arctic in its overarching 'America First' strategy. It was not – as had been the case during the Obama administration – predominantly concerned with climate change adaptation and mitigation. Trump's Arctic strategy included an increased focus on military presence in various domains, particularly in the North Atlantic and European Arctic regions (Office of the Under Secretary of Defense for Policy 2019).

One significant driver of US attention to Arctic security in an Alaskan context has been concerns about homeland security and homeland defence. Alaska is currently home to nine military installations, including Joint Base Elmendorf-Richardson, Eielson Air Force Base, Clear Space Base, Fort Wainwright, Fort Greely, and Coast Guard Stations Kodiak, Valdez, and Juneau. Over the last decade, the US military has elevated the importance of Alaska by investing in personnel, assets, and infrastructure. These include, for example, the deployment of fifty-four fifth-generation F-35 aircraft to Eielson Air Force Base, the renaming of US Army Alaska to the 11th Airborne Division ('Arctic Angels'), and over USD 350 million in funding to upgrade military infrastructure across the state from 2023 to 2024 (The Associated Press 2022; Fernandez 2023).

Moreover, the US Coast Guard – part of the Department of Homeland Security (but part of the Department of the Navy in wartime) – concentrates on homeland security challenges with a particular Alaskan component, including emergency response, law enforcement, and fishing inspections. The US Coast Guard released an Implementation Plan in 2023 for the 2019 Arctic Strategic Outlook. These documents note that 'Russia and China's persistent challenges to the rules-based international order around the globe cause concern of similar infringement to the continued peaceful stability of the Arctic region' but also that 'US Coast Guard and the Russian Border Guard have a history of practical cooperation and [we] should endeavour to maintain that collaboration' (US Coast Guard 2019: 10). Today, US–Russian Coast Guard joint exercises have been paused but the two sides maintain essential areas of cooperation in search and rescue and emergency response.

Some analysts have argued that the USA has yet to make additional significant investments in the North in terms of civilian infrastructure, military capabilities, and warning systems (Conley et al. 2020: 4–6). For example, inadequate land (e.g., road network and construction) and maritime (deepwater port and search and rescue vessel) infrastructures pose challenges for Alaskan residents and local-, state-, and national-level officials. The US military has also encountered difficulties in procuring cold-weather gear and funding cold-weather operations training, challenging its operational capability (Science Daily 2021).

The US Coast Guard has struggled for years to acquire the funds to construct three icebreaker 'polar security cutters' (US Coast Guard 2024). However, two are currently being designed, with the first ships due for service in the 2030s. These developments come after US Coast Guard calls for new assets for over a decade (Humpert 2024a). Moreover, by end of 2025, the Trump administration signed an agreement with Finland over the delivery of up to eleven medium class icebreakers ('Arctic Security Cutters') to be constructed in Helsinki and Texas (Humpert 2025).

In the federal US system, getting priority for Arctic security issues is naturally in constant competition with a range of other issues. The Alaskan representation in Washington, DC has had a focus on Arctic security issues over the last two decades. In 2017, Senator Sullivan notably called on President Trump to pay attention to Russian military posturing in the vicinity of Alaska: 'What has been our national security strategy in the Arctic? Well, I think until recently, from the US perspective, from the [Pentagon] perspective, it really hasn't existed' (Gramer 2017). Both Senators Sullivan and Murkowski expressed alarm over Russian and Chinese military vessels operating together near the Aleutian Islands in 2023. 'This is a stark reminder of Alaska's proximity to both China and Russia, as well as the essential role our state plays in our national defence and territorial sovereignty. Incursions like this are why we are working so hard to secure funding and resources to expand our military's capacity and capabilities in Alaska', said Murkowski of the 2023 occurrence (Murkowski 2023a).

Another relevant aspect in an Alaskan/homeland context is the relationship the USA has with Canada in the North. The security relations between these states are intertwined, almost to the extent that they are difficult to distinguish (at least in the case of Canada). As outlined in the previous chapter, NORAD provides the framework for cooperation, although the interlinkages also permeate many other security spheres such as coast guard cooperation, surveillance, and special operations (Government of Canada 2014).

However, in 1985, an incident occurred when the US Coast Guard icebreaker *Polar Sea* transited the Northwest passage from Greenland to Alaska on a resupply mission without seeking permission from the Canadian government. While the Canadian government had been informed of the voyage, it had severely underestimated the public outcry that was to ensue (Lackenbauer 2010). The 1988 Agreement on Arctic Cooperation was drafted in response and 'pledges that all navigation by US icebreakers within waters claimed by Canada to be internal will be undertaken with the consent of the Government of Canada' (Canada–United States 1988). This represented, in the words of then-Prime Minister Brian Mulroney 'a practical solution that is consistent with the requirements of Canadian sovereignty in the Arctic' (Canada–USA 1989). Taken together, these Arctic disputes have contributed to what Whitney Lackenbauer terms 'sovereignty anxiety', that is the perception that Canada is struggling to uphold its sovereignty in the Arctic; thus, it is prone to security threats in the region (Lackenbauer 2011b; Østhagen and Lackenbauer 2023).[1]

[1] For more on this, see, for example, Pharand 2007; Byers and Lalonde 2009; D. M. Johnston 2002; Lajeunesse 2008; Lackenbauer 2010; Moe and Stokke 2019; Moe 2014; Østreng 1999.

Also, as previously outlined, the *only* maritime boundary dispute that remains in the Arctic concerns delineation in the Beaufort Sea between Canada and the USA. The Beaufort Sea is the shallow portion of the Arctic Ocean located between Alaska's and Canada's High Arctic islands just north of the Mackenzie River Delta. The dispute over the location of the boundary began in 1976, when the USA protested against the line that Canada was using when issuing oil and gas concessions (McDorman 2009: 184). The dispute centres on the wording of the treaty concluded between Russia and Great Britain in 1825 (Great Britain–Russia 1825). The USA took on Russia's treaty rights when it purchased Alaska in 1867, and Canada acquired Britain's rights in 1880. In the late 1970s, Canada and the USA unsuccessfully sought to resolve the Beaufort Sea dispute.

Beginning in the early 2000s, another collaborative attempt with a renewed Arctic focus was made (Boswell 2008; Griffiths 2010). This was a partnership born of necessity as both countries required a complete scientific picture of the seabed to determine the geographic extent of their sovereign rights to an extended continental shelf more than 200 NM from shore and neither of them had two icebreakers that were capable of the task. As argued by James Baker and Michael Byers (2012), this collaborative mapping may also have opened the door to a resolution of the boundary dispute by showing that the continental shelf in the Beaufort Sea might stretch to 350 NM or even farther from the shore. In March 2010, the Canadian government signalled its desire to 'work with other northern countries to settle boundary disagreements' (Government of Canada 2010). However, discussions were suspended at some point in 2011 after the two countries decided that they would need more scientific information on the existence and location of hydrocarbon reserves before negotiating a boundary (Byers and Østhagen 2017). World oil prices dropped sharply in 2014; in 2015, Shell shut down its activity north of Alaska, without having made noteworthy findings. Then, in December 2016, both the Canadian and US sides of the Beaufort Sea were put off limits for further oil and gas development as a result of a moratorium announced by the Obama administration and the new Trudeau government (Reuters 2016).

Since the USA has not ratified the Law of the Sea Convention, despite, for example, Senator Murkowski for years arguing that it should (Murkowski 2023c), it cannot submit claims for an extended continental shelf to the CLCS. Instead, in 2023, it unilaterally claimed an extended continental shelf in the Bering Sea and the Chukchi Sea (US Department of State 2023), sparking negative reactions, with Russia threatening to withdraw from the Law of the Sea (Todorov 2024), and some scholars questioning the effects the act might have on general adherence to the Law of the Sea (Solski 2024). At the same time, the move also seemed to spark renewed interest in negotiations on the maritime boundary dispute with Canada, as a joint task force was created in October 2024 (Edvardsen and Molid Martinussen 2024b). However, with Donald Trump returning to the White House at the end of January 2025, efforts to find joint solutions with Canada in the North seem to be off the agenda.

Russia and Supporting European Allies

There is a difference between what the USA engages in domestically or in its immediate abroad surrounding Alaska and how its Arctic interests have materialised in the European part of the Arctic. As examined in the previous chapter, the USA, having recognised the geostrategic importance of Iceland during World War II, established the Iceland Defence Force in 1951 and signed a bilateral defence pact in 1961 that permitted US military personnel and basing rights in Iceland. The Barents Sea was eventually deemed integral for the Soviet nuclear triad (Breemer 1989), which prompted a continued US presence ranging from participation in exercises to military operations and storage of military equipment in Norway (Skogrand and Tamnes 2001).

When the Cold War ended, that part of the Arctic also lost some of its strategic significance for the USA. Russia, experiencing socio-economic distress after the dissolution of the Soviet Union, could neither afford to sustain a high number of military personnel in the Arctic, nor properly maintain its nuclear-armed ballistic submarines (Bowermaster 1993). Russia therefore gradually diminished its security footprint in the Arctic as new security issues, such as the wars in Chechnya, came to prominence. Russia's reduced military presence in the Arctic may also explain the US shift away from Arctic security capabilities and relative absence from the region in the first decade of the post-Cold War era (Huebert 2013b). However, as Arctic states' foreign and security policies became tied to the 'Arctic spirit of cooperation' (Devyatkin 2023), the USA also engaged in these issues – at least on a foreign policy level. As the Arctic Council continued to grow in political prominence, then-Secretary of State Hillary Clinton became the first US Secretary to attend a Ministerial Meeting, in Nuuk, Greenland in 2011 (Arctic Council 2011b).

Following the Russian annexation of Crimea in 2014, the USA launched the European Deterrence Initiative (first named Reassurance) within three months. This initiative aimed to increase the US military presence in Europe, thereby strengthening its capacity, visibility, and preparedness to discourage further Russian aggression. Significant investments were made in European military infrastructure, with a focus on nations along Russia's periphery such as Norway, Poland, and the Baltic states. The overarching goal was to deter Russia while reassuring NATO members. In 2018, the US Navy reinstated its Second Fleet, specifically designated for naval strike operations in the North Atlantic (Maritime Executive 2019). This revival not only bolstered the region's defence but also served as a formidable deterrent with an augmented fleet of high-quality vessels. Prioritising forward deterrence and control of transatlantic sea lanes of communication re-emerged as crucial aspects of US security concerns.

The tense security relations that followed between the West and Russia led the USA to also look north and counter the Russian presence there by increasing its military exercises and/or maritime security operations in the Barents Sea. For example, during the Trident Juncture exercise in 2018, the largest military exercise in Northern Europe in decades, USS *Harry S. Truman* became the first aircraft carrier of a Western state to sail above the Arctic Circle since the end of the Cold War (Eckstein 2018). As outlined in the previous chapter on Norway, B1 bombers, Seawolf-class submarines, and US Marines have all been increasingly active in Norway and the European Arctic as part of military exercises or as 'rotating forces' (Nilsen 2021a; Marines 2020; Trevithick 2020).

The USA also entered into the Supplementary Defense Cooperation Agreement with Norway in 2021 and subsequently with Finland and Sweden in 2023. The agreements allow US forces to establish military strongholds on the participating countries' territories. By aiming to enhance force readiness and reduce reaction time, the US posture in Northern Europe signified a shorter strategic warning time between US and Russian forces in potential stand-offs, crises, or armed conflicts. This was interpreted as a response to the Russian annexation of

Crimea in 2014 and the resumption of perceiving Russia as a potential threat in the North Atlantic (Østhagen 2021a).

This situation has arisen not only because the Nordic or North Atlantic states have requested help in deterring Russia, but also because of US national interests: the forces of the Russian Northern Fleet are described as a potential threat to the USA (Devyatkin 2019). The activity of the US Sixth Fleet in the Barents Sea, the reactivation of the US Second Fleet in 2018, and increased US participation in NATO exercises in Norway since 2014 – such as Cold Response and Trident Juncture – are examples of the USA's ability and willingness to engage in security activities in parts of the Arctic as required (Wegge 2023). However, this stands somewhat in contrast to engagement in Alaska, as described previously. US engagement in the European Arctic is influenced by the presence of Russia's military presence and strategic weapons systems on the Kola Peninsula.

China and Strategic Competition

It is at the same time apparent that another primary driver for US security strategy and presence in the Arctic is China. China's vocal interest in the Arctic has elicited security reactions particularly from the USA. During the Trump presidency, a 2019 DoD report to Congress concerning the Arctic avoided the terms 'climate change' and 'global warming' in favour of 'changing physical environment' and made over a dozen references to China (Office of the Under Secretary of Defense for Policy 2019). Similarly, the *Strategic Outlook for the Arctic* published by the US Navy in 2019 mentioned 'environmental change', such as thinning ice, but paid more attention to the geopolitical situation and great power competition (US Navy 2019).

An example of this China-obsession came with the Pompeo speech in Finland (US Department of State 2019). The speech was seen as undermining the efforts of all Arctic states – including the USA and Russia – to exalt the track record of mutually beneficial cooperation in the region (Devyatkin 2023). There seems to have been a concern that China's involvement in the Arctic is a means to gain access to regional resources and new shipping lanes, but that the long-term aim may be to add a military component to its Arctic presence.

This dynamic, which can arguably be interpreted is a consequence of the wider systemic rivalry between the USA and China, has prompted the USA's desire to contain China's global expansion as China grows its worldwide military and diplomatic power (Sørensen 2019). This seems to have had less to do with aggressive Chinese activity *in* the Arctic and more to do with a wider policy of restricting China's global expansion in as many domains as possible as a consequence of the emerging bipolar system. Meanwhile, the Pentagon warned in 2019 that Chinese 'civilian research could support a strengthened Chinese military presence in the Arctic Ocean, which could include deploying submarines to the region as a deterrent against nuclear attacks' (Office of the Under Secretary of Defense for Policy 2019).

Both the first Trump and the Biden administration thus paid great attention to China and Russia's influences on Arctic security and their strategic rhetoric assertively challenges the USA's perceived competitors in the Arctic region (The White House 2022b). Although the rhetoric was slightly toned down, and the opposition to climate change statements was gone, the Biden administration (2021–2025) did not diverge from this new US security approach in the North. The Biden administration released a National Strategy for the Arctic Region in 2022 (updating the 2013 one) as well as an Implementation Plan for the Strategy in 2023. The 2022 National Strategy is advanced across four pillars: 'Security, Climate Change and Environmental Protection, Sustainable Economic Development, and International Cooperation and Governance.' It also warns about China's 'dual-use research with intelligence or military applications in the Arctic' (The White House 2022c).

Russia's invasion of Ukraine is mentioned in the National Strategy as the reason behind fractured international cooperation and Russia's Arctic militarisation is seen as a catalyst for NATO unity. However, the Strategy places great importance on how China 'seeks to increase its influence in the Arctic through an expanded slate of economic, diplomatic, scientific, and military activities' and expresses alarm over Chinese icebreaker and naval vessels. The National Strategy also notes the 'risks of further militarization or unintended conflict, including those resulting from geopolitical tensions with Russia or the People's Republic of China' (The White House 2022c).

Further, in 2024, the US Department of Defense released its fourth Arctic Strategy (the others came in 2013, 2016, and 2019), stating that the 'Arctic serves as an avenue for power projection to Europe and is vital to the defense of Atlantic sea lines of communication between

North America and Europe' (US Department of Defense 2024a, 2); and particularly focusing on the role of China and Russia:

> PRC and Russian activities in the Arctic – including their growing cooperation – the enlargement of NATO, and the increasing effects of climate change herald a new, more dynamic Arctic security environment. These changes, as well as the growing cooperation between Russia and the PRC, have the potential to alter the Arctic's stability and threat picture. (US Department of Defense 2024b)

US Arctic officials have thus increasingly named China (as well as China's confluence with Russia) as security risks to the USA in the Arctic. Former State Department Arctic Coordinator DeHart stated that Chinese investments in the Arctic were 'definitely a concern' and should be 'looked at through a national security lens' (AFSA 2021). Iris Ferguson, former Deputy Assistant Secretary of Defense for Arctic and Global Resilience, stated, 'We're seeing increased geopolitical activity by Russia, as well as China, in the region' (Garamone 2022). Ted Stevens Center for Arctic Security Studies Director Randy 'Church' Kee noted that 'increased activities in the Arctic from both China and Russia represent a set of security and defense challenges that potentially hold US national interests at some level of risk' (Kee 2021).[2] Finally, in the Pentagon's 2024 Arctic Strategy, a clear assertion of the fears related to China is stated:

> Although the vast majority of the Arctic is under the jurisdiction of sovereign states, the PRC seeks to promote the Arctic region as a 'global commons' in order to shift Arctic governance in its favor. The PRC's 2018 Arctic Policy claims non-Arctic states should contribute to the region's 'shared future for mankind' due to the Arctic's global significance. Its 'Polar Silk Road' has been used to gain a footing in the Arctic by pursuing investments in infrastructure and natural resources, including in the territory of NATO allies. (US Department of Defense 2024b)

Others warn about China's long-term geoeconomic interests and actions, calling for a different way of conceptualising 'threats' from China in the Arctic (Brady 2017: 254–67; Doshi, Dale-Huang, and Gaoqi 2021: 10–13; Sun 2020). Stephanie Pezard et al. concluded that the actions and related effects of Chinese actors have been rather

[2] The Ted Stevens Center is a regional centre of the US Department of Defense.

limited. At the same time, China is increasingly attempting to gain influence in various branches of Arctic economic activity. From a US perspective, the scholars recommended several pathways to dealing with China in the Arctic: cooperation with Russia, increased involvement in Greenland, and fostering local awareness of the sensitive nature of investments and involvement (Pezard et al. 2022: 85–92).

The problem here, however, is the lack of specificity concerning the 'China threat'. Which particular part of the Arctic and/or which activity is in focus is seldom outlined. At the same time, it did become clear from 2020 onwards that China *was* indeed interested in showing an Arctic presence by participating in military exercises primarily organised by Russia in the Bering Sea/Strait domain, which also – from 2021 – includes a Chinese signalling component off the coast of Alaska.

Writ large, Russian military exercises interfering with Alaskan fishers have caused concern over the violation of US rights, as well as the fear of conflict escalation between the fishers and the Russian Navy (M. Baker 2020; The Associated Press 2020b). Then, in both 2021 and 2022, the US Coast Guard encountered Chinese and Russian warships operating together off the coast of the Aleutian Islands (Bye 2022; Mahadzir 2023). After a 2023 Russian–Chinese naval exercise near Alaska, US news outlets stoked concern about the event, calling it an 'aggressive manoeuvre' and a 'close call' (Armbruster 2023), also showcasing how Alaskan security issues had received wider US attention. Moreover, for the first time, in the summer of 2024 Chinese and Russian bombers conducted joint operations in the Alaska Air Defense Identification Zone, albeit without violating US airspace (H. Williams, Bingen, and MacKenzie 2024).

In September 2024, coast guard vessels from both countries conducted joint operations in the same region (Xiao 2024). The degree to which a Chinese–Russian Arctic partnership exists remains unclear, but the potential for the states to collaborate influences US Arctic security policy and fuels federal attention on Arctic security issues. We return to these dynamics in the next chapter, looking at the Bering Sea security subregional complex. Here it is still worth noting that we cannot discount how this increased Chinese physical presence (albeit from non-existent or low levels) in the North Pacific off the coast of Alaska seems to have further triggered US policymakers into focusing on China in the Arctic. As Trump returned to the White House in January 2025, the Arctic seemed to be high on the administration's agenda.

The Greenland Drama

As mentioned, during Donald Trump's first term as president (2017–2021), someone must have notified him about great power rivalry in the Arctic. Suddenly, he raised the Arctic on the agenda of American foreign and security policy. The second Trump administration continued with this, particularly with an emphasis on the need to acquire more control over Greenland.

In December 2024, incoming President Donald Trump reignited his ambition for US control over Greenland. On 22 December, Trump publicly declared American ownership of Greenland an 'absolute necessity' for national security and global freedom, coinciding with his appointment of Ken Howery as ambassador to Denmark (Armstrong 2024). This echoed his earlier, unsuccessful 2019 bid to purchase the island. Greenland's leadership swiftly rejected the notion. Greenlandic Prime Minister Mute Egede reaffirmed the island's sovereignty, stating, 'Greenland is ours. We are not for sale and will never be for sale.' Denmark's Prime Minister Mette Frederiksen also dismissed the idea as 'absurd', emphasising openness to cooperation but not territorial negotiation.

Despite diplomatic pushbacks, Trump's efforts intensified. In early 2025, he asserted that the USA would 'go as far as we have to go' to gain control of Greenland (Brennan 2025). Reports emerged in August 2025 alleging covert US influence operations in Greenland. Danish broadcaster DR revealed that three Americans linked to Trump were attempting to sway Greenlanders towards secession from Denmark, compiling lists of pro-Trump supporters and exploiting internal tensions (Kirby 2025). Danish Foreign Minister Lars Løkke Rasmussen summoned the US chargé d'affaires to protest what he called unacceptable interference in the Kingdom's internal affairs. Prime Minister Frederiksen echoed the sentiment, warning against undermining Greenlandic democracy. While Vice President JD Vance's March 2025 visit to Greenland was officially diplomatic, Danish officials viewed it as a pressure tactic. The controversy escalated into a rare diplomatic rebuke, signalling Denmark's alarm over US ambitions.

As described in Chapter 3, Greenland is a separate country, but not a separate state. Originally, the Norse settlements on Greenland were part of Norway and thus also became part of Denmark-Norway from 1380. When Norway became independent in 1814, Greenland,

Iceland, and the Faroe Islands remained under Danish control. Today, Greenland is still part of the Kingdom of Denmark (*Rigsfællesskabet*) but has extensive autonomy and can decide on most policy areas without interference from Copenhagen. For example, Greenland left the EU in 1985, of which Denmark is still a part.

Greenland is also important for the United States. Geographically, the world's largest island is part of the North American continent and thus a piece of the North American defence system. The United States has therefore maintained its military base Pituffik (formerly Thule) in the very north-east of the island since 1951. Today, the base is part of the US Space Force and important for monitoring rocket and space activity. The island could also become an important link in maintaining supply lines from the US east coast to northern Europe. In addition, Greenland has critical minerals that can be very valuable. It is therefore perhaps not completely surprising that a populist US administration muses about control, arguing that Denmark has not looked after either the people or the resource potential of the world's largest island. Greenland is, after all, part of the North American continent, and American presidents have expressed interest in the island a number of times in the past (in 1867, 1910, 1946, and 1955).

From a Greenlandic perspective, however, there are other interests at stake. The dream of independence is alive and well but not shared by everyone. Independence also raises a number of questions. Who will cover the large deficit needed to run services for the 57,000 citizens on the island? Who will be responsible for defence? Relations between the authorities in Copenhagen and Nuuk have not always been good and discussions about independence from the Kingdom have arisen periodically (Jacobsen and Lyngholm Lindbjerg 2024). This has become more acute as interest in the Arctic has increased. US interference in this domestic political issue is fuelling the fire. It is still unlikely that Greenland will become part of the USA. However, should a form of independence become an option, closer cooperation, both economically and militarily, with both Canada and the USA could be a natural development.

However, despite both the strategic relevance and economic potential associated with Greenland, the Trump administration (and analysts in Trump's wake) arguably overestimates the geopolitical importance of Greenland. The administration has mused about Greenland's strategic position for shipping and resource development and for countering both Russia and China in the Arctic. However, if you are to take the 'shortcut' between Europe and Asia via the Arctic Ocean then Greenland is far from the route you are using. Kirkenes, in North Norway, on the other hand, is centrally located. Should sea transport increase further across the Arctic Ocean, Greenland is unlikely to become a central part of that activity (unlike, for example, Svalbard). If you on the other hand are interested in extracting oil, gas, or minerals, then some mining projects in Greenland may be attractive, but there are other places where it is perhaps easier and cheaper to extract the same resources. The idea of Greenland as a 'treasure chest' of mineral resources is both inaccurate and ignores the fact that the Greenlandic population has often been critical of large-scale mining.

The USA has also thrown in the pretext that they must control the island for their own security because China has shown interest there, while Russia is 'only' 1,000 kilometres away as the crow flies. But being mindful of geography and geopolitics, per the fundamental point in this book, this argument also does not hold up under scrutiny. There are no Chinese advances in Greenland. And Greenland is not particularly more strategically important for dealing with Russia in the North than other parts of the Arctic. If the focus is on Russia, both Alaska and Northern Norway are more important and face more immediate security challenges. The biggest challenge seen through American eyes would be ballistic missiles or bombers from Russia, but these are handled via the American Pituffik base and NORAD (with Canada). Greenland is therefore only a small, and not the most central, part of the larger geopolitical picture in the Arctic.

Despite these nuances and the immediate and continued backlash to Trump's statements from not only Greenlandic and Danish officials, but also other European allies, the administration does not seem to have given up on the notion of more control on or over Greenland. Trump's interest in Greenland is linked to his, and the USA's, increased interest in the Arctic, as outlined here. Despite the fact that the USA is a superpower, its presence and capabilities in its part of the Arctic – northern Alaska – have not been a priority. The country has only one fully functional icebreaker, while China, which is not near the Arctic Ocean, has four. China's growing interest in the Arctic in particular has led to a new American outlook of the North.

The interest in Greenland is also a clear sign of the US's foreign policy doctrine under Trump's second administration: 'America First'.

Not as a doctrine that considers what goes against America's long-term strategic interests, but as a doctrine that speaks to Trump's voters and fan base. These people are perhaps unlikely to care about melting ice in the Arctic or the relationship with Danish authorities in Copenhagen. On the other hand, they respond positively to the image of a strong leader who promotes American interests first. Relatedly, it is clear that the Trump administration wants to put pressure on European countries to force a new normal where Europe takes responsibility for its own security and carries more of the economic burden. As Vice President Vance said in his meeting with Ukrainian President Zelensky in February 2025: 'You have to be thankful' (Maher and Liptak 2025). Thus, issues such as the Kingdom of Denmark having to pay more for the defence of Greenland become an important signpost.

However, what is being discussed is the territory of another state and assumedly also a close ally. The Trump administration's approach thus seems counterproductive. By sending his son Eric on a four-hour 'vacation trip' to Nuuk before he was inaugurated as president, sending Vice President Vance to the Pituffik Space Base in March, and at the same time threatening to use force to get what he wants, Trump created a negative backlash in both Nuuk and Copenhagen. The suggestion of using pressurising tools such as economic sanctions or even military presence is even more serious. If, in the worst-case scenario, the Trump administration was to actually use military force and extract an agreement that benefits the USA, it would undermine both the NATO alliance and the world order that the USA benefits from. An annexation would severely damage the foundation for cooperation between Europe and the USA. It would also likely create strong internal backlash from the American political system and the American Armed Forces.

In essence, the Trump debacle concerning Greenland underlines two things. First, it marks how the USA has fully taken up its role as an Arctic country and increasingly wants to engage in Arctic security issues (both in practice and rhetorically). Second, this debacle also highlights what seems to be a symptom of a larger trend in the increased interest in Arctic geopolitics and security, namely, to ignore the particularities of Arctic geography to the peril of policy decisions.

If China truly was the dominant worry for the Trump administration in an Arctic security context, more attention should arguably have been placed on Alaska and increased Chinese–Russian activity in that vicinity. As Trump invited Russian President Putin to Anchorage, Alaska for a hastily put together summit to discuss an end to the Ukraine war, five Chinese research vessels were operating in international waters off Alaska (Chan 2025). The message was clear: China is able to operate in distant waters close to the American homeland and pose security challenges in new arenas. This, however, had little to do with Greenland, located more than 4,000 kilometres away on the other side of the Arctic.

A Reluctant Arctic Actor No More

It is apparent that a shift in the US federal government's Arctic security interests and engagement occurred from 2017 to 2018. Statements from top politicians, new strategies launched by various branches of the Armed Forces, and investments – albeit modest – in Arctic capabilities in Alaska all materialised around this time. This is in contrast to the decade before, when there was much discussion about the Arctic from environmental and resource perspectives, but limited engagement in terms of security initiatives and investments. Ergo, the USA was dubbed a 'reluctant Arctic actor' (Huebert 2013b), probably in no small part due to the physical and 'mental' distance between the lower forty-eight and the state of Alaska.

What caused this shift? So far, this chapter has outlined aspects of US history in terms of Arctic security, the range of different state and federal actors engaged in Arctic security policy decision-making, and various scales of US security interests and engagement over the last two decades. From these accounts, it is clear that Russia is the primary antagonist for the USA in an Arctic regional context. This was the case during the Cold War, and Russia's military, economic, and political resurgence in the Arctic since around 2007 has also been the primary driver for Alaskan and homeland security concerns. Senators from that state, alongside various sections of the Armed Forces, have focused on the possible threat emanating from just across the Bering Strait as well as from Russia's strategic arsenal located on the Kola Peninsula.

Despite regional security relations across the circumpolar region having remained amicable in recent decades, increased tensions since 2014, especially in the North Atlantic, prompted the USA to revisit its security strategy for the Arctic. US naval activity and increased US participation in NATO exercises in Norway demonstrate the USA's increased security presence in the Nordic NATO context. This security shift occurred after relations between Russia and the other Arctic states deteriorated due to the 2014 Ukraine crisis. US engagement came about both as a realisation of the need to manage a possible emerging threat from Russia in the North Atlantic and because NATO allies – especially Norway – have been calling for increased US engagement in European Arctic defence (Heier 2018).

Still, Russia's modernisation of its Arctic military capacities (in line with its politically stated goals) started before 2014. In 2008, Norway launched what it called a Core Area Initiative in NATO, in order to get the alliance – and primarily the USA – to focus more on traditional questions of territorial defence, due to increasing concerns over Russian military build-up in the European Arctic. Russian strategic investments and activity did not, however, spur the USA into publishing a range of Arctic security documents, statements, and strategies. Around the same time, the USA *left* Keflavik air base in Iceland, deactivated the Second Fleet with responsibilities for the Atlantic, and – under President Obama – Hillary Clinton and Sergey Lavrov literally pressed the reset button on relations. Similarly, Alaskan calls for focus on Arctic *security* issues were not met until they acquired a Chinese flair. Such was the case after the Russian–Chinese naval exercises near Alaska from 2021 to 2023 and the Alaskan senators' subsequent reactions.

However, as showcased, China's increased engagement in the Arctic due to its heightened global prominence and cooperation with Russia has been a central, if not a primary, factor in increased US attention on the Arctic. It is perhaps telling that in the Pentagon's 2024 Arctic Strategy, the first actor it focuses on in a separate heading is China, not Russia (US Department of Defense 2024b). However, some have cautioned about the 'limited value of the Arctic as a realm of Sino-American competition' (Lajeunesse and Choi 2022). From a China-centric perspective, 'the level of access and target density [in the Arctic] simply do not exist' (Lajeunesse and Choi 2022: 1065). This is consistent with the relatively low level of Chinese strategic engagement with the Arctic. However, it is clear that China represents a long-term strategic challenge for the USA in the Arctic, as in other parts of the world. From a US-centric perspective, China's limited presence and growing interest in the Arctic is enough to warrant a reluctant-no-more US Arctic security strategy.

In other words, China's engagement in the Arctic has been a central factor underpinning increased US security engagement in the Arctic – especially when compared to US strategy vis-à-vis Russia. The effects of the war in Ukraine impact Chinese–Russian cooperation in the Arctic and Western perceptions of their partnership (Hsiung 2022). Rhetoric has increased in strategic documents and public statements about dangerous Russian–Chinese confluence and how China's interests do not necessarily align with the interests of Western Arctic states (Clark 2024).

This falls in line with other strategic studies of US foreign policy shifts, focusing on the global or structural level as a decisive factor in understanding US policies (Homolar 2011). In contrast, perhaps, to the other Arctic states, the USA – whether it wants to be or not – is entangled in a systemic outlook that forces certain priorities. That is not to say that there is no agency involved. The actions taken by the first Trump administration and its officials in focusing explicitly on China in the Arctic were examples of agency by US actors, shifting focus away from cooperation on 'soft' policy issues to 'hard' security matters in an Arctic context.

At the same time, Russia's full-scale invasion of Ukraine in 2022 shifted attention in the Arctic towards Russia. Putin's decision to send troops into Ukraine in 2022 led to the seven other Arctic states condemning Russia's military actions and the suspension of regional cooperation in various forums, such as the Arctic Council. In 2022, the White House announced the termination of scientific cooperation with Russian government-affiliated institutions (The White House 2022a). Further sanctions and suspensions of dialogue were imposed on Russia. The accession of Finland and Sweden to NATO means that US security engagement in the European Arctic is unlikely to be reduced in years to come – in contrast to the rest of Europe and despite general calls from the USA for European partners to share more of the regional security burden.

The US security footprint, namely what it spends funding on and where it trains its Arctic-positioned troops, is conversely, heavily skewed towards Russia-focused deterrence activity, especially in the North Atlantic and European Arctic. That, however, might change, due to increased Sino–Russian military activity just off the coast of Alaska (Tingstad, Pezard, and Shokh 2024b). The 2022 war in Ukraine further severely affected Arctic security, but it too can be seen in the context of intensifying USA–China competition and has led to a new level of Chinese–Russian collaboration (Chen 2023).

In sum, despite a change in US Arctic security concerns after 2014 in the context of the Ukraine crisis, what truly seems to have spurred the USA into a new Arctic modus operandi – at least in terms of policy statements and speeches – is increased global competition with China. That does not discount the other drivers examined in this chapter: the role of Alaska in US security considerations; security and safety concerns in the Bering Sea/Strait region; the pull factor of European NATO allies arguing for a more determined presence to deter Russia in the European Arctic; the push factor in terms of US security concerns linked with Russian Arctic rearmament and the nuclear triad; and the decline in cooperation given increasingly tense security relations with Russia.

Still, the USA is a different actor altogether to the other seven Arctic states. Its global posture and enduring status as the world's only superpower imply a different range and breadth of security concerns. Former President Obama and security analysts have referred to Russia as a regional power (Borger 2014; O'Brien 2023). China, however, prompts a different response from the USA, starting with Obama's 'Pivot to Asia'. Arctic security concerns would no doubt have become more important even without the rise of China. However, one cannot understand statements such as President Trump's Greenland purchase proposal, former State Secretary Pompeo's 2019 Rovaniemi speech, or how the Biden administration maintained some of Trump's security rhetoric concerning the Arctic without taking into account the Chinese dimension.

Looking at the previous chapters and what has been outlined in this one, if we take the case of Norway, its foreign and security policy decisions in the High North are, of course, influenced by developments at the other levels of analysis (Østhagen 2021a). The balance of power and tensions between the USA, China, and Russia at the international level have direct consequences for Norwegian practices, especially as it relates to Russia–NATO dynamics. Also, at the regional Arctic level, the various circumpolar agreements and dynamics within the Arctic region – such as the Arctic Council's work, the Ilulissat declarations of 2008 and 2018, and discussions of how to improve regional security relations – have an impact on High North security dynamics and Norway's immediate foreign policy. Initiatives and developments on the Arctic level can spur actions or impact the Norway–Russia bilateral relationship (for example, the 2010 boundary agreement). However, North Atlantic security dynamics are only to a limited extent impacted by USA–China relations and related global struggles. These dynamics materialise more prominently in the Bering Sea/Strait region.

Similarly, looking at the USA and its Arctic security practices, although it obviously has a global outlook on security interests and engagements, in an Arctic context the USA is also restricted by geography, as geography defines its scope for action. This action revolves around the Bering Sea and Bering Strait, as a domain where increased Russian–Chinese activity is spurring concern and a security response. In the same domain, the USA and Russia did – up until 2022 – initiate bilateral efforts to manage increased Bering Strait traffic and improve coast guard information sharing and emergency response. And it was not accidental that Trump and Putin met in Alaska in 2025, as they mused restarting economic cooperation in that part of the Arctic (Nichols and Ayaydin 2025).

In essence, describing the national level is more complex than can be explained solely by dynamics at another level (Østhagen and Rottem 2023). Clearly, a state's deliberations on the development of foreign policy consist of more than just a balance of power and/or common interests. The relationships in the North between Norway and Russia and the USA and Russia are examples of activity at the national level that is not necessarily characterised by either a systemic balance of power or regional cooperation but includes a bit of both. The national perspective taking into account the immediate security practices that emanate from the near abroad seems a more instructive approach to uncover the developments of Arctic geopolitics, than broad-sweeping regional generalisations.

What this and the previous chapter have done is thus to showcase the differences between two Arctic coastal states, both of which have a considerable population in the Arctic and have placed emphasis on their northern regions in various policy documents and security initiatives over the last decade. Still, all states are not equal. The differences between Norway's and the USA's Arctic security interests, approaches, and tangible engagements vary according to size, reach, and power. Norway has been working to get the USA engaged in *its* near abroad to manage an increasingly antagonistic Russia. The USA does not need Norway to reciprocate, despite Norwegian troops flying to Alaska to participate in special forces exercises (US Department of War 2025).

At the same time, geography, regional constructs, and regional security complexes do come into play. Alaska and its role in US security concerns, as well as the geographic domain that it lies in, infer certain security dynamics that the USA increasingly cannot ignore. As such, the USA is more similar to Norway than one might initially assume. Even the world's only superpower cannot escape the dictates of geography and the concerns that arise from the geopolitical alliance between Russia and China in that specific domain. Exactly *how* we can theorise these trends, and the relationships between geography, security, and the Arctic, or subarctic, regions, this book now turns to.

7 *Arctic Security Subregions*

This chapter will uncover more of Arctic geopolitical dynamics by doing several things. It will lay out the argument for why, in fact, it makes sense to divide the Arctic into subregional security complexes (SRSCs) when we discuss security and geopolitics in the North. Thereafter, this chapter will take a dive into the two SRSCs that in fact make up security relations and dynamics in an Arctic setting, namely the Barents SRSC and the Beringia SRSC. This is arguably where most Arctic security interactions and practices have materialised over the last decades, but these domains are only slightly, if at all, interlinked.

Centring SRSCs

This book has so far established that when we study Arctic security dynamics, the circumpolar lens is often too broad and fails to capture the complexities and nuances of the dynamics that take place in parts of the Arctic. As particularly apparent in the previous chapter, examining US and Norwegian security practices, when we study and describe Arctic security relations, the national perspective taking into account the immediate 'near abroad' is often the more fruitful approach to uncover what is actually happening in terms of security and defence issues. What emerges from such a distinction is the emphasis on, and role of, geography. It is clear that the defining features that impact state security practices are size of the spatial domain in question, distance between states, and operability.

Thus, only a few Arctic issues relating to security described in Chapters 3–6 have a pan-Arctic or circumpolar effect, such as diplomacy and cooperation in pan-Arctic forums and the threat from Russian strategic forces. Instead, subregions, or SRSCs, are where security issues and practices compound. Arctic security dynamics and practices are in other words primarily grounded in Arctic SRSCs, not in the whole circumpolar region (see Figure 7.1).

Figure 7.1 Map of the Arctic, with the two subregions highlighted by circles. Note: The location of the circles is not accidental. The European/Barents subregion is located further North than the Pacific one, basically due to different climatic conditions defining the operating areas and in turn where security practices take place. Circles added by author. By CIA World Fact Book, Public Domain, www.cia.gov/the-world-factbook/static/97828fc01510ce36884bf7e07a508436/arctic_phy.pdf.

Buzan and Wæver also highlight the importance of what they term *subcomplexes*: 'Subcomplexes represent distinctive patterns of security interdependence that are nonetheless caught up in a wider pattern that defines the RSC as a whole.'[1] The idea of subcomplexes concerning the Arctic has been mooted before. Åtland hinted at this in 2007, with his examination of the European Arctic as a possible separate regional security complex. Karsten Friis (2021) did the same in his article 'Analyzing Security Subregions: Forces of Push, Pull, and Resistance in Nordic Defense Cooperation'. However, both scholars do not relate these subregions to the wider notions of an Arctic region, or security region. Moreover, Wilfrid Greaves (2019), in his article 'Arctic Break Up: Climate Change, Geopolitics, and the Fragmenting Arctic Security Region', takes a different approach, arguing that the Arctic is in fact breaking up into subregions due to both climatic and security changes. The premise was thus that the Arctic, for a brief period in time, existed as a distinct regional complex on its own, but that is now changing: 'The result will be a circumpolar region that is less distinctly 'Arctic' than in the past, as the exceptional and cooperative nature of recent Arctic politics is replaced by adjacent security sub-regions characterized by different combinations of great power influence, economic nationalism and investment, environmental change, and ongoing human insecurity.' (Greaves 2019)

This builds on the premise that there was an Arctic regional security complex, namely that the Arctic states' security practices were truly intertwined, and that the region did serve – even for a brief time – as an outlet of these. Indeed, perhaps for some states such as Canada and Norway in the period from around 2005 until 2014, the spatial-political outlet for security interactions and initiatives truly was the Arctic (although they emphasised different Arctics).[2] From 2005 onwards, the Norwegian government argued that the Arctic was Norway's 'primary strategic foreign policy priority' (Stoltenberg et al. 2005: 6). Similarly, Canadian Prime Minister Stephen Harper elevated

[1] 'Subcomplexes have essentially the same definition as RSCs, the difference being that a subcomplex is firmly embedded within a larger RSC. Subcomplexes represent distinctive patterns of security interdependence that are nonetheless caught up in a wider pattern that defines the RSC as a whole … Subcomplexes are not a necessary feature of RSCs, but they are not uncommon either, especially where the number of states in an RSC is relatively large. The device of subcomplexes eliminates most of what might otherwise occur as disturbing cases of overlapping membership between RSCs: e.g., if the Gulf and the Levant were seen as separate RSCs, Iraq would be a member of both but, with these as subcomplexes, Iraq can be a member both of the Gulf subcomplex and of the wider Middle Eastern one' (Buzan and Wæver 2003: 51–52).

[2] See Østhagen, Sharp, and Hilde 2018.

the Arctic as a national priority when in office from 2006 (Dodds 2011; Lackenbauer 2013b).

However, even then there was an explicit divergence between practices and defence and foreign policy interests, both domestically in each of those states, and between them (Østhagen, Sharp, and Hilde 2018). Undoubtedly the Arctic (what the two states meant by that – High North and the North in their respective contexts) was central to both states and still is. But as explained in Chapters 4–6, these different Arctics did not, and still do not, sufficiently overlap to constitute a coherent region or even subregion – at least not in terms of security practices. The Arctic does not become a security region just because state leaders emphasise the importance of their respective parts of the Arctic. However, we can instead divide the Arctic into several distinct, albeit slightly overlapping, SRSCs when examining security relations. The two main ones of interest here are the Barents SRSC (or the North Atlantic SRSC) and the Beringia SRSC (or the North Pacific SRSC). This adds another layer of complexity and nuance to the three-level analysis so far employed in this book (see Figure 7.2).

What marks these SRSCs is that they are the spatial parts of the Arctic where states' security interactions are more intense than in other parts. The geographical proximity of states is the first key variable. The second is security interactions driven by the presence of military forces for both conventional and strategic purposes. Therefore,

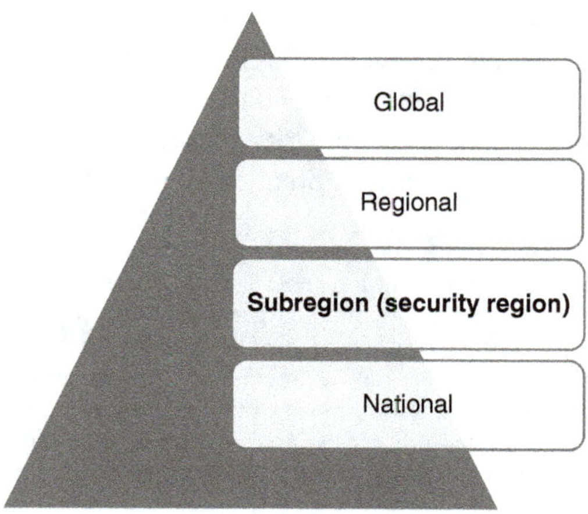

Figure 7.2 A simple division into four levels makes it easier to delineate the various dynamics of the Arctic from each other, with emphasis added on the additional 'subregion'.

although we might also think of the North-West Atlantic domain (Greenland, Canada, Iceland) as a third SRSC in terms of state interactions or in terms of resource management and ocean governance, it lacks the security component that is present in the other two highlighted areas. The Kingdom of Denmark, Canada, and Iceland are allies under the NATO umbrella and collaborate closely on security issues – and despite Trump's ideas about control over Greenland, we are currently not witnessing sufficient security practices in and around Greenland to warrant a whole new subregional complex emerging.

Another candidate for a SRSC would be the territory and maritime domain in the Russian Central Arctic and Far East, stretching from Novaya Zemlya to the Chukchi Sea. This would incorporate vast parts of the Russian Arctic and the Russian maritime domain towards the North Pole. As we explored in the previous chapter, this is also a region of strategic and military importance. However, there are limited security interactions amongst states in this part of the Arctic. This is due to the remoteness and distance from nearby Arctic states, whether it is from Norway and Finland to the West, the USA to the East, or Canada and Greenland across the North Pole. We should also note that the 'High Arctic' – the area surrounding the North Pole – does not qualify as a SRSC in terms of security simply because there is limited state presence and activity there, and the proximity to both states' territory and security interests is too great.

From a state security perspective, it is arguably therefore most relevant to examine the two SRSCs with the highest levels of security dynamics and related practices, the Barents SRSC and the Beringia SRSC. These two parts of the Arctic are defined by a maritime domain/sea (Barents and Bering Seas), where Russia – as the main security concern for the other seven Arctic states – is forced to interact with one or more of the other Arctic states due to geographic proximity. Moreover, the presence of the Russian strategic fleets – the Pacific Fleet and the Northern Fleet – deems both domains crucial in Russian strategic thinking and thus also requires a response from NATO countries. At the same time, the two domains are markedly different along a range of parameters.

The Barents SRSC

The Barents Sea is located north of the Norwegian and Russian mainland, just off the Arctic Ocean itself (see Figure 7.3). Its boundaries are set by the Svalbard archipelago (Norway) to the west, Franz Josef

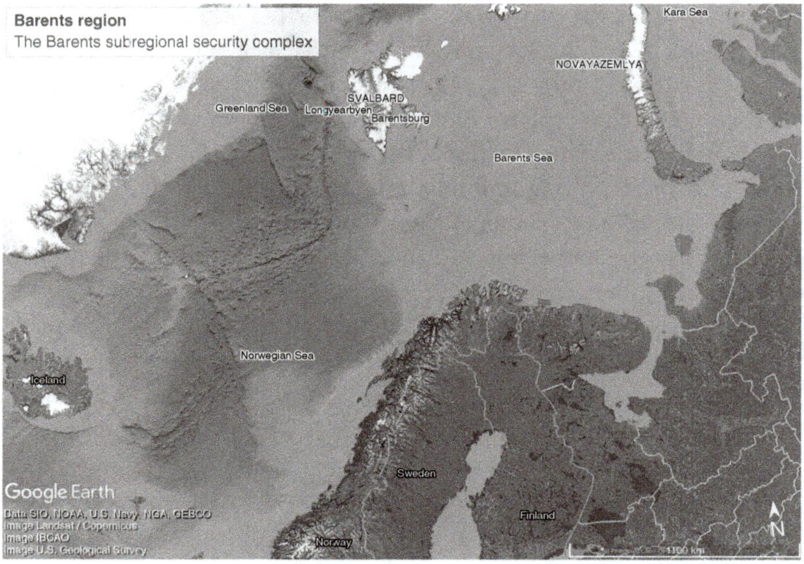

Figure 7.3 Image of the Barents subregion. Made with Google Earth.

Land (Russia) to the north, Novaya Zemlya (Russia) to the east, and the European continent (mainland Norway and Russia) to the south. The Sea is named after the Dutch explorer Willem Barentsz, who ventured northwards in search of a sea route to Asia via the northern hemisphere in the late sixteenth century.

When looking at the North Atlantic/Barents Sea, the defining actor is obviously Russia. The Barents Sea holds strategic and economic importance for Russia for several reasons: it provides access to the Arctic Ocean as well as the Atlantic Ocean writ large as its northwestern ports remain ice-free even during wintertime. As described in Chapter 4, the strategic military importance of the Kola Peninsula and this part of the Arctic cannot be underestimated for the Russian state. The other primary state in this context – in addition to Russia – is Norway. Historically, being the only northern NATO member bordering the USSR, and with its vast coastline exposed towards the Barents Sea and the Arctic, Norway plays a particularly central role in that subregion's security dynamics.

Iceland and Greenland are also relevant; however, their limited population and lack of military assets do not infer great regional clout vis-à-vis Russia and they are located at a relative distance. Finland

and Sweden are relevant too – especially after their decision to join the NATO alliance in 2022 – although a limited northern population, lack of infrastructure, and political prioritisation of the North infers less immediate significance.

In the Barents Sea domain, the change in the defence and security posture started in 2007, as Russia resumed Cold War patterns of military activity on the Kola Peninsula and then engaged in conflict with Georgia in 2008. Norway responded by, for example, launching the Core Area Initiative in NATO in 2008 and indeed had already – from 2006 – initiated the Cold Response military exercise in the North. This is the start of the contemporary tit-for-tat response and military build-up that has continued in this maritime and land domain, primarily driven by Russian military expansion. When, on 24 February 2022, Russia launched its full-scale invasion of Ukraine, previously held Western assumptions about Russian calculations and the Arctic, and especially the Barents Sea, as an exceptional or 'low-tension' space were questioned. In many ways these dynamics peaked in 2024 when Sweden was accepted as a NATO member, deeming the whole subregion a NATO–Russia-defined spatial domain.

The Svalbard Focal Point

The archipelago of Svalbard stands out as a specific geographic location for the importance of geography and security-related dynamics – or geopolitics – in the Barents Sea domain. A range of other cases exist, such as the land border between Norway and Russia, or the extensive 1,340 km long now-NATO border between Finland and Russia. However, Svalbard is a particularly relevant example of geopolitics, security, and state concerns, all in a SRSC. As a fibre-optic cable running from Svalbard to the Norwegian mainland was disrupted just weeks before the Russian invasion of Ukraine (Fredriksen et al. 2022), fears of hybrid activity perpetrated by Russia rose across the Arctic region, and Svalbard became a focal point for some of the renewed public concern about 'Arctic geopolitics'.[3]

[3] The Norwegian public broadcaster NRK highlighted the archipelago's particularities and conflict potential (Fredriksen et al. 2022; Finnset, Berg, and Rostad 2022; Wormdal 2022; B. M. Pettersen et al. 2023), while international broadcasters such as the *BBC* (BBC News 2023), *The Guardian* (Bryant 2023), *Le Parisien* (Brugeron 2022), *USA Today* (USA Today 2023), *Deutsche Welle*

Traditional geopolitical analyses focus on the relationship between geography and conflict-driven politics amongst states. In this instance, the mere presence of Russian – and previously Soviet – communities on Svalbard can serve as a case study for geopolitical rationales as manifested in human activity and physical infrastructure (Houck 2022). Moreover, as activity has increased in the near-Arctic area because of changing geophysical conditions as well as states being eager to showcase their Arctic economic interests and research capabilities, Svalbard has become an increasingly popular topic in discussions of power politics and their manifestations – through Arctic policy documents, research stations, or geoeconomic actions.

Over time, as Norway has implemented stricter or more coordinated regulations, applying to everything from tourism to research, and limited certain types of activities that do not fall under the scope of the 1920 Treaty, some Treaty signatories have voiced discontent over specific issues (J. H. Jørgensen 2010; Pedersen 2008a). Such complaints have come primarily from the USSR, and later Russia, as the only state with a sizeable population in their own communities on Svalbard (ИноСМИ 2017). These complaints have concerned Russian companies not being allowed to use helicopters for tourism; what Russia has claimed as being extensive environmental regulations and the creation of national parks; and questions concerning the alleged use of the Svalbard satellite station for military purposes (Ministry of Foreign Affairs of the Russian Federation 2020; Moe and Jensen 2020).[4] The Norwegian government's response has been to focus on how Norwegian policies regarding Svalbard and the Russian population do not in fact breach the Treaty (Moe and Jensen 2020; High North News 2016; Wormdal 2022; B. M. Pettersen et al. 2023).

In particular, Russia has complained about Norway allegedly using Svalbard for military purposes, in breach of Article 9 of the Treaty

(Nahmen 2023), and *Al Jazeera* (Al Jazeera 2022) issued lengthy features about Svalbard. A pinnacle of these unspecified concerns came in May 2024, when the owners of the last private property on Svalbard – constituting 0.1 per cent of the land – tried to raise global interest in buying that property by stating its 'geopolitical significance' and possible Chinese interest in it, albeit without further specifying what, whom, or why (Rappaport 2024; Guy 2024; France24 2024).

[4] For a detailed description of the history of such complaints and the rationale behind them, see Jørgen Holten Jørgensen's 'Russisk svalbardpolitikk (Russian Svalbard Policy)' (J. H. Jørgensen 2010).

(Tamnes 2020). The Norwegian Coast Guard regularly docks in Longyearbyen to resupply, and the Norwegian Navy sends a frigate to Svalbard once a year to highlight Norwegian sovereignty and capacities in the area. Russia sees this as a challenge to the Svalbard Treaty – although, as already discussed, the Treaty does not hinder Norway having a military presence on or around the archipelago as long as the purpose is not 'warlike' (Finnset, Berg, and Rostad 2022). This points to the crucial difference between being a 'demilitarised' zone and adhering to the actual wording of Article 9 in the Treaty.[5] Russian sensitivities to the question of military activity on Svalbard relate, to some extent, to the proximity of Svalbard to the Northern Fleet on the Kola Peninsula and its strategic position for defending Arctic territory and projecting power in the GIUK (Greenland, Iceland, and the United Kingdom) and Bear (after Bear island) gaps – a point I return to below.

Similar complaints have been raised by Russia concerning the Norwegian satellite station located on Svalbard – which is the world's largest commercial ground station (KSAT 2021) – where Russia has questioned whether the data gathered are used for 'warlike purposes' (B. M. Pettersen, Schaubert, and Eriksen 2022). This point is difficult to prove or refute, as satellite data concerning weather and climate might be used for a range of purposes and spread to a wide range of users. Still, in 2022, Norwegian authorities denied such use of Svalbard's satellites to a US company and a Turkish company, both of which have links to the military. Some argued that this was due to heightened Norwegian sensitivities related to Article 9 (Wormdal 2022).

Beyond opposition to Norwegian policies and research coordination, it is clear that, since Russia's full-scale invasion of Ukraine in February 2022, the Russian authorities seem more intent on reinvigorating their Svalbard presence for both symbolic and strategic purposes (Haugli and Bjøru 2023a).[6] In the aftermath, and in tandem with tense relations between Norway and Russia over a range of

[5] 'Norway undertakes not to create nor to allow the establishment of any naval base in the territories specified in Article 1 and not to construct any fortification in the said territories, which may never be used for warlike purposes' (Svalbard Treaty 1920).

[6] See also Mathieu Landriault (2024) on how the Russian media narrative changed after 2022, becoming less concerned with highlighting that Svalbard is a part of Norway.

issues – including shipments to Svalbard in the summer of 2022 – Russia declared its wide-ranging plans for a new research station on the archipelago (Edvardsen 2023a). This turned into the larger plan that Trust Arktikugol announced to revive the abandoned settlement of Pyramiden (Nilsen 2023c). What is to come of this plan is yet to be seen. The Norwegian Intelligence Service has argued that Trust Arktikugol 'will increasingly become dependent on government transfers' while also seeking to expand its activities in other areas beyond coal mining (Norwegian Intelligence Service 2023: 19). This is likely what is motivating Trust Arktikugol to make increasingly lofty and sweeping statements about future plans and investments, alongside an increase in applications to the Governor of Svalbard for the right to initiate new activities (Edvardsen 2023b).

Moreover, in May 2023, the Russian consul general in Barentsburg led a parade that involved a helicopter and dozens of vehicles waving Russian flags to mark the anniversary of the USSR's defeat of Nazi Germany (Staalesen 2023a). A few months later, Trust Arktikugol unlawfully erected a cross outside of Pyramiden, which was interpreted as in support of the Russian war in Ukraine (Nilsen 2023b). These actions are probably as much aimed at decision makers in Moscow, to raise awareness of Russia's Svalbard positioning, as they are meant as a challenge to the Norwegian authorities. At the same time, this must be seen as part of a larger Russian strategy that aims to both strengthen its Svalbard presence and increase pressure on Norwegian authorities, perhaps in order to force them to overreact or overreach. This is also about upholding a form of 'strategic uncertainty' and an escalatory approach that hopes to provoke Norway to violate the rights of third-party nationals (Østhagen, Jørgensen, and Moe 2020).

Still, in response to such symbolic actions and concerns over possible tension and escalation in Svalbard, Norwegian authorities – through the Governor of Svalbard and the Norwegian Police Security Service – have increased both their presence and their openness regarding their presence (Haugli and Bjøru 2023a). While these actions and complaints from Treaty signatories do not directly challenge Norwegian sovereignty over Svalbard, the sum of complaints could amount to a larger challenge to how Norway adheres to and interprets the Treaty. This also ties into concerns that some have voiced after 2022 about the Norwegian authorities losing control over foreign activity on the islands (Haugli and Bjøru 2023b).[7] In addition, the concern has been that if Russia wanted to escalate a conflict while retaining some form of plausible deniability, it could initiate actions based on these complaints to undermine Norwegian sovereignty.

Russia looms large in all the geopolitical dimensions highlighted. Military use of the archipelago is relevant only in a scenario with a NATO–Russia conflict. Other possible adversaries in a large-scale conflict (e.g., China) are simply too far removed from Svalbard, underscoring the importance of geography in this context. There are two aspects of these security dynamics and practices that may intensify further, linking to the wider SRSC. The first is the issue of access to resources, and possible attempts by vessels from various states to claim their rights at sea, as seen with the EU in the snow crab case (Østhagen and Raspotnik 2018; Tiller and Nyman 2017).[8] The fact that snow crabs are defined by all relevant parties as sedentary species also means that how they are managed sets a precedence for other shelf resources such as hydrocarbons and seabed minerals. The attention to oil and gas development on the shelf around Svalbard has somewhat

[7] Relatedly, an issue that attracted considerable attention and critique from parts of Longyearbyen's population was the Norwegian Government's decision in mid 2022 to add a requirement to who could vote and stand for elections for the Longyearbyen Community Council (Norwegian Ministry of Justice and Public Security 2022a). The government stated, 'Persons elected to the Council must have a good knowledge of the objectives of Svalbard policy and the special framework conditions for Svalbard. Consequently, a requirement that foreign nationals have at least three years of residence in a municipality on the mainland to be able to vote and stand for election in Longyearbyen has been introduced' (Norwegian Ministry of Justice and Public Security 2022b: 47–48). For an academic analysis of the community development in Longyearbyen, see Sokolíčková 2022.

[8] In recent years, however, the EU represented by the European Commission has been the most active proponent of the view that Norway must adhere to the Svalbard Treaty's non-discrimination provisions concerning activities at sea. This issue was raised on the political agenda in Brussels through a dispute between Norway and the EU over the right to fish for snow crabs around 2015 (Østhagen and Raspotnik 2018; Raspotnik and Østhagen 2019). It received further attention with a dispute over cod quotas from 2020 to 2021 that came about as a result of Brexit (Moe and Jensen 2020; Moens and Galindo 2021). The EU even went so far as to award itself licences to catch snow crabs, breaking with both its own position that Norway still has jurisdiction to manage economic activity in the area, and with coastal state rights under the Law of the Sea (Moe and Jensen 2020). That case ended up in the Supreme Court of Norway, where the ruling confirmed the Norwegian position on the issue (Supreme Court of Norway 2023).

subsided since the heyday of Norwegian Arctic oil and gas interest around 2007–2010, although that might change in the future.

Some have speculated whether the role of China – with the world's largest fishing fleet – and potential Chinese claims to equal access to fishing rights could come into play concerning fisheries in the Arctic (Hoel 2023; Jonassen 2023). Liu speculates that Svalbard and the surrounding FPZ could become a focal point for this interest, although China is also wary of causing a rift with Norway over the issue, as China seeks cooperative relations with Norway and portrays itself as a benign actor in Arctic affairs (N. Liu 2021). As mentioned, in China's Arctic Policy from 2018, the Spitsbergen (Svalbard) Treaty is referred to six times to underscore China's Arctic rights, interest, and history, stating that 'China maintains that all activities to explore and utilize the Arctic should abide by treaties such as the UNCLOS and the Spitsbergen Treaty as well as general international law' (State Council of the People's Republic of China 2018).

The second aspect in this SRSC that might be a cause for concern is a possible escalation of incidents between Russian vessels and the Norwegian Coast Guard that could get caught up in wider security concerns (Østhagen 2018). Russian fishers are – per the administrative fisheries cooperation – allowed to fish in the FPZ, but the Norwegian Coast Guard is responsible for fisheries inspections and possible fines and/or arrests. However, the Russian state makes a point of not recognising Norwegian jurisdiction to inspect and arrest in these waters (A.-K. Jørgensen and Østhagen 2020). This has led to tense situations where Russia has threatened the use of military force to 'protect' the rights of its fishers (Østhagen 2018).[9] The tenser the wider security relations between Norway and Russia, the greater are fears that such small-scale incidents at sea involving the Norwegian authorities and Russian fishers could escalate out of proportion (Østhagen 2023d). It should, however, be noted that on a day-to-day basis, the Norwegian Coast Guard regularly inspects Russian fishing vessels without any incidents or escalation as Russia de facto accepts Norwegian jurisdiction, even if the captains of the vessels refuse to sign the inspection documents afterwards (Hønneland 1999).

[9] Most notably the *Elektron* incident in 2005 and the Gorbatsjov/Pokramovitsj incident the same year (Lepperød 2005; Tjønn 2005). For analysis of the former case, see Inderberg 2007; Bruusgaard 2006; Åtland and Bruusgaard 2009; Tiller 2010; Fermann and Inderberg 2015; Østhagen 2018.

Complicating the matter for Norway is the fact that both fishing vessels and research vessels from Russia have access rights to Norwegian waters in the Barents Sea that are difficult to curtail. That fishers with a quota may fish in all of the Barents Sea constitutes a core pillar of the successful co-management scheme of fisheries cooperation between Norway and Russia (Hønneland and Jørgensen 2015; Hønneland 2014; Stokke 2022; A.-K. Jørgensen 2022). Norwegian politicians have remained wary of disrupting this regime, even after Russia's invasion of Ukraine in 2022 (Oma Vyvial 2023; Hønneland and Jørgensen 2025).

As this Svalbard discussion showcases, these issues are part of the Barents Sea security domain which is important for Russian power projection. Russia will in all likelihood be intent on rebuilding its Arctic force posture and capabilities that have diminished through fighting in Ukraine and in response to Finland and Sweden joining NATO. As put by the Norwegian Intelligence Service, 'Svalbard has military strategic significance for Russia, and the Kremlin will prioritise presence also in 2023, despite failing economic rationales'[10] (Norwegian Intelligence Service 2023, 19). At the same time, calling it 'NATO's Achilles Heel', as James Wither did (2018), might be taking this argument too far and overestimating Svalbard's strategic position in Russian military thinking. With a focus on security dynamics, equally central are the Barents Sea dynamics evolving as increased economic activity in this part of the Arctic takes place, most notably fisheries, shipping, and – possibly – seabed mining (Hovelsrud, Kaltenborn, and Olsen 2020; Ylvisåker 2020; Hvinden and Østhagen 2024).

Finally, a related and relevant question is whether Russian actions (or those of any other state) in the Barents Sea might trigger a collective NATO response, despite diverging legal interpretations about sovereign rights in this maritime domain – as discussed above. However, this is less relevant when discussing Svalbard's strategic position, as there is limited strategic advantage in operating *from* the FPZ specifically. It is only holding Svalbard territory as a forward operating base for North Atlantic operations that would make sense in that scenario – and even that is, as mentioned, questionable due to the limited benefits of such an operation.

However, there are more lenses through which to understand Russian actions and interests in Svalbard, or the wider Barents SRSC,

[10] Author's translation.

beyond the security-oriented or military ones that look only at defence strategies and possible hotspots for escalation. As the only other state with communities on Svalbard, and with interests in economic activities ranging from coal mining to tourism and fisheries, Russia also benefits from maintaining the status quo. The current situation on Svalbard favours Russian economic interests, as well as the wider Russian desire to ensure that regional relations in the Barents Sea remain politically stable. Scholars have thus concluded that undermining the Svalbard regime or disregarding the Treaty itself have so far not been in Russia's interests (Todorov 2020a, 2020b; Moe and Jensen 2020; A.-K. Jørgensen and Moe 2023).

The Changing Nature of the Barents SRCS

In turn, what does this walkthrough of contemporary security challenges, concerns, and dynamics regarding the Svalbard archipelago tell us about the Barents SRSC? It showcases how a subregional space and related security dynamics – in this case primarily between Norway and Russia – can cluster around specific spatial objects or domains. Svalbard is unique in the sense that although it is Norwegian territory, it has a somewhat shared history between Norway and Russia, it is located close to strategic Russian forces, and it is the spatial focal point for much of the specific security concerns and tensions between Norway and Russia. Moreover, Svalbard – due to the Svalbard Treaty and its unique set-up – has led to interest from a range of non-regional actors and might – with the increased interest of countries such as India and Turkey – become a central security element in the European Arctic and thus also wider Arctic security relations.

The art of balancing Norway vis-à-vis Russia writ large has had to exercise in the Barents SRSC is about more than just deterrence and reassurance regarding foreign base policy, nuclear weapons, and military exercise activity. The balance has been about securing allied support for deterrence, while at the same time Norway's policymakers have worked to preserve everything from Arctic cooperation forums to assessing the possibility of new measures vis-à-vis Russia to reduce the security policy tension in the North (Østhagen 2023b).[11]

[11] These in themselves are not necessarily opposites, and the idea that there is an ideal balance point is perhaps artificial (e.g., Bjur 2021).

Decades of cooperation that started during the 1990s, primarily through the Barents Euro-Arctic Council and related mechanisms between Norway and Russia, but also including Finland and Sweden, were meant to enable more stable subregional relations and foster stability and even 'peace' (Onsager and Stuvøy 2022; Hønneland 2012b; Leland et al. 2008).

Despite its considerable military and foreign policy capacity, Norway is still relatively limited in scope when it comes to a global security outlook and non-regional entanglements. As the discussions in Chapter 5 highlighted, various governments since 2005 have pinpointed the High North as the country's primary security interest. Regardless of that rhetorical exercise, if we look at Norway's military positioning, expenditure, and focus, they tell the same story. Almost all of Norway's security concerns and practices relate to its High North domain and relations with Russia in the North. In a geographical and spatial sense, this entails the shared 197 km land border with Russia, the maritime boundary and related maritime space in the Barents Sea, the surrounding airspace, and the Svalbard archipelago.

In turn, it is not the 'Arctic' but the immediate surroundings that have been the determinant for Norway's security practices. As put by the Norwegian foreign, and later, prime minister: 'Our geography defines who we are as a nation. The Barents Sea, the Norwegian Sea, the North Sea, Skagerrak, the coastlines and islands, the continental shelf, the ocean floor, the ocean depths' (Gahr Støre 2023a).[12] Had Svalbard been located in the South, in the North Sea between Norway and say the UK, its status as a geopolitical focal point would have been limited. For Russia, however, the Barents Sea does not constitute the

[12] The full section referred to Norway's size as a 'small' country:

> During my period as Minister of Foreign Affairs, I decided I would never refer to Norway as a 'small country.' Because what does it mean to call something 'small'? Well, with only 5.4 million people, we are certainly at the lower end of the population scale of the world's 200 countries. In terms of land area, though, we move up to a position of roughly 60th in the world. And if we add the ocean and the ocean floor – which are also a part of Norway – we need to multiply our land mass by seven, and then we suddenly shoot up to 17th place. A small country? Hardly. Moreover: We have the world's second longest coastline. Including our islands, it is 100,915 kilometres long. Our geography defines who we are as a nation. The Barents Sea, the Norwegian Sea, the North Sea, Skagerrak, the coastlines and islands, the continental shelf, the ocean floor, the ocean depths (Gahr Støre 2023a).

primary security concern for the state, or 'define' what the Russian 'nation' is (e.g., Hønneland 2016).

Although difficult to rank given the size and multiregional positioning of Russia as the world's largest country spanning eleven time zones, it seems plausible to argue that the Barents Sea – despite its centrality for Russia's strategic second-strike capability and nuclear triad – is conceived in a somewhat similar way to the other geographic maritime domains where Russia directly encounters a NATO state; namely the Baltic domain, the Black Sea domain, and – to some extent – the Bering Sea domain (see next section). That does not mean that this SRSC is not highly relevant for Russian security practices, but it also showcases the importance of geography and taking into consideration the different, overlapping, and/or extended conceptions of subregions and security practices across different geographic domains.

Finally, an interesting dimension of this subregion will be the role of Finland and Sweden in the future. With fewer inhabitants in their Arctic territories, and less immediate emphasis on the High North (or the Arctic) in their security and foreign policy endeavours, focus has – at least in terms of security practices – primarily been on what we can conceive of as a Baltic SRSC (Kopra and Mikkola 2025; Khorrami and Moyer 2025). Questions remain what Russian responses to an additional 1,340 km long NATO border with Finland, partially in the Arctic, will mean for Russian conventional military postures on the Kola Peninsula and in turn the Finnish and Swedish responses. Moreover, seeing Finland-Norway-Sweden as a joint continuum might drag focus and attention from the two NATO newcomers northwards, although this remains to be seen. Here it is equally important to analyse what the different state actors do in terms of security postures, exercises, and new base structures, as much as what the various state officials emphasise rhetorically.

The Beringia SRSC

Moving to the Beringia subregion, it is defined as the land and maritime area bounded on the west by the Lena River in Russia; on the east by the Mackenzie River in Canada; on the north by seventy-two degrees north latitude in the Chukchi Sea; and on the south by the tip of the Kamchatka Peninsula (National Park Service 2023). The Bering Sea is named after Vitus Jonassen Bering, a Danish explorer in

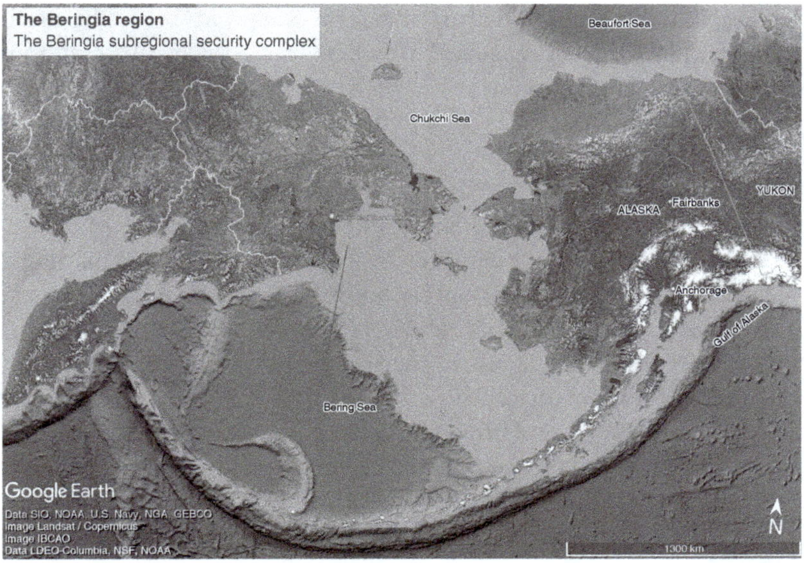

Figure 7.4 Image of the Beringia subregion. Made with Google Earth.

the service of the Russian Navy, who set out to explore the maritime region towards the Bering Strait in 1728. It connects to the Arctic Ocean through the Bering Strait, a narrow passage that separates the Chukchi Peninsula of Russia from the Seward Peninsula of Alaska. The USA and Russia are separated only by 4 kilometres/2.5 miles of water between the Big Diomede (Russia) and Little Diomede (USA) islands (see Figure 7.4).

In terms of security dynamics, the level of activity in this part of the Arctic (or subarctic, since the Arctic Circle crosses just north of the Bering Strait) has been lower than the security developments seen in the Barents Sea domain. Although the USA has maintained a force presence in Alaska, it is only in recent years that the state has attracted considerable attention related to security threats. Whereas Svalbard serves as an illustration of a geopolitical dispute that might erupt further, in the Bering Sea/Strait the primary security issue in near-to-medium term is the increased Chinese–Russian security collaboration off the coast of Alaska and the related possible responses from the USA. This might further come into play given related concerns over the impact on Alaskan fishing activity and fears over the disruption of the Bering Sea/Strait maritime boundary and fisheries arrangements.

When the USA purchased Alaska from Russia in 1867, the maritime boundary in the Bering Sea/Strait area was defined, albeit without clearly determined map lines and two interpretations of the line itself, namely the rhomb line (Russia) and the geodetic line (USA) (Kaczynski 2007: 2–3; Byers 2013). In 1990, the USSR and the USA agreed to settle the rather extensive maritime boundary in the Bering Sea/Strait; this was known as the Baker–Shevardnadze agreement (United States–Union of Soviet Socialist Republics 1990; Verville 1993).[13] This agreement stretched through the Bering Strait between Alaska and Russia and extended into the Arctic Ocean to the north, and the Bering Sea to the south.[14] The US Congress ratified the agreement in 1990; however, after the collapse of the USSR at the end of 1991, the new Russian Duma never ratified it. Although this boundary treaty is not in force, both sides have formally respected its terms (Verville 1993: 454; R. W. Smith 1994; Schofield 2015). At times, Russian politicians and fishermen have criticised the agreement, arguing that it disfavours Russian fishing interests in the Bering Sea and that Russia ceded more of the formerly disputed area than the USA (Kaczynski 2007: 3).

China's Expanding Presence

The Pacific Fleet's base in Viliuchinsk/Petropavlovsk, on the eastern seaboard of the Kamchatka Peninsula, is eastern Russia's closest permanent naval base to Arctic waters. It is home to Pacific Fleet nuclear-powered submarines, in addition to some surface ships. Naval operations in the Bering Strait and the Chukchi Sea by the Pacific Fleet are becoming common, adding to Russia's Arctic presence alongside the Northern Fleet located in the European Arctic. Starting around 2020, Russian military activity in this maritime domain has increased.

13 Named after the two foreign ministers from the United States of America and the Union of Soviet Socialist Republics who negotiated and signed the agreement.

14 The agreement was based on the line defining the western limit of the area covered by the 1867 Alaska purchase (Russia–United States 1867). Depending on the delineation of outer continental shelf limits beyond their EEZ limits, the boundary line relevant to the Arctic Ocean is a straight line heading due north from a specified point in the Bering Straits 'as far as permitted under international law' and thus to their 200 NM limits and potentially further seaward in the Central Arctic Ocean (United States–Union of Soviet Socialist Republics 1990: para. 3 (1)).

Russian military exercises interfering with Alaskan fishers have caused concern over the violation of US rights, as well as the fear of escalation between the fishers and the Russian Navy (M. Baker 2020; The Associated Press 2020a).

For the first time, in the summer of 2024 Chinese and Russian bombers conducted joint operations in the Alaska Air Defense Identification Zone, albeit without violating US airspace (H. Williams, Bingen, and MacKenzie 2024). In September 2024, coast guard vessels from both countries conducted joint operations in the same region (Xiao 2024). This is in stark contrast to Russia's scepticism about Chinese naval activities a few years ago. When four Chinese vessels passed the Tsugaru Strait and circled Japan for the first time in October 2008, Russia was concerned about China's possible advance into the Arctic Ocean (Dutton 2009; Kotani 2019). In 2010, the Commander-in-Chief of the Russian Navy, Admiral Vysotsky, warned that Russia should closely watch China, which is 'advancing their interests very intensively, in every possible way' in the Arctic (Faulconbridge 2010). In 2013, the Chinese research icebreaker *Xue Long* attempted to enter the Sea of Okhotsk without the courtesy of notifying Russia beforehand, which resulted in a warning shot being fired.

Notably, in August 2015, the second phase of their Joint Sea 2015 exercise took place in the Sea of Japan near Vladivostok, involving over twenty ships from both nations (Gady 2015). Afterwards, five Chinese ships appeared in the Bering Sea off Alaska. Chinese military commentators suggested that this tested China's far sea capabilities and aimed to enhance navigational security and strategic presence in the Arctic (CCTV 2015). China's first Arctic operation following a Russian naval exercise highlighted Russian acceptance of their presence and was seen as a political gesture.

The scale of Russian–Chinese exercises grew significantly after Russia's invasion of Ukraine in February 2022. In September 2022, the Chinese Navy participated in the Russian Vostok 2022 exercise in the Sea of Okhotsk, marking its first use of army, navy, and air force elements in a joint exercise (X. Liu and Guo 2022). Following Vostok 2022, Russian and Chinese navies conducted their second joint patrol in the Pacific, with Nanchang visiting the Bering Sea (X. Liu 2022). Their presence, monitored by the US Coast Guard, included joint patrols in the North Pacific. In July 2023, China organised a first North/Joint-2023 exercise by the PLA Northern Theatre, with Russia as an invitee. Following that exercise, Russia and China conducted their third joint patrol in July 2023 (China.com 2023). During the Russian-China Joint Patrol in 2023, Russian and Chinese navy vessels passed the Sea of Japan, La Pérouse Strait, the Sea of Okhotsk, and Kamchatka Strait, sailing by Alaska and the Aleutian Islands, which triggered US security concerns (Mahadzir 2023; Murkowski 2023b).

The following year, in July 2024, Chinese and Russian navies held the Joint Sea-2024 exercises in the South China Sea, focusing on anti-submarine, air defence, and anti-missile drills. They then conducted a joint Pacific patrol from south of Jeju Island through the Osumi Strait to the South China Sea. Moreover, on 25 July 2024, a joint aerial strategic patrol took place over the Bering Sea involving Russian and Chinese bombers entering the Alaska Air Defense Zone and drawing US attention (The Moscow Times 2024). Furthermore, in September 2024, Russia led its largest post-Soviet naval exercises, Ocean-2024, across several seas, highlighting its nuclear capabilities in the Arctic, with Chinese participation (Dyner 2024). Later that month, China organised the North/Joint-2024 exercise, lasting nearly two weeks and featuring advanced drills with major Chinese and Russian vessels (X. Liu and Guo 2024). These naval exercises were, in other words, part of an emerging trend where Russia and China have jointly increased their operations in the North Pacific theatre and – increasingly – off the coast of Alaska, even including an airborne component in 2024.

There is also another element to this, namely the softer coast guard element. In April 2023, Russian and Chinese Coast Guards signed a cooperation agreement in the city of Murmansk, in the European Arctic, and China was invited to observe the Russian Arctic Patrol 2023 exercise (Nilsen 2023a). Joint coast guard or naval operations are relatively straightforward ways for both countries to demonstrate their collaboration and send a message to the USA that it should not assume dominance in the Bering Sea. Notably, Chinese Coast Guard ships entered the Arctic Ocean for the first time in autumn 2024. For China, joint patrols in the Arctic were meant as a reciprocal response to increased US presence in the South China Sea and also to demonstrate its civilian capacities in an Arctic context (S. H. Choi 2024). Russian Coast Guard ships conducted joint exercises and patrols in the North Pacific in September and October 2024, which was a direct consequence of the agreement signed in 2023 (TASS 2024).

The Chinese Coast Guard is a paramilitary organ not formally integrated into the Chinese Navy.

As such, China views the coast guard ships as a low-level signalling tool, albeit one that increasingly deploys in order to convey power and ability to operate (S. H. Choi 2024). This is becoming evident in the South and East China Seas, where the Coast Guard upholds and reinforces disputed Chinese claims and positions over everything from reef ownership to access to fisheries (Midford and Østhagen 2024). In the words of a Chinese military commentator, Professor Shao Yonglin, 'the Coast Guard and the Navy are completely different' because 'the Coast Guard patrols and enforces laws at sea' while 'the navy plays a role from the perspective of national security' (CCTV 2024b). At the same time, Chinese media celebrated this as 'not only break[ing] through your first island chain, but also sail[ing] thousands of nautical miles all the way to the Arctic Ocean after breaking through the island chain'.

These events and this increased activity started before, but increased after, Russia's 2022 invasion of Ukraine. They have also taken place in the context of tense wider China–USA relations and when it is apparent that the asymmetry between Russia and China is shifting in favour of the latter. Still, the Bering Strait and North Pacific (Beringia) spatial domains remain distant from China and are considered to be unfamiliar operational environments. Russia plays a significant role in these drills. By inviting China closer to areas considered important for its defence and by increasing interoperability with China, the Moscow regime aims to strengthen its relations with China, while also assessing China's growing naval capabilities (Conley et al. 2024). Simultaneously, the signalling component here is central: Russia is obviously highlighting its ability and willingness to protect its interests in the Arctic, linked to, for example, discouraging the USA from conducting Freedom of Navigation Operations along the Northern Sea Route (Gosnell and Saunes 2024).

For China, the drills are also an opportunity to demonstrate their power projection capabilities, while at the same time supporting Russia. Additionally, China seems to want to highlight a reciprocal response regarding the South China Sea, where the USA conducts regular Freedom of Navigation Operations. For instance, the sighting of Chinese Navy vessels in the Bering Sea in 2015 was intended to serve as a signal, providing a reciprocal response to the US involvement in the South China Sea. Although the US reaction was calm, somewhat damping the intended signalling effect, China sees its operations in the Bering region as tit for tat (Pincus 2020).

Chinese media military analysts continue to stress that areas such as the Bering Sea are far away for China and areas that China does not understand very well. As Richard Weitz (2021) observes, Russia and China seem to take turns holding maritime exercises in regions where each of them seeks to project power, including the South China Sea, Senkaku/Diaoyu Islands, and the Baltic Sea. Following this logic, the Eastern Arctic is a region where Russia is more interested in power projection. What is common among the Joint Sea exercises is that they were always conducted in several theatres. According to the China Power China-Russia Joint Military Exercises database (CSIS 2024), Joint Sea exercises have been conducted since 2012 and have not shied away from areas with maritime disputes.

Simultaneously, China is expanding its interests in the Eastern Arctic region by, for example, organising its own military exercises in the region, the North/Joint exercises. This is a significant change. China and Russia are also continuing to make slow progress on increasing interoperability. Even in 2022, there were reports of Russian cadets learning Chinese in order to improve communications (Kimura 2022). Furthermore, Russia and China understand that the two navies have 'a win-win effect of making up for each other's weaknesses' (CCTV 2024a). China also continues its capacity building, including construction of yet another civilian icebreaker, which is about to start in 2025. China's icebreaker fleet consists mainly of research icebreakers with limited military significance.

An Emerging SRSC

Chinese encroachments facilitated by an isolated Russia might complicate the wider Arctic security landscape in the longer term. Looking at what these developments mean for the Beringia SRSC, the relationship between Russia and the USA in this domain is obviously complex, shaped by a mix of cooperation and competition. As with the case of the Barents SRSC, geography matters. The fact that you can – although they do not share a border on land – walk between Russia and the USA, or between Asia and North America, at certain times of year, speaks to the need to manage neighbourly relations linked to everything from fisheries to maritime traffic. In this geographic domain, the two primary actors are undoubtedly Russia and the USA, although what has changed in the last decade is the involvement of China. This in turn spurs more attention to the subregion by US decision makers who would arguably not be otherwise concerned with Alaskan security issues (see Chapter 6).

Despite the political tensions between Russia and the USA – especially after 2014 – there is a history of collaboration over fisheries management, environmental protection, and maritime safety in this region. The Bering Sea is significant for both countries economically, especially in terms of fishing.[15] In the period 2021–2023, total pollock catches were in the region of 1.7–1.8 million tonnes, making it the largest single-stock fishery in the world; 66–78 per cent was taken in the US EEZ, the rest in the Russian EEZ (NOAA 2024b). Consequently, there have been forms of cooperation between Russian and US authorities, especially on the practical level between the US Coast Guard and the FSB Border Guard, even after 2014 (Østhagen 2016).

For example, from around the 2010s, the USA focused on developing a well-functioning cooperative relationship with the Russian Coast Guard, as traffic in the Arctic was, and is, on the rise. Most of the traffic increase in these waters derives from Arctic shipping, or shipping to destinations in the Arctic itself, as a means of resupplying isolated coastal communities or for bringing natural resources to markets further south. Based on the desire to implement the multilateral agreements signed under the auspices of the Arctic Council in 2011 and 2013, the regional US Coast Guard authority from District 17, Juneau, Alaska regularly met with their Russian counterparts in

[15] The Bering Sea is home to a number of marine species of significant economic value, such as cod, pollock, salmon, king and snow crab, seal, and whale. In the Bering Sea at large, and in the Donut Hole in particular, many of the fish stocks have been subject to overfishing, in particular the Alaskan pollock. In the 1980s, the expansion of overfishing prompted negotiations between the former USSR and the USA, and in 1988 an agreement on mutual fisheries relations was signed between the two parties. The agreement established a consultative committee on fisheries in the Bering Sea. Constituting the most lucrative fisheries in North America, the pollock stock was severely decimated in the early 1990s, after a decade of overfishing in the Donut Hole. An agreement on a temporary moratorium with stringent enforcement measures on pollock was reached between the various national stakeholders in 1994, including Japan, China, Poland, and South Korea.

Vladivostok. This continued despite Russia's annexation of Crimea and initiating of conflict in Ukraine in 2014, but contact and interaction have been scaled back and – after 2022 – limited to only ad hoc contact as needed.

Although both countries have continued operating emergency services in the Bering Strait, and emergency cooperation has continued, an increased Russian military presence and shifting stocks due to climate change have worried some US fishers. Alaska's lucrative snow crab harvest was cancelled in 2023 for the first time ever due to stock collapse, which scientists warn will become increasingly common with climate change. As many species seek cooler, deeper waters, they are likely to migrate beyond historical boundaries, with Russia already exploiting new fishing locations within its EEZ. This is likely to put further pressures on efforts to reduce overfishing, or even lead to poaching and illegal, unreported, and unregulated fishing (IUU) in this SRSC (Wenger 2022). China's increased role here, also with a coast guard presence, could lead to more activity by Chinese trawlers and in turn spark conflict. The linkages between resource interests and the activities of the Chinese Coast Guard, in addition to the joint exercises between China and Russia, makes volatile fisheries relations a particularly relevant geopolitical component of this SRSC (Evans and Østhagen 2023).

Furthermore, with the return of Donald Trump as US President in 2025, there were several signals coming from the White House concerning the possibility of increased collaboration between the USA and Russia in the Bering Sea/Strait domain. These hinted at resuming dialogue on for example enabling shipping through the Strait, or finding other joint ventures in the Arctic, as – assumedly – a carrot in negotiations trying to bring some form of ceasefire in Ukraine. A few months later, in August 2025, Trump and Putin met in Anchorage, Alaska, to discuss the war in Ukraine, but also ways to enable Arctic-specific collaboration going forward. The meeting did not lead to anything, but again signals not only the relevance of geographic proximity in US–Russia relations but also the distinctness of these dynamics in contrast to those in the Barents SRSC.

In essence, what a study of both this SRSC and the previous one showcases is the importance of studying the specific issues and developments that take place a level below the regional (Arctic) and the global but that constitutes more than the 'national'. US foreign, security, and defence policies are not defined by the interactions it has with Russia and – increasingly – China in the North Pacific. It is probably not within the top five global security concerns that the USA must deal with. Neither is it likely to make the top five for China. For Russia it might. However, this domain still takes up considerable time, resources, and leads to – crucially – security practices by all the mentioned actors. The point is not to quantify which geographic domain holds the most sway in states' security considerations. The point is that states participate in multiple geographical security domains, at a smaller scale than 'North America', 'Pacific', 'Europe', or 'Arctic'.

The dynamics described here in the Bering Sea are linked to everything from fisheries management to how increased Russian–Chinese military collaboration impacts US security practices, which in turn impacts Russian and Chinese security practices, creating dynamics that have a clear regional or subregional characteristic. These dynamics are different from USA–China relations in the South China Sea, or USA–Russia relations in Northern Europe (i.e., the Barents SRSC). Therefore, they must be analysed and understood in order to improve our understanding of state behaviour in international relations, as well as the limitations posed by geography and how states practice their security dynamics in such domains.

That is not to say that Beringia SRSC dynamics are defined solely by their geographic attributes. A more accessible Bering Strait leading into the Chukchi Sea and the Arctic Ocean will not necessarily lead to more Chinese activity that triggers a US military response. Moving fish stocks in the North Pacific might lead to cooperative and innovative mechanisms to manage their fluctuations, instead of competition. And, as we return to in the next chapter, the proximity of Russia and the USA across the Bering Strait might act as an impetus for collaboration on softer issues, instead of spurring rivalry (as already showcased by the Trump overtures in Alaska in 2025). Geography matters in defining the frames and boundaries for SRSCs, which in turn are influenced by, but also influence, state action in that domain. However, it is the linkages between security concerns, geography, and the regional composition of actors and interests that make up a spatial domain in which security dynamics coalesce.

8 *A Subregional Approach to Security and Geography*

This chapter summarises the findings from across the previous chapters and attempts to discuss both the 'region' status of the Arctic and what the Arctic is a case of in terms of security dynamics, regionality, and security regions. First, it evaluates the relevance and usefulness of security regions theories – the RSCT in particular – in unpacking Arctic security relations. Before turning to what the Arctic *is* a case of, and how we can better understand the circumpolar regional security dynamics, it is worthwhile considering why the Arctic has increasingly been described as an RSC or even just as a coherent region in terms of security issues. Thereafter it takes a look at what these findings mean for the study of security regions and RSCs more broadly. It concludes with an emphasis on state-centric small-scale SRSCs, which can roughly be conceptualised as circles of overlapping spatial and security domains (i.e., a Venn diagram), before musing about the increasing relevance of maritime 'regions' in security considerations.

Security Regions and the Poor Arctic Fit

Is the Arctic a region? The short answer is most likely yes, because that is the way it is described and referred to in everything from encyclopaedias to historical documents around the world. Whether it is or not also depends on whom you ask and your object of reference. A natural scientist studying changes to sea ice texture or the movements of the beluga whales would draw a line zigzagging across the top of the globe, defining the Arctic region according to climatic criteria. If you ask a political scientist, on the other hand, the answer to whether the Arctic constitutes a region would depend on whether that scientist studies Arctic political developments such as the Arctic Council (then the answer tends to be yes) or traditional security questions such as NATO–Russia relations (then the answer tends to be no, although that might have changed in recent years depending on where that person sits).

If, however, you ask a foreign minister from one of the eight Arctic states they would tend to define the Arctic as a region in order to achieve certain political goals. These goals seemed to, at least up until Russia's full-scale invasion of Ukraine in 2022, revolve around regional cooperation, expanding governance mechanisms, and portraying the Arctic as an 'exceptional' or sheltered part of the world in terms of state security interactions. Others point to a shared identity and culture among the Arctic states – Canada, the Kingdom of Denmark, Finland, Iceland, Norway, Russia, Sweden, and the United States. Their geographical proximity and common Arctic interests, along with contributions from Indigenous Peoples, enhance the region's 'northern' identity (e.g., Penikett 2017).

This book does not intend to give a definite answer concerning whether or not the Arctic is a region because that answer inevitably varies. Instead, for the purpose of being able to answer another question which is at the heart of so-called geopolitics – What is the chance of conflict between states in the Arctic? – the focus in this book has been on security dynamics amongst the Arctic states (and non-Arctic states) and whether – as it often is portrayed – the Arctic is a *security* region.[1] As outlined in Chapter 2, the concept of security regions – merging security dynamics with a regional outlook (as defined, at least partially, by geography) – as well as securitising acts across that spatial domain, provides a theoretical approach from which to examine the contemporary Arctic and approximate an answer to the conflict question.

Moreover, despite not being listed in the seminal framework for analysis published by Barry Buzan and Ole Wæver in 2003, the Arctic is increasingly described as a security region or an RSC. In a review of that 2003-book, Gunhild Hoogensen Gjørv argued that the framework's omission of the Arctic space was problematic, as it was an example of a spatial domain that did not fit with neatly categorising the world into security regions (Hoogensen 2005). More recently, arguing for the existence of an Arctic RSC has gained prominence as scholars and policymakers examine the region's somewhat distinct security dynamics (Exner-Pirot 2013; Tüysüzoğlu 2021; Sadurski 2024).

The argument has been that states within the Arctic region exhibit a substantial degree of security interdependence. Non-traditional security issues such as climate change, resource development, environmental

[1] As well as the EU, as *sui generis* actor (Raspotnik and Østhagen 2019).

concerns, and new shipping routes present shared security challenges necessitating cooperation, or at least interaction, among Arctic states (Padrtová 2017; Sadurski 2024). In addition, the argument is that security dynamics in the Arctic are converging, influenced by factors such as climate change, melting sea ice, and the potential for increased economic activity, which in turn has the potential to create a RSC (Lanteigne 2016), or already has done (Greaves 2019). In turn, scholars argue that the RSCT helps analyse the region's security environment, highlighting the interplay of cooperation and confrontation among Arctic states (Padrtová 2017).

There are two dimensions to these analyses: the emphasis on the foreign policy narrative as extensively described in Chapter 3, and the extended or alternative conceptions of security beyond the state or military-oriented one. These two elements at times merge, as Arctic states (primarily the foreign ministries), or other actors such as Indigenous Groups, the Arctic Council, Arctic conferences, or even Arctic scholars create narratives about Arctic issues and (non-military) security concerns.

There is no doubt that environmental concerns on both a local level (say, coastal erosion in Kotzebue, Alaska) and on a global level (say, the impact of Arctic climate change on the Indian monsoon) are serious and real security concerns for, in this instance, people in Kotzebue or India. Similarly, human security concerns are becoming more acute in the Arctic, linked to everything from loss of traditional livelihoods in, for example, Northern Finland and Northern Canada, to safety concerns arising from having cruise ships carrying 4,000 passengers sailing around Svalbard when there is no emergency response system in place for that scale. Along any of these parameters (or sets of non-state-centric security issues), one could conceive of a security complex or security region. This was the premise of Heather Exner-Pirot's (2013) article 'What Is the Arctic a Case Of? The Arctic as a Regional Environmental Security Complex and the Implications For Policy'.

However, Exner-Pirot's approach is perhaps more equivalent to region-building or regionalist approaches,[2] than to an analysis of whether security practices – at least with a state-oriented perspective – truly overlap across the Arctic spatial domain. The focus is not on

[2] As also Jacobsen, Wæver, and Pram Gad highlight (2024c, 21).

defining whether the Arctic is a spatial domain where security issues are intertwined, but rather how, and why, Arctic actors are expanding their joint efforts in dealing with environmental concerns and issues.

Emphasising certain pan-Arctic issues in a circumpolar context, such as developing a 'Polar Code' for Arctic shipping (Rayfuse 2014; Ø. Jensen 2016), or efforts to reduce black carbon emissions (Arctic Council 2015; Exner-Pirot 2016), are undoubtedly relevant in region-building approaches when trying to tackle shared challenges across the circumpolar area that are indeed only relevant in a circumpolar context. There is also no doubt that all Arctic states – even Russia – have eventually come to recognise over the last three decades that their interests align and that there are positive-sum elements to be gained from cooperation, alongside the mere effects of cooperation in and of itself on creating relations, trust, and confidence. All the Arctic states' foreign policy statements started to emphasise the Arctic as a unique and special place for international affairs, governed by cooperation and the Arctic Council's role as the primary forum for Arctic matters (Heininen *et al.* 2020). In other words, region-building has been taking place, and still is – to some extent – even after the limited contact between Russia and the other seven states.

Also, there is perhaps no doubt (anymore) that the overarching and most menacing security threat to society, and thus also to states as the constructs they are, is climate change writ large. The emergencies, famine, domestic unrest, migration, health issues, and economic catastrophe that follows from a 2100-perspective will put tremendous pressure on state structures, and will in many instances across the globe, and perhaps also in the Arctic, lead to the reconfiguration and even collapse of some states in their current form. Still, the question is: Do Arctic states *practice* this concern, or in the language of Buzan and Wæver, do they engage in securitisation of this (or these) issue(s), to the extent that this security 'problem' is interlinked across the Arctic (but does not extend beyond)? Arguably not.[3] If the premise for a security region is that it comes into being by securitising acts amongst actors (here states) within more or less the same geographic space,

[3] See Gerard Toal's work *Oceans Rise Empires Fall: Why Geopolitics Hastens Climate Catastrophe* for more on this discrepancy between immediate security concerns and the long-term climate concern (Toal 2024).

there also needs to be some level of urgency, or at least hierarchy and priority, concerning the issues that are securitised.

Although environmental and climate change issues dominate much of the Arctic conversation in both the big Arctic conferences, in Arctic states' policy documents, and in Arctic Council deliberations, these issues are not always framed as region-specific, nor are they always framed as *security* issues. Moreover, the climate change dimension is inherently non-regional, in the sense that the Arctic states themselves emphasise that 'emissions causing the Arctic to warm faster do not originate in the Arctic – they are the result of collective emissions of industries and energy systems in the rest of the world' (Varteressian 2024). That is also the argument used by the likes of China, the EU, India, and the UK when engaging in Arctic issues, for example, the EU arguing that 'Climate action is of particular importance to the Arctic, given the immense knock-on effects of Arctic warming. The legislative proposals under the European Green Deal (EGD), will be at the heart of the EU's Arctic engagement' (European Commission 2021).[4]

We might desire that environmental or human security concerns, which are found in a pan-Arctic context, dominate the eight states' security outlook and interactions. This is not the same as saying that the geographic region that defines (climate) security matters for the eight states that are partially located above the Arctic Circle is the whole circumpolar area. In turn, this is also not the same as constituting central acts in a security region or a RSC.

Arctic state's foreign ministers – at least up until 2022 – could talk all they wanted about the Arctic as a region of shared security concerns or interests, or an exceptional space for state-to-state and people-to-people cooperation. However, we are not talking about a RSC if the ministers of defence or chiefs of defence from the same states simultaneously contradicted such notions by highlighting their considerable divergence in how they viewed the Russian threat in the Arctic and related responses. Bluntly put, it is not enough to only listen to what the mouth is saying, we also need to see what the hands (or, in this case, the military) are doing.

[4] The EU is in fact the one actor that seems to place particular emphasis on the threat emanating from climate change, 'Climate change is the most comprehensive threat the Arctic is facing and has reached an unprecedented crisis point' (European Commission 2021).

Examining states' security-oriented actions in terms of resource expenditure, activity, and – gradually since 2014 – political statements, paints a different picture. This has been outlined in this book by looking at, for example, where NATO member state-led, or NATO-led, military exercises have taken place; where Russia has been modernising and investing in military capabilities; or how Norway and the USA (despite being the 'superpower') focus on different parts of the Arctic when they emphasise security concerns or develop military or security strategies for their respective Arctic territories.[5]

A state's concern with its security, namely its survival, focuses first and foremost on its immediate geographic surroundings and the nearby states. The quote from Canadian military official Riviere that any land invasion would become a rescue operation (Fridtjof Nansen Institute 2024a) perhaps best illustrates the divergence between security conceptions in an Arctic context.[6] Canada does face airborne, and perhaps naval, concerns emanating from the Arctic parts of Russia (Lasserre and Têtu 2016), but arguably not more, and probably less, than the UK does (Depledge, Kennedy-Pipe, and Roger 2019; Depledge and Østhagen 2021) – a non-Arctic state.

An equivalent example would be the idea of a European security region, where states' security practices and concerns cannot in any meaningful way be disentangled from each other. However, when looking at the security practices of some states in Europe, for example Spain, Greece, or Finland, these showcase the opposite. Whereas Spain is the primary example of a European NATO-member unwilling to reach a 3.5 per cent defence spending target, Finland is currently revamping its armed forces to a degree not seen for decades. Greece is similarly focused on military spending, but due to a completely different set of security issues emanating not from a shared border with Russia (as in the case with Finland), but its proximity to the instability emanating from parts of the Middle East. As with the Arctic example, 'Europe' as a label for a security region is much too wide and in lack of specificity, despite rhetoric coming from the political leaders from all the three mentioned countries arguing for a joint European approach and joint security interests.

[5] Although the USA has been operating in both domains, it tends to talk about *its* Arctic when using the term 'Arctic', at least in recent years.

[6] 'I challenge anyone to have a land invasion of the Canadian North … it would end up being a search and rescue operation.'

In essence, what descriptions of 'the Arctic' or 'Europe' in terms of security and geopolitics tend to ignore are the fundamental lessons of geography and of accuracy concerning the space in question. Whether the discussions are on pan-Arctic/region-wide security and politics,[7] on specific actors in the Arctic,[8] or framed as handbooks exploring aspects of this region,[9] the works tend to lack specificity regarding *which* part of the related area is in question. Instead, the Arctic, or Europe for that matter, is made use of more as a concept or an idea that frames the scholarly work in question and attracts attention, whereas the actual object of study is not really the Arctic, but one part of this rather immense territorial and ocean space.[10]

The problem in much of the work leaning on notions of the Arctic as a coherent space for security matters is that state security interests and defence issues are decentralised. Arctic-wide studies are, for the most part, concerned with the 'softer' dimensions of security, or security politics. Descriptions and discourses of contemporary or near-contemporary political, climatic, and/or economic events and dynamics are laid out. The examples or key pieces of evidence used are, to a great extent, the same as those outlined here – at least in Chapter 3: the Russian seabed flag planting, the Ilulissat declaration, the emergence of the Arctic Council as a primary forum with observers and a permanent secretariat, efforts to tackle shared concerns such as the Central Arctic Ocean agreement, and the ups and downs when it comes to the notion of an Arctic exceptionalism. Military concerns, investments, and activities do not figure prominently, if at all, in these analyses.

However, state security in a strict sense means *also* examining the role of the armed forces, the military, and the actions, statements, and descriptions of the ministries or departments of defence. These tend to get limited attention in studies looking at circumpolar security

[7] E.g., Kraska 2011; Emmerson 2010; Greaves and Lackenbauer 2021; Depledge and Lackenbauer 2021; Tamnes and Offerdal 2014; Zellen 2013; Hoogensen Gjørv *et al.* 2014; Mineev, Bourmistrov, and Mellemvik 2022; Goodsite and Swanström 2023; Østhagen 2020a.

[8] E.g., Østhagen 2023c; Raspotnik 2018; Lackenbauer 2011a; Jian, n.d.

[9] E.g., Hoogensen Gjørv, Lanteigne, and Sam-Aggrey 2020; Nuttall, Christensen, and Siegert 2018; J. Weber 2020; Finger and Rekvig 2022.

[10] This author is guilty of the same, especially when choosing a title for a seminar or presentation. The 'Arctic' seems to hold a certain fascination and intrigue, which in turn makes it difficult to avoid it in more popularised contexts.

dynamics especially when attempting to see whether the RSC framework, or the security region category writ large, fits.

Although heavily inspired by the RSCT,[11] the approach in this book, described in Chapter 2, adds a couple of tweaks or limitations to the RSC model. These tweaks are based on the attempt here at linking works on *state* security, historic and contemporary studies of the influence of *geography* on international affairs, and wider *regionalist* and security regions debates. A key deviation from the RSCT is an exclusive focus on states and their security practices – defined as threat perceptions, interests, policies, and actions. Another is being explicitly attentive to how the security dynamics that we inherently are interested in uncovering and understanding operate at different levels of analysis that in turn influence each other; these are constrained by both constructed and material factors including – especially – the geographic dimension.

For example, when looking at the global or international dynamics that have a direct security impact on the Arctic, unpacking NATO–Russia interactions in the European parts of the Arctic hinges on the concrete steps taken by Russia and NATO member states to ensure their own security in that geographic domain. Although this is in essence the core of security (for states) – military exercises, positioning, and operations to either deter, or prepare for, outright conflict – these developments and examples are seldom part of descriptions of the Arctic in the works highlighting the fit with the RSCT, or works that describe 'Arctic geopolitics' as a regional affair either implicitly or explicitly. Military security has become decentralised in favour of other security approaches to the extent that it does not even figure in the analyses anymore, at least not in the pan-Arctic studies. Paradoxically, Arctic states are simultaneously investing in military assets at levels not seen since the end of the Cold War.

[11] To repeat, the core premise of the RSCT approach is that an RSC comprises 'a set of units whose major processes of securitisation, desecuritisation, or both are so interlinked that their security problems cannot reasonably be analysed or resolved apart from one another' (Buzan and Wæver 2003: 44). One problem, if evaluating the Arctic against the traditional RSCT, as pointed out by Jacobsen, Wæver, and Pram Gad (2024c: 22), is that in a technical sense, 'the Arctic is and was not an RSC because it is neither the primary security context for the super or great powers in the region, nor is it sufficiently marginal for the overarching superpower security dynamics to be primary for any lesser actors'.

On the other hand, scholars that *do* work on military and defence questions seldom lean on the RSCT or other comprehensive theories of regions, region-building, and security complexes to explain the *multitude* of Arctic security dynamics.[12] For them, such a scope is arguably too broad to be able to produce relevant insights for a more traditional security-oriented audience, as the link to useful studies of potentially practical relevance becomes paramount. As put by David Auerswald (2022), 'All security is local'.[13] Examples of military studies related to the Arctic, or more accurately specific parts of the Arctic, abound. Several have been utilised here, for example, Depledge (2020) examining where states perform military exercises; Åtland, Nilsen, and Pedersen (2022, 2024) exploring the change in Russian NOTAMs in the Barents Sea; Wegge *et al.* (2023) and Auerswald (2022) on national defence policies; Kjellén (2022) on the changes to the Russian Northern Fleet; Lackenbauer (2013a) on the Canadian Rangers; or Lajeunesse and Choi (2022) on the presence of Chinese submarines in parts of the Arctic.

These works heavily inform the chapters here looking at Arctic security practices where the military component of security practices is apparent. What such works sometimes lack, however, is linking, for example, China's Arctic rather generic and toothless policy initiative from 2018 to increased US military attention given to Arctic security, which in turn might lead to military exercises in North Norway and then spur Russia to perform symbolic acts on Svalbard. In fairness, the goal of these studies is not to provide a holistic examination of security dynamics across all of the Arctic spatial domain. Still, for the purpose of this book, we cannot conceive of Arctic geopolitics and answer the question of whether or not the Arctic is a coherent security region, without taking into account both (1) the 'softer' or overarching political developments that tend to be absent from military or security studies, *and* (2) the changes and developments in states' military approaches and activities related to (parts of) the Arctic.

Is, then, the Arctic a security region in terms of state security practices (e.g., stated concerns, interests, approaches, focus, threat perceptions, and securitising acts), where interactions amongst the units

[12] An exception here is, for example, Kristian Åtland's (2007) study looking at how the European Arctic fits with the RSCT and securitisation approaches.

[13] For more on the local-security nexus, see Iselin Németh Winther (2024).

(states) cannot be analysed without taking into account the other units? As already outlined here, the short answer is no. For that, the Arctic is too vast *geographically* and the *security practices* of the states that make up the Arctic are only marginally aligned. Naturally, Finnish, Norwegian, Swedish, and partially Russian security practices are intertwined and entangled to the extent that they cannot meaningfully be analysed independently of each other. They also share the same immediate geographic space. The same goes for Canadian and Greenlandic (Danish) security being interlinked with the USA. But these dynamics do not sufficiently overlap and influence each other. Through the examination in this book of the different levels of security dynamics taking place on a regional level, global level, and national level, it is clear that the central security issues are either grounded at a global level or at a national level.

In essence, the author of this book concluded already in 2021 that 'the Northern European and North American security domains are only marginally aligned' (Østhagen 2021b: 68). That still holds true in terms of defence issues and state security practices. However, what that analysis got wrong, or lacked, was a limited appreciation of the interlinkages between security regions and regionalism,[14] between material objects and practices, and between socially constructed interactions and institutions. It is the *combination* of these dynamics, across levels, and defined by geography that is what we must try to grasp.

If we take the securitisation approach seriously and attempt to unpack which issues dominate the discourse (and actions) of states within a regional space, the impact of forums that are the product of region-building – such as the Arctic Council – must also be taken into account. For example, it can be argued that up until 2014 Arctic NATO member states were wary of overstating their security concerns about Russia in an Arctic setting because they did not want to negatively impact other forms of softer Arctic cooperation and they wanted uphold the discourse and ideas surrounding 'exceptionalism'. Foreign ministries and defence ministries *do* talk to each other, at least sometimes. Although that approach was, to some extent, a charade – masking an increasing divergence between military investments and exercises and foreign policy statements – we must still include it

[14] Another deficiency in that article was an overly reductionist and simplified utilisation of the RSCT framework.

when unpacking regional dynamics concerning Arctic states' security. Ignoring it, and only looking at – for example – defence strategies and military operations, also does not capture the full picture.

The problem is how to measure, describe, or distinguish between words and actions while not assuming that the security traits of a region, or a state, are something inherent just 'waiting to be discovered' (Allen, Massey, and Cochrane 1998: 2). It is *also* important to take into account that there are certain characteristics or geographic facts – such as the existence of an ocean – that are difficult to ignore when discussing the same issues. In other words, we must pay attention to both state discourse *and* security actions by the same states, considering the apparent divergence between the foreign and defence ministries in particular, when examining security practices.

If we consider the two dimensions arguing against conceiving of the Arctic as a security region – geography and state practices – and combine them with the regionalisation or region-building efforts as outlined throughout this book, a slightly different picture emerges. As has been showcased, political and region-building efforts – which the Arctic is an example of – can have an impact on state practices concerning security. And as more critical and nuanced conceptions of geography and geopolitics have taught us, the way geographic characteristics define politics and in turn are being used to define politics, is not a constant, but can change. It is the interplay between these different things that ultimately becomes the totality of regional security dynamics.

To highlight this interplay, take, for example, the Norway–Russia maritime dispute that was resolved in 2010, two years after the Norwegian and Russian foreign ministers shared vodka and watched a football match aboard an icebreaker in the port of Murmansk. For more than four decades, from 1977 until 2010, one central security concern for Norwegian decision makers in Oslo was how to avoid possible Russian challenges to, and escalations over, the unclear jurisdiction and sovereign rights in the relatively large disputed area in the Barents Sea (Stabrun 2009). However, in 2010, the two states agreed to an extensive maritime boundary, hailed as a 'new era' in their relations (V. L. Pettersen 2010), removing a significant liability in Norway's security and defence policy. Part of the reason behind this agreement was the broader Arctic 'spirit of cooperation' which emerged in the period 2005–2010, coinciding with Gahr Støre as the Norwegian foreign minister building personal relations with Lavrov as the Russian foreign minister (Gahr Støre 2010a; Lavrov and Gahr Støre 2010; Gahr Støre 2008: 45–55).

Those relations, and the broader context of Russia's desire to affirm the primacy of the Law of the Sea regime in the Arctic and 'tidy up its spatial fringes' (Moe, Fjærtoft, and Øverland 2011: 158),[15] impacted the ability of the two states to reach an agreement. Today, such an agreement would be impossible due to the tense security relations between Norway and Russia, but at the time, the wider intangible region-building approaches and the dynamism of circumpolar affairs had a concrete impact on security concerns and practices for the related Arctic states. This example highlights the interplay of the geographically bounded state security concerns, the way they were interpreted and changed with the 2010 agreement, and how that agreement was a result of both bilateral-specific efforts *and* regional Arctic-wide developments.

However, that is not the same as arguing that Norway's or Russia's security practices are tied up primarily in a circumpolar or pan-Arctic context, namely that the 'Arctic region' is the determining outlet for security interactions and practices concerning those two states writ large. Instead, these security practices take place at a much smaller and geographically bounded space, that is the Barents Sea regional domain – at least for Norway. Russia, on the other hand, operates in a range of different smaller security subregions ranging from the Barents SRSC to the Beringia SRSC and everything in between.

In other words, although the Arctic does not constitute a coherent RSC or a security region, Arctic-wide issues and developments still matter for Arctic states and their security practices. However, if we want to get more specific about the linkages between geography, state security, and regional practices, it makes more sense to scale down to a subregional dimension and examine parts of the Arctic and how they do indeed constitute spatial locales of particular relevance to the security of the participatory states.

There Is No Arctic 'Security Region', but Why Does That Matter?

It seems reasonable to conclude that Arctic states' security practices are not primarily defined by, or operate across, the whole circumpolar

[15] See also Solstad 2012; Holsbø 2011; Østhagen 2022: 78.

region. They did not before Russia's full-scale invasion of Ukraine in 2022, and they have not since. One could argue that US security practices do, given the USA's centrality in both North American, North Pacific (to the extent this is different from the former), and European Arctic security issues, but this has little to do with the regional overlap of practices and threats, and everything to do with the USA's position as a global superpower.

Looking to the security practices by the other Arctic states over the last two decades, even including the regional 'superpower' Russia, and the defence posture that follows, the Northern European and North American security regions are only marginally aligned, if even at all. Security concerns related to Russia as perceived by the other seven Arctic states also do not sufficiently converge, nor does Russian actions and security practices. Canada and Finland, for example, have varying threat perceptions and operate in different security geographies. Russia also does not treat the Arctic seven as one coherent block, despite narratives from the Russian Ministry of Foreign Affairs since 2022.[16] Instead, Russia sees its vast Arctic coastline as segmented into different theatres of operation, the most prominent ones being the Barents/North Atlantic one and the Beringia/North Pacific one, which align with the two SRSCs discussed above. This contradicts arguments that the Arctic is a current or emerging security region – it is simply too vast and inaccessible to fit the various definitions of a spatial domain where security practices sufficiently overlap and intertwine.

Given that there are in fact many 'Arctics' with varying characteristics and differing security dynamics, why does the idea of the Arctic as a political space for understanding state security practices persist? In some ways, the Arctic states (specifically their foreign ministries) have become victims of their own success. The deliberate branding of the Arctic as one coherent space where all eight states could engage on soft diplomacy issues such as dealing with local effects of climate change and producing good scientific data, with even Russia deeming circumpolar affairs as 'exceptional' (Devyatkin 2023), must partly be seen as a response to international attention and pressure as described in both Chapters 3 and 4. This was especially related to actors such as China and the EU wanting influence over Arctic governance processes

[16] E.g., Wilhelmsen and Gjerde 2018; Wilhelmsen and Hjermann 2022.

that Arctic states preferred to manage themselves. Narratives matter, including in foreign policy.

At the same time, there was probably a genuine belief amongst Arctic foreign ministers and their diplomats that the subregional military and security trends that were developing could be overcome, or at least be mitigated, by pan-Arctic cooperation. In other words, a belief existed that region-building could eventually make defence security concerns less relevant (as in the case of the EU). The Arctic Council, as the foremost expression of this belief, was even nominated for the Nobel Peace Prize in 2019 and 2022 (Jonassen 2022). In a way, it seems almost like the Arctic states' foreign ministries and diplomats were trying to avoid facing the defence developments that took place in various parts of the Arctic by upholding the notion of the Arctic as a political region, when security practices instead highlighted a different reality.

Even in 2008 – when Arctic coastal states foreign ministers gathered in Greenland and declared the primacy of regional cooperation (The Ilulissat Declaration 2008) – an apparent divergence had emerged between the foreign policy rhetoric and aspirations concerning cooperation and exceptionalism and state security practices by the same actors. Arctic states had already begun investing in, and preparing for, more military activity and possible tension in the northern *subregions*. Despite highlighting shared interests in regional cooperation, upholding the Law of the Sea regime, and promoting climate and research action, the same states (albeit different parts of the state organisational structures) were preparing for more military activity in their immediate geographic spaces.

Finally, perhaps there was, and still is, a lack of knowledge about 'the other' side, namely what actually occurs in dynamics between Russia and Norway (and other NATO states) in the North Atlantic, or what the other parts of the Arctic look like. Few Canadians have set foot in the Canadian North, let alone Norwegians or Swedes. Probably more Norwegians have been to Malaga or Lisbon, than to Kirkenes or Longyearbyen – northern towns in the same country a mere two- or three-hour flight away. As Greaves (2019) argues citing Ken Coates *et al.* (2008: 55): 'The Arctic's inaccessible terrain, vast distances, cold weather, and sea ice also helped deter military aggression and prevent some inter-state conflicts, such as allaying concerns of a Soviet ground invasion of northern Canada during the Cold War.'[17] This, however, is not actually the case in the *European* Arctic. Here, instead, Greaves is talking about the Canadian Arctic, albeit extrapolating to make it relevant for the whole Arctic region. Fundamentally, it is in this transition from one Arctic experience to the wider Arctic region that some of the errors occur, at least in terms of describing defence and state security practices.

Instead, when unpacking regional – in this case Arctic – security issues and dynamics, it helps to be specific about what the object of examination is. If we ask what the security concerns, interests, and actions by Arctic states amount to, the answer depends on which state and what territory is in play. As shown here, the Arctic security concerns of Norway and the USA are widely different. Or, to make the comparison less lopsided; the same goes for Canada and Finland. This in turn speaks to the difficulties of concluding that the Arctic truly is a coherent security region, not only in terms of security practices, but also in the way we conceive of and describe the North.

However, in contrast to the pan-regional perspective, or the global perspective, the approach argued for here does not lend itself to quick generalisation about 'a new Cold War' or an 'Arctic exceptionalism'. In this approach, a statement such as 'China has strong military interests in the Arctic, including trying to get a nuclear-armed submarine up into Arctic waters' (Tatlow 2024b) lacks relevance and even validity, because it lacks specificity about geography: Which Arctic waters, where, and why? The devil, or in this case a Chinese submarine, is in the detail.

Why and how do these conclusions matter? The need for accuracy when discussing geopolitics in specific parts of the world is not only an academic endeavour; it also enables us to better understand the chance of conflict erupting or taking place in the Arctic. For that enquiry, the discussion here over the region label holds certain lessons. The fundamental one is that the answer depends on (1) *which part* of the Arctic we are discussing, and (2) *what issue* we are looking at.

For example, the likelihood of conflict in the Barents SRSC could be linked to both unintended and intended escalation in the Barents Sea,

[17] Further: 'The deterrent effect of the harsh northern climate remains relevant to national security, with the chief of Canada's defence staff citing it as recently as 2010 as part of his lack of concern over the need for conventional defence in the Arctic' (Greaves 2019).

on the border between Russia and Finland or Norway in the North, or on or around the Svalbard archipelago between Norway and Russia. The likelihood of escalation is still relatively low but has increased considerably since Russia's full-scale invasion of Ukraine in 2022, linked to miscalculation and deliberate disruptive acts by Russia. None of this is necessarily resource- or climate-linked or has anything to do with China.

In the Beringia SRSC, the chance of an outright standoff between US naval or coast guard forces and a joint Chinese–Russian naval exercise that disrupts Alaskan fishers has also risen considerably, especially as security dynamics are likely to continue on current trajectories.[18] The second US Trump presidency from January 2025 adds a further disruptive and unpredictable element to these security dynamics, given that it was the first Trump administration that started using also the Arctic as an arena to challenge China's global ambitions and posture.

Some, on the other hand, argue that focusing on the Pacific-side of Arctic security issues – as highlighted in the US Department of Defense's Arctic Strategy from 2024 – is a mistake by the USA, as the real threat lies in the European Arctic, emanating from Russia (Bekkevold and Hilde 2024). However, that ignores the fundamental point made in this book, namely the importance of these SRSCs and the evolving security dynamics in them, going beyond the *general* Arctic connotations that in essence fail to accurately describe the specific concerns across all the different parts of that vast space. This book also argues against evaluating which Arctic SRSC faces the most immediate security threats: security dynamics are inherently different in different subregions and might change rather rapidly (as both the Barents and Beringia SRSCs show).

Furthermore, the framework employed in this book holds relevance for understanding other issues that tend to attract attention to geopolitics in the Arctic. Looking at, for example, territorial claims, it is fundamental to reiterate that no territorial disputes remain in the Arctic region. No land border between any of the eight Arctic states is in dispute and no Arctic state is currently claiming another Arctic

[18] Although fisheries disruption is also taking place in the Barents Sea context with Russian military exercises (Nilsen 2024a). In an unrelated story of a more benign character, a Norwegian fisher 'caught' a US nuclear-powered submarine in its net (Nilsen 2024b).

state's territory for historic reasons or to protect an ethnic minority.[19] However, a dispute *is* likely to emerge over extended continental shelf rights in the Arctic Ocean between Canada, the Kingdom of Denmark, and Russia (and to some extent the USA).

Yet if we take the subregional approach seriously, it becomes apparent that the SRSC that these claims take place in – the geographic domain around the North Pole – holds limited relevance when discussing state security practices. The distances to population centres, military base structures, or just any territory where states can operate from, alongside the limited – if any – commercial or strategic value of this space, means that this potential dispute will hold limited consequences, if any, for state security practices. The dispute might lead to symbolic posturing, diplomatic efforts, and/or pandering to domestic audiences with statements about 'protecting' Arctic interests and the like (e.g., Staalesen 2023b). That is, however, nowhere near the level and intensity of the security dynamics we see developing in the Barents and Bering SRSCs with military operations, exercises, and political escalation.[20]

Similarly, other Arctic issues can be better understood through the SRSC lens presented here. One is NATO's Arctic role. As discussed in Chapter 4, with seven Arctic states as members of the NATO alliance, there are increasing discussions about either a joint NATO Arctic command or a NATO Arctic policy (Mottola 2023; Jouan 2024; Babikian and Nesheiwat 2024; Odgaard 2024). The primary assumption presupposed by an apparent need for a NATO Arctic command, or even a joint NATO Arctic policy, is that the Arctic is a region where NATO security interests are more or less coherent. As shown, this is arguably not the case. Arctic NATO members do not operate in the same geographic space, nor agree on NATO's potential roles in the North.

[19] Furthermore, no island is in dispute (the last minor island dispute was settled between Canada and the Kingdom of Denmark in 2022). The closest situations to a dispute over sovereignty concern the legal status of shipping through Canadian and Russian territorial waters and domestic discussions in some Arctic countries about Indigenous/local rights and land ownership. At sea, seven out of eight maritime boundaries are agreed. Of these, only one remains: between Canada and the United States in the Beaufort Sea – and this might be settled in the near future. This is in stark contrast to other maritime domains around the world encircled by states (the Mediterranean, the South China Sea, and the Caribbean).

[20] It is also important to note that the seabed is not the same as 'territory'.

NATO is also not a unitary actor that coordinates the Arctic states' security practices, despite Russian rhetoric to the contrary.

The need for accuracy when discussing security affairs and the role of the NATO defence alliance in the Arctic is thus central. Developments in the North Pacific present growing security concerns for the US, but NATO's role in that area is limited, whereas US–Canadian collaboration is necessary. Conversely, NATO plays a crucial role in the security of the Nordic states in the European Arctic. This is not due to decisions from Brussels HQ or because NATO members such as Portugal or Greece are identified as essential security contributors. Instead, NATO serves as a framework through which the USA, along with certain northern states such as the UK, France, and Germany, engages with Nordic security issues, beyond their individual interests in mitigating threats from the Russian Kola Peninsula. One might be critical of military exercise activity in the Arctic, or what policy choices Arctic states make when it comes to their security alignments and guarantees. Yet to shorthand it all under a 'NATO' bracket not only fails to capture the complexities of what NATO actually is and does, but it also falls into the Russian rhetorical trap of describing everything done by Western states as a coordinated plot by NATO.

Based on these premises, beyond establishing an 'Arctic Command' (Mottola 2023; Palmer and Gosnell 2024), there are still some things that NATO *can* do more of when it comes to the Arctic. This in turn underscores the relevance of the subregional approach that this book argues for. Misinformation and lack of knowledge abound within the NATO organisation, and amongst NATO member states, when it comes to the Arctic. Even NATO's own 'brain' (Allied Command Transformation) highlights that 'The Arctic region carries significant strategic interest: maritime routes, resource access, climate conservation, and territorial claims, the High North is an important priority for NATO' (NATO 2023b). Discounting the vagueness concerning lumping unspecified resources, routes, and 'territorial claims' together, the fact that the Arctic and the High North are used interchangeably, showcases a lack of understanding by the authors of that article. As already highlighted, there is a clear and purposeful distinction between the Arctic and the High North, proposed by NATO member states. That there is one, and why, is probably not fully understood across the thirty-two NATO member states, nor by everyone in NATO Brussels HQ.

While it is clear that Canada and the USA through NORAD and USNORTHCOM must align their Arctic-related security awareness with JFC Norfolk (which all the Nordic states have joined),[21] this does not necessarily call for the establishment of a joint Arctic Command. In fact, given that the security concerns of the European Arctic are interlinked with what happens in the Baltic region, Norfolk coordination with NATO Command JFC Brunssum, responsible for the Baltics and NATO's Eastern flank, is equally vital. In other words, the subregions help pinpoint the divergence between Arctic NATO member countries, and also showcase the interlinkages with, for example, the conceived Baltic SRSC – stepping beyond vague statements about 'the Arctic'.

A third example and a looming issue in Arctic security affairs concerns China's role and presence. Again, as has been showcased here with the difference in focus from Norway and the USA, looking at SRSCs better captures the nuances of China's Arctic endeavour. In a Barents SRSC, China does play a role as it continues to enhance its cooperation with Russia, which in turn has subregional expressions such as the BRICS plans for a Svalbard research station, or Chinese interests in investing in European Arctic infrastructure for its Belt and Road Initiative. However, the security practices that emanate from China and its interactions with the Arctic states in that space are limited, despite calls to investigate what China is doing on Svalbard (G. Johnsen 2024), or fears of general Chinese 'hybrid' operations in the area.

In the Beringia SRSC, however, as described in the previous chapters, other dynamics are at play. To what extent these Chinese operations (still primarily jointly with Russia) will increase in scope and effect, in turn also eliciting reactions from the USA, is unclear. Geography comes into play, as China can utilise a perceived strategic link from the East China Sea all the way to the Sea of Okhotsk and the Bering Sea in order to showcase capabilities, but also signal to the USA if there are signs of escalation concerning Taiwan or even the Philippines and activity in the South China Sea. This is thus a crucial geographic space to watch as, or if, security relations between China and the USA deteriorate further.

Being mindful of the different Arctic SRSCs can also provide accuracy to our understanding of other actors external to the region and

[21] Others suggest an 'Arctic Security Initiative (ASI)' in the framework of NATO (Conley and Arts 2023).

their related interests. For example, the EU has for decades been developing both an Arctic policy and a narrative about its Arctic relevance. However, as argued by Raspotnik and Stepien (2023), more spatial awareness in its policy approaches might be a more fruitful and less friction-causing way to apply its EU policy. Focusing on close-to-home, namely the European Arctic where two of its member states are located above the Arctic Circle, might prove more effectful for *its* policy. But it might also provide a lens for analysts to better understand what influences EU-Arctic-policymaking in Brussels and its related implications.

The same can be said for India. India is the world's largest state in terms of population numbers and could start rivalling China along a range of economic, political, and security parameters in the span of a few decades. For India, as an aspiring Arctic actor, some uncertainty abounds when it comes to Arctic security and geopolitics, however. Decision makers in New Delhi must conceive of India's Arctic footprint along different issue areas, levels of political dynamics, and specific geographic parts of the Arctic. Its role in terms of investments, research activity, and even a military footprint will possibly be different in the North Pacific and Beringia SRSC, than in the European Arctic and Barents SRSC.

Finally, the subregional approach also helps us when studying the linkages between climate change, the Arctic, and security dynamics. There is undoubtedly an overarching link between the Arctic and climate change, as the region signifies everything from global warming to humanity's relationship with nature. And climate change will assumedly only lead to more direct state security concerns in the future, also in the Arctic. Too often however, climate change and its various Arctic effects are used to highlight the relevance and importance of the region, but the actual causal linkages with the object of study are not clearly spelled out. In many instances, climate change is often a contextual factor with limited direct bearing on the question at hand, at least when we examine geopolitics and (military) security relations in the Arctic.[22]

Put differently: Is ice melting a factor behind Russia's military build-up? Is ice melting a factor leading to concerns about resource

[22] Others have highlighted this as well, when showcasing that in fact, at least in the Barents SRSC, climate change does not actually have a significant impact on states' security posture there (Knutsen and Pedersen 2024: 175).

governance? Is ice melting prompting Chinese interests? Or, conversely, would a reversal of the melting of sea ice in the Arctic lead to a reduction in Russia's military investments? Would we see fewer military exercises in the North Atlantic and/or North Pacific? Would China be less eager to invest in LNG facilities? Would the EU be less interested in developing a role in the region? The answer to these questions is not a straightforward yes or no, but how we use climate change and particularly ice melting as a contextual or causal factor is often not made clear in Arctic security studies.

Being specific, again, about which part or which country in the Arctic we are discussing, and with regard to what issue, is crucial here. In many instances, although climate change is obviously important in the Arctic – both spurring attention and being a pretext for actors to get involved – it is not necessarily the primary driver of the state security practices and related geopolitical tensions we see emerging in the different parts of the Arctic. In other instances, such as when it comes to spurring geopolitical interest from the United Arab Emirates, or challenging current regional fisheries management organisations in areas in or close to the Arctic, climate change is an obvious driver of the issue at hand. Subregions can again be a more relevant lens, or frame, to explore the issue of climate change and its linkages to security practices (i.e., *geopolitics*).

The argument here is not that these Arctic SRCSs are bound to end up as more conflictual spaces, nor that they have inherent dynamics that cannot be superseded by regional constructs, impact of world events, or changes in the actors themselves. The argument is instead that we need to take another look at the map when we attempt to ascribe certain security dynamics across vast geographical spaces. Sometimes, such dynamics do transcend the limitations of immediate geography. Most of the time, however, they do not.

Lessons for the Study of Security Regions

Returning to Buzan and Wæver, separating the world into RSCs and focusing on the map of Eurasia in Figure 8.1 – which a considerable part of the Arctic spatial domain belongs to – can help illustrate the point. In it, it seems apparent that what has been highlighted as a separate SRSC in the European Arctic is just an extension of a wider Western European RSC. As argued by Greaves (2019), the Barents

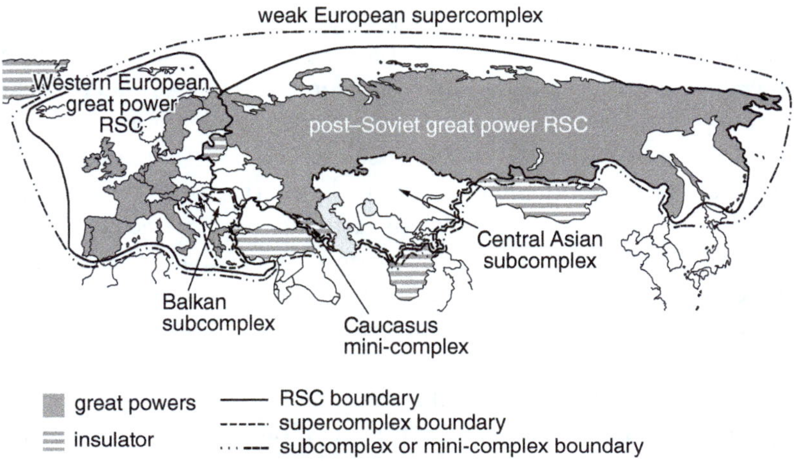

Figure 8.1 Map of the European RSCs. From Buzan and Wæver (2003: 350). Reproduced with permission of the Licensor through PLSclear.

SRSC or the European Arctic is 'simply the northern zone of the broader European RSC'.

However, from the review and discussion so far in this book, it seems plausible to argue that Spain's security, or 'military security', practices have limited consequences, if any, for Poland, Norway, or Estonia. Perhaps in other fields certain actions and practices might have a wider impact, such as a joint EU policy to manage migrant flows from North Africa or putting in place a Western sanctions regime targeting Russia. However, by grouping all such practices under the larger umbrella of securitisation and securitising acts, the boundaries of the respective regions become blurry and in essence subject to the boundaries of those very acts – which in and of themselves often hinge not on geographic or spatial concerns, but on ideas of regionality and political cooperation. In turn, 'the Arctic' or 'Western Europe' says little about where states in *that* space operate in terms of security practices.

The core premise of this book is that if we want to comprehend what states actually do, where they send their armed forces, and where they feel insecure and even threatened, most of those practices revolve around geographic proximity to other states and shared borders and boundaries. There is a reason why Switzerland can allow itself to spend only 0.76 per cent of its GDP on defence and remain 'neutral' in security affairs (O'Neill 2024), whereas Estonia, Norway, and Bulgaria have all joined NATO and spend 3.43 per cent, 2.2 per cent, and 2.18 per cent respectively on defence (Falkenek 2024). Or, in an Arctic case, why Canada – at least before the return of US President Trump in 2025 – was spending 1.37 per cent, struggling to reach the 2 per cent target (Penney 2024). These are highly simplified examples – of course states' considerations of security, defence spending, and neutrality hinges on a range of factors, from historic context to domestic politics. Still, geography matters – more than is often taken into consideration when discussing international security.

There are naturally also a whole range of security issues that can heavily influence states' security posture and security practices but are not necessarily conditioned by geography and *other* states. The French military engagement in its former colonies across Africa, Asia, and South America is one example. Another is the considerable wave of terrorism attacks across European countries such as Belgium, France, Spain, and the UK in the 2000s and 2010s, as a non-state non-territorial security concern that dominated much of state security practices and led to actions in everywhere from Afghanistan to Syria and Iraq.

The argument here is not that these issues are irrelevant. Instead, the argument is that we also cannot ignore the fundamental importance of the security practices that revolve around geographic proximity between states. Despite idealistic notions of an end to territorial conflict and border wars in the 1990s, today we see that is still where a lot, if not most, of states' security posture, resources, and concerns materialise.

The Arctic SRSCs make clear the continued centrality of geography in determining states' security practices. The USA, despite being the global superpower, cannot ignore the confines of the North Pacific and the Bering Sea. In that domain, the USA must engage with Russia in some way or another, as it also did during the Cold War. Changes in climate, ice conditions, fish stock movement and abundance, as well as technology, impact that domain, making Russia–USA security interactions brittle. The addition of China there also emphasises the (relative) proximity between various SRSCs. As highlighted in Figure 8.2, the Bering SRSC is located relatively close to a SRSC that can be conceived around the Kuril Islands and the Sea of Okhotsk, which in turn interlinks with a SRSC that could be conceived around Japan and South Korea. China has also argued that part of its increased

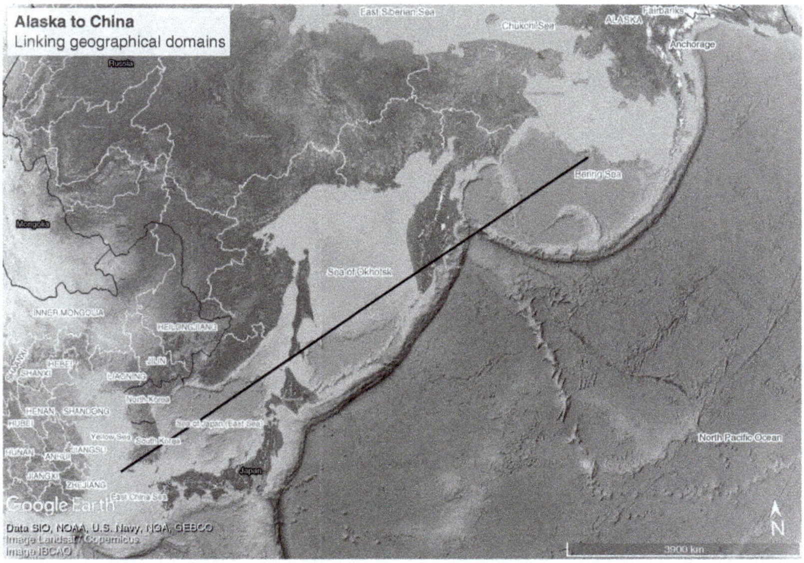

Figure 8.2 A line drawn from Alaska to China. Made with Google Earth.

naval activity off the coast of Alaska must be seen in response to the US presence in the East and South China Seas. This is not an example of decoupling geography and security concerns; it is instead an example of seeing various geographic domains as interlinked.

This linkage underscores how the concept of SRSCs must be seen as overlapping geographically bounded spatial domains, which have their own security dynamics based on proximity between states and state practices but must nevertheless also be conceived of as linking with other nearby SRSCs. Similarly, in the European context, the Barents SRSC has direct overlap and interlinkages with the Baltic SRSC, although these retain their distinctly separate dynamics. The central focal points of these SRSCs are geographic spots where states intersect and interact with each other, either over borders on land or boundaries at sea, and the immediate surrounding areas.

Thus, states typically find themselves in several SRSCs at once, with varying degrees of intensity concerning their security practices and dynamics. It is therefore relevant to envision a Venn diagram of overlapping security SRSCs, as displayed, albeit rudimentarily, in Figure 8.3. These Venn diagram circles are a starting point for analysis. What in turn might prove fruitful is identifying the exact limitations

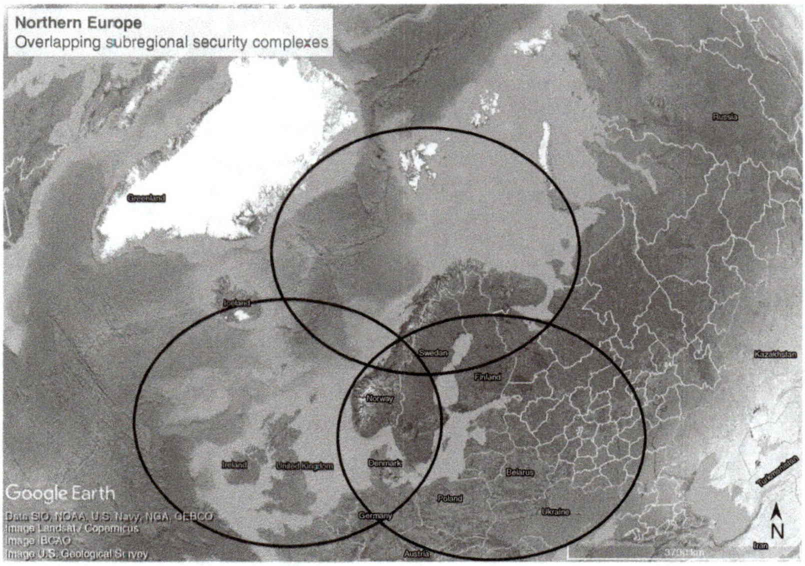

Figure 8.3 A Northern Europe of overlapping SRSCs. Made with Google Earth.

and boundaries of each, and also how, and why, some overlap, but others perhaps do not.

Latvian security practices, for example, have considerable bearing on Polish and Belarussian security practices but limited influence on Norwegian ones. But a large-scale security issue or escalation in the Baltic domain will also impact Russian and Norwegian calculations in the Barents Sea. Spanish–Moroccan security interactions and practices, however, have little bearing on either of these domains, unless the security issue at hand takes on scaled-up characteristics (such as Spain invoking article four or five of the North Atlantic Treaty) which inherently then decouples the issue from what is in focus here – the dynamics emanating from constraints and state interactions in geographical subregions.

From this examination of the Arctic SRSCs, some key lessons for theorising about regions and security practice thus emerge. This does not only concern the RSCT, but will inevitably refer to that theoretical framework, given its centrality in unpacking security regions. The approach here is not to evaluate the RSCT or make use of it to explain the Arctic in the twenty-first century. Instead, the approach here has been to lean on the conceptualisations and ideas of the RSCT, as well as other approaches to security regions, region-building, geopolitics, while retaining a specific focus on the state and its security practices.

First, the subregions that emerge have centred on explaining state security practices where they coalesce, ignoring other subregional domains not dominated by security considerations, and also shedding the shackles of the pan-region-wide scope. Although the framework in this book is sympathetic to trying to unpack 'patterns of enmity and amity', including the 'mixture of history, politics, and material conditions' (Buzan and Wæver 2003: 47), this notion is arguably too broad in scope.

Instead, Buzan's state-centric definition of a RSC from 1983 might be a better fit (Buzan 1983: 106): 'a group of states whose primary security concerns link together sufficiently closely that their national securities cannot reasonably be considered apart from one another'. The emphasis is thus less on how international or systemic structures define the parameters for state security at a given regional level, and more about how the regions are defined by states and their security practices (so-called bottom-up), although these processes are obviously reconstitutive. It therefore becomes essential to include the domestic, or national, level and security perspective, and give more weight to geography and just how quickly states lose their ability to project force without great expense. Security practices of states are consequently something local, seldom something global.

This does not mean that the geography of regions is an absolute limitation that locks the surrounding states into predetermined patterns of behaviour. We also cannot predict the trajectory of these dynamics and relations going forward based on geography alone. This is where this book markedly deviates from traditional analyses of geopolitics. For example, Chinese–Russian relations in the Bering Sea/North Pacific showcase how we must understand the *interlinkages* between geography (Russia and the USA in shared maritime space in the Bering Sea) and other dimensions of security amongst states, namely security dynamics at various levels or scales, and the relevance of the intangible elements in Arctic 'regionalism'.

The increase in joint exercises by China and Russia, in 2024 taking on both an airborne and a coast guard element alongside the naval one, must be understood as a direct consequence of increased general cooperation between the states in general after 2014 which has a limited geographic rationale. In 2014, Western states sanctioned Russia and cut various forms of military collaboration in response to Russian actions in Ukraine, while they had – led by the USA – been pivoting towards containing China from around 2010 (e.g., Canrong 2016). The increase in scale of collaboration between Russia and China must be linked to further deterioration in relations with the West after Russia's invasion of Ukraine in 2022. What happens on the global or systemic level naturally will impact subregional expressions of security practices.

Furthermore, the symbolic elements of operating in the Arctic, as exemplified by the Chinese Coast Guard's fanfare about reaching the Arctic Ocean in 2024 (McCarthy 2024), as well as the increased symbolic elements of Russian–Chinese cooperation in the Arctic (e.g., Stensdal and Heggelund 2024), are drivers of Chinese security practices in this subregion. That in turn spurs US attention to Arctic security issues (as exemplified with the DoD Arctic Security Strategy from 2024). There is, in other words, an interplay between political and region-building conceptions of the Arctic, how certain parts of the Arctic such as the Beringia SRSC become expressions of global security dynamics, and the limitations and opportunities defined by geography in those very same spatial domains. This interplay is what constitutes the regional security dynamics (or an RSC) taking place in much smaller geographical domains – the SRSCs – than what is normally used to refer to regions, namely North American security dynamics or Pacific security dynamics.

Second, a consequence of recentring the state and downscaling the scope is that security regions can overlap, or states can participate in several simultaneously by virtue of their geography. Great powers (i.e., states with relatively more recognised capacity than their peers) operate,[23] as highlighted by much of the security regions literature, in several security regions by virtue of either overlay – when they dominate the region to the extent that no local patterns

[23] See Iver B. Neumann (2008) for an extensive discussion on different ways of understanding what a great power is: 'Weber's realist criteria – material resources and the ability to project power … Durkheim's moral criteria – setting an example for how a state should be in the world', and – what Neumann argues for – a 'governance' account which 'takes cognizance of the materiality of the discourse' and 'is predicated on the intersubjective character of recognition'.

exist – or penetration – when they align with actors in the region and assert influence. However, great powers also participate in SRSCs by virtue of their geography and can arguably belong to several simultaneously.

Although Buzan and Wæver (2003: 46) appreciate this through their notions of overlay and penetration – that is, '[s]maller states will usually find themselves locked into an RSC with their neighbours, great powers will typically penetrate several adjacent regions, and superpowers will range over the whole planet' – they arguably neglect that states can belong to several regions at once – at least when operating with scaled-down SRSCs as the primary focus. Moreover, most states are small powers (e.g., Baldacchino and Wivel 2020). Studying their security concerns, overlaps, and regional linkages provides a more accurate approach to international studies, given that the emphasis is on understanding state behaviour and actions. Thus, this framework does not operate under the assumption that the SRSC boundaries are firm and mutually exclusive. SRSCs bleed into each other and overlap. Scaling down regions provides a more accurate and empirically oriented picture of security dynamics and – crucially – security concerns of the participating *states*.

Third, this also links to a point concerning the notion of security regions as the 'primary' spatial or regional domain for security practices. In the approach here, they do not have to be. In fact, defining what the primary spatial domain for any given state is opens a Pandora's box of subjective evaluations and fluctuations depending on how you measure security interests and practices, where geographic spatial domains are delineated, and temporality. This is not to say that certain SRSCs do not hold more clout than others in defining the security practices of participating states, but to highlight one as primary might be a futile process of limited and rapidly expiring analytical value.

For example, in the case of Norway, it is clear from both political rhetoric and security practices ranging from military investments to exercises that the High North, or the Barents SRSC, is the most dominant spatial domain for that state's security practices. However, Norway cannot discount other nearby SRSCs (see Figure 8.3), such as what we could conceive of as a North Sea/UK-centred SRSC, or the nearby Baltic SRSC. These hold importance in Norwegian security practices, albeit not with the same significance as the Barents one.

The USA, in the case of the Beringia SRSC, is in a different position. For that state, when focusing on its geographic space and near abroad, other more immediate security concerns and in turn subsequent practices can most likely be identified along parts of the border with Mexico, or historically, in its relations with Cuba across the Straits of Florida, or even Canada to the north-east. Developments on the border between the USA and Mexico around the Gulf of Mexico have little bearing on the Beringia SRSC however. The point here is that states often take part in several SRSCs by virtue of their geography and their relations with the bordering or proximate states, and making the existence of a subregion contingent on being the primary one for all participating states is a futile effort.

In summary, where this study deviates from traditional approaches to security regions and RSCs, is the scale of the 'regional' and where to place the emphasis as a starting point of analysis. This is due to a focus on the state and its security practices as the primary referent object. If state security practices are the starting point, the RSC becomes something much smaller and immediate, defined (and limited) by geography. Furthermore, subregional dynamics must be conceived in relation to the international system, the broader region, *and* the national dynamics. The question is to what degree a SRSC is influenced by dynamics at those various levels, and how in turn the SRSC enables and limits state security practices. Finally, a focus on SRSCs not only opens up the possibility of, but hinges on, these spatial security domains being at least partially overlapping. In turn, this should provide for a more accurate assessment of the linkages between security, geography, and the regional spatial domain.

Maritime Regions as a New Frontier?

Lastly, a few words about maritime space and oceans are pertinent here, given that the Arctic comprises one relatively large maritime space, and that the security subregions identified and examined are also framed around seas. The approach put forth here does not posit that subregions must coalesce around maritime space. However, there are some interesting trends that this book also fits into when it comes to the study of maritime regions, security dynamics, and geopolitics.

Several trends are making (coastal) states increasingly concerned with maritime space, sovereign rights, and notions of upholding sovereignty at sea.[24] Whether this is due to drivers related to climatic changes, resource potential, reimagined social relations with the ocean (J. Anderson and Peters 2014; Steinberg 2001), or increased polarisation in global governance discussions depends on both the maritime domain in question and on which actor we examine. In any case, the result is the same, namely that enclosed or semi-enclosed maritime domains such as the Arctic Ocean, the Black Sea, the Caribbean Sea, the Mediterranean Sea, or the South and East China Seas are receiving increased attention as spatial containers for all those mentioned drivers.

As Christian Bueger, Elizabeth Mendenhall, and Rebecca Strating put it in their edited volume *The Politics of Global Ocean Regions* (2025, 1), 'despite their fundamental importance to international relations, our understanding of how ocean spaces are politically organised remains surprisingly limited'. Jacobsen, Wæver, and Pram Gad (2024a, 337) reflect, in their work on Greenland in Arctic security, that the RSCT might hold relevance in understanding security dynamics across other regional maritime domains, although they argue for retaining a focus on land as the domain where security interactions are the most intense.

Based on the examination of the Arctic, and its security subregions, in this book, the approach remains agnostic to the centrality of the oceans, or the maritime domain, in terms of security dynamics. There is no doubt that territory, and related border conflicts and disputes, has been, and continues to constitute, the primary security issue for many states. When states have unclear or disputed borders, that domain is a continuous source of friction. The conflict between China and India in the Himalayas, the conflict between Colombia and Venezuela along their Caribbean-facing border, or the conflict over where to delineate territory in Western Sahara exemplify the contemporariness of conflict surrounding borders on land (Tir and Diehl 2002; A. Rosen 2015).

Inherently, it is the ability of states to interact across geographical space that makes border regions particularly contentious. Land is one of the most valuable assets for a state, both in terms of function and in terms of symbolism. Therefore, so-called patterns of enmity and amity

[24] Which in and of itself could be seen as a contradiction, given that states do not have – beyond twelve nautical miles – sovereignty at sea, but instead only sovereign rights.

are dominated by border issues, or historic relations across those borders, such as that between Latvia and Russia, or Mexico and the USA. Even where borders are settled and have been so for a long time, they can be a source of contention, especially when relations between bordering states change due to issues that may or may not be related to the border at all. However, more and more border disputes *are* settled (Schenoni *et al.* 2020).

That in turn leaves maritime domains, where states are in relative proximity with each other, and might even dispute maritime boundaries, jurisdiction, and sovereign rights (Østhagen 2021c). The various maritime parts of the Arctic are examples of this, although the Arctic Ocean writ large is actually an outlier in a global context since most maritime disputes are settled there. An interesting question is whether there are certain characteristics with state interactions in the maritime domain that differ markedly from those characteristics on land (Nemeth *et al.* 2014). The oceans have generally been seen as the antithesis of land – opposite and inherently different – or as a void (Steinberg 2001). Maritime space and its value for states have been defined as inherently functional. That may make it seem tempting to equate the value and importance of maritime space (and subsequent disputes over such space) to the functional value of that space (D. M. Johnston 1988).

However, in some instances, it might be said that maritime disputes are coming to resemble traditional territorial conflicts on land (Østhagen 2021c; Mendenhall 2018). Especially in bilateral relations that are already fraught, as with Colombia–Venezuela or Norway–Russia (over Svalbard), maritime disputes may prove to hold latent conflict potential. However, the maritime domain has certain characteristics that nevertheless keeps it separate from the terrestrial domain. John Mearsheimer (2001: 84) deemed it 'the stopping power of water'. There are geographical barriers that hinder prolonged interactions between the actors concerned. Maritime boundaries and sovereign rights are also a construct of international law, and (coastal) states seem to depend on the Law of the Sea regime while also wanting to apply the regime to their own advantage (Mendenhall 2019). Also, what we are discussing regarding states and maritime space is sovereign rights to resources in the water column or on the seabed, not *exclusive* rights to the entire maritime 'territory' in question.

At the same time, as this focus on SRSCs brings forth, the immediate geography where state security practices coalesce might revolve around a maritime domain. Whereas a border on land provides a fixed and often well-guarded point of contact, which – as in the case of Norway/Finland and Russia in the Arctic – leads to a relative stasis in relations and activity over that space, the maritime domain is a much more 'fluid' space. States cannot deny passage through their EEZs or even hinder military activity such as large-scale exercises; they may only deny actors' access to marine resources and apply environmental regulations in their maritime zones.

Also, as the Barents SRSC showcases, the access of Russian research vessels to the Norwegian EEZ, the FPZ, and the shelf rests on UN Convention on the Law of the Sea Article 246 (United Nations 1982), where 'the coastal State should normally grant its consent' except in a few specific circumstances (Woker *et al.* 2020: 2). Thus, the burden of proof concerning Russian vessels conducting illegal activities in Norwegian waters, including in the FPZ, lies with the Norwegian authorities – which is diplomatically and administratively strenuous. If tensions were to mount even further between NATO and Russia at large, it is also likely that Russia could decide to engage in so-called hybrid warfare in or around Svalbard. Concern in Norway over the activities of Russian trawlers and research vessels in Norwegian waters is already prevalent (Juul Stensrud and Østhagen 2024).

Such security concerns, which in turn spur security practices and impact the general security dynamics between actors within that specific geographic space, are not relevant across all ocean space. More than 70 per cent of the globe is ocean. Instead, it is where states are located in geographic proximity to each other across restricted maritime space that interactions might be of relevance to security practices. Enclosed or semi-enclosed maritime spaces, or spaces where state boundaries at sea congregate, are particularly interesting in this regard. Not all such spaces are necessarily dominated by security practices of the nearby states, but they have the potential to be. The most relevant spaces around the world seem to be the Arabian Sea, the Arafura Sea, the Bay of Bengal, the Black Sea, the Caribbean Sea, the Caspian Sea, the East and South China Seas, the Gulf of Guinea, the Sea of Japan, the Persian Gulf, and the Red Sea/Gulf of Aden, in addition to the Barents and Bering Seas.

As showcased in both the Barents and the Bering Seas, uncertainty over sovereign rights, jurisdiction, under-the-threshold operations, and even military exercises close to territorial waters of the opposing state lead to heightened security concerns, in turn impacting (and constituting) the Barents and Beringia SRSCs. The maritime space thus becomes a spatial domain that states utilise in order to signal and, possibly, escalate tension, without engaging in direct conflict. Whereas escalation on land along a fixed border might entail direct involvement of military units, at sea softer means can be deployed, such as coast guard vessels firing water cannons, or even a fishing trawler that 'accidently' damages seabed infrastructure (Ondraskova, Berzina, and Arts 2024).

At the same time, due to the fluidity of the maritime space, as well as the characteristics of ocean-based resources and activities, there is also room for collaboration. Whereas bilateral cooperation with a neighbouring state across a land border might be difficult to initiate if security relations are tense, certain maritime tasks, such as search and rescue, environmental protection and response, and management of transboundary marine resources, not only suggest collaboration, but are sometimes dependent on it. Moreover, for delimitation in the maritime domain, both states may have valid legal claims to a given area, in which case it becomes a matter of 'reasonable sacrifice such as would make possible a division of the area of overlap' (Weil 1989: 91–92) or even sharing – as with oil and gas resources or a joint fisheries zone. The Barents and the Bering Seas are again examples of this, where both Norway and the USA initiated cooperation with the USSR in the 1970s, and later expanded this with Russia in the 1990s and 2000s, in order to manage everything from overfishing to increased maritime traffic.

As Michael Byers terms it (2019), there are certain spatial domains that are 'cold, dark, and dangerous', which in turn might prompt states to cooperate with each other for positive-sum benefits, regardless of negative security interactions elsewhere or over other issues. Although the 'cold' and 'dark' elements may vary depending on which sea or ocean we examine, the maritime domain in general might in fact be exactly such a space that allows low-level forms of cooperation amongst states that may find it difficult to engage in other forms of cooperation, geographically bounded or not. Not only is the Arctic Ocean (or the various SRSCs) an example of this, we can, for example, find similar examples in Southeast Asia (Beckman *et al.* 2013; Midford 2015; Simon 2012; Mishra 2017).

Perhaps there are certain elements when it comes to the ocean that makes the potential for low-level disputes, as well as low-level forms of collaboration, relatively high. These concluding remarks of this chapter therefore suggest looking not only at security subregions in general, but also interactions related to security and security practices in these smaller or geographically bounded maritime spaces.

9 Conclusion

The Benefits of the Arctic Illusion

Returning to the story from Chapter 1 and the icebreaker in the port of Murmansk in June 2008, Russia ended up losing the football match against Spain. After drinking vodka with Norwegian Foreign Minister Gahr Støre, things more or less went downhill for Russian Foreign Minister Sergey Lavrov. Still, from 2008 to 2013, there was relative optimism and, at times, exaggeration regarding Arctic resource potential, while Arctic states – and primarily Russia – simultaneously invested in Arctic military capabilities. There was also a considerable interest in the 'Arctic' that stemmed from limited understanding of regional variances and complexities.

From 2014 onwards, a decline in relations with Russia and a drop in global commodity markets shifted the narrative away from overly optimistic projections, leading to a sense of Arctic disillusionment.[1] However, from 2018, events such as China's first Arctic policy, the first Trump administration's vocal stance on Arctic security issues, and increasingly aggressive military activity from Russia began to attract a different kind of global interest. Russia was not invited to participate in the Euro 2024 tournament due to its illegal invasion of Ukraine in 2022, which must have been a low point, at least in terms of football. In the context of Arctic geopolitics, tensions had already been mounting between Russia and the other Arctic states before 2014.

Russia's invasion of Ukraine in 2022 further underscored these trends and added an immediacy to concerns about potential conflict escalation in the Arctic. Both Russia and NATO states are demonstrating their military capabilities in this region, raising worries about potential escalation and outright conflict, for example, related to the

[1] The Norwegian conservative commentator Asle Toje called Norway's High North endeavour a *nachspiel*, that is an afterparty where everyone in fact had lost interest and wanted to go home (Toje 2015).

Svalbard archipelago. This is undoubtedly Arctic *geopolitics*. But does it concern all of the Arctic?

The Arctic spans all time zones, eight countries, and three continents. Its territory is six times as large as the European Union. Just the Arctic Ocean is six times larger than the Mediterranean Sea. Due to the Gulf Stream, the European Arctic coasts and the waters of North-West Russia are completely ice-free even during the coldest months of the year, setting them apart from the ice-covered waters of North Alaska and Canada. Drifting sea ice, darkness, and low temperatures are no doubt common traits in the Arctic region, but the climatic conditions vary across it – and they are changing faster than anywhere else on the planet.

Furthermore, there are considerable differences between parts of the Arctic in terms of levels of development, culture, infrastructure, and resources. The way of life in the university city of Tromsø is different from the industrial and naval city of Murmansk, and these cities are relatively close to one another – a mere 13 hour and 1,000 km drive. If one also considers the differences between the local Inuit communities in north-west Greenland and the coastal communities on the east coast of Alaska, it is apparent that the Arctic is by no means one coherent region on a range of, if not most, parameters.

Still, the security interests of Arctic states are increasingly seen as intertwined. The Arctic is portrayed either as a region where great power rivalries abound and military conflict is likely, or – before 2022 – as a part of the world defined by positive-sum interests expressed through various cooperative mechanisms. The Arctic is subject to descriptions of being an arena for growing great power rivalry, and as a spatial domain where military activity is increasing and conflict might be unavoidable due to the clashing interests of both Arctic and non-Arctic states. The Arctic is thus implicitly or explicitly portrayed as a 'security region': a spatial domain in which the security concerns and interests of states are interlinked and overlapping. Such portrayals, which tend to focus on state-oriented or military security, seldom make use of the concepts upon which they rely, in this case the concept of security regions or regional security complexes.

The region label is frequently used, but without a proper analysis of what this label means when we pair it with state security practices. As explored throughout this book, the idea that up until 2022 the Arctic was a unique region, and that its cooperative traits and lack of internal dispute led to some shielding from security affairs, did and does not correspond with the security practices of Arctic states even during that time. However, describing the Arctic as an arena for geopolitics, interstate conflict, and intertwined security dynamics also does not fit both previous and current realities.

Is the Arctic truly a 'region' if we take a closer look at the eight Arctic states and the various Arctic subregions? Do the security interests and practices of states overlap and intertwine within this defined geographic space that comprises 4 per cent of the globe?

The answer to these questions necessitates a conceptualisation of the linkages between security, geography, and regions in international relations. This book has analysed this interplay in the context of the Arctic in the twenty-first century, with a focus on state security practices, the divergence between foreign and defence policies, and the role of regional cooperation. In turn, this means leaning on notions such as geopolitics (i.e., the relationship between security and geography in international affairs), regionalism (i.e., the social construction of regions), and regional security complexes (i.e., theories attempting to define the relationship between geography, security, and the regional).

This book has explored the interconnection of security interests and practices across various geographic scales, focusing on how geography and regionalism impact security. To understand the complexities of Arctic security interactions, this book has advocated for a multilevel analysis encompassing regional, global, and national dynamics, before also adjusting our scope and looking at the subregional. This matters not only to get a better picture of the security threats and conflict potential in the Arctic. It also matters in order to improve the way we conceive of regional security relations in other regions around the world.

Leaning on the regional security complex theory (RSCT) by Buzan and Wæver from 2003, this diverges from the RSCT by narrowing the (geographic) focus, while also centring the state and notions of state security. Moreover, this book argues that subregional security complexes (SRSCs) do not need to be the primary spatial domain defining security practices of participating states as long an impact can be discerned. Security regions, or more accurately, subregions, are in turn overlapping. As such, this is less of a grand theory of regions and security interactions in the world, and more of a way to operationalise how to unpack how a state's security practices are influenced by geography, that is, its near abroad and immediate spatial environment.

A key feature of the argument put forth in this book is about questioning the role of geographic proximity vis-à-vis other relevant factors when gauging security concerns and interests. This work reinforces two ongoing, or returning if you will, trends in international studies (whether traditional international relations, geopolitics, or security studies), namely placing importance on geographic traits defining state actions, and focusing on state or military security. What is novel is linking the various strands of regional or territorial studies with debates on geopolitics and international relations, while also giving weight to processes of regionalisation and region-building.

The Arctic is the ideal case study: because it has increasingly been referred to as a region in terms of security and geopolitics, or even as a regional security complex; and because there are in fact multiple subregions that are markedly different from, but still interlink with, wider security dynamics related to the Arctic at global, circumpolar, and national levels. Furthermore, the Arctic highlights the clustering of security dynamics around enclosed maritime spaces. The book concludes that the idea of an 'Arctic region' in security terms remains relevant not as a factual description, but as an aspirational framework for cooperation and confidence-building.

In essence this work is about taking geography (i.e., a version of geopolitics) seriously, but nuancing and qualifying it, while also leaning on the work done by Buzan and Wæver and all those that have followed in their footsteps with the RSCT. And perhaps most importantly, it is about better understanding the Arctic region(s) and security developments and conflict potential in that geographic space.

As showcased here, 'Arctic geopolitics' actually takes place in Arctic subregional security complexes (SRSCs), specifically the Barents SRSC and the Beringia SRSC – named after the Barents Sea and the Beringia Sea/land space. In those domains, conflict potential is undoubtedly real, although we must be specific about where and over what. Here we must unpack the way these subregions figure in systemic competition, relevant national strategic security policies, as well as broader regional dynamics. Separating between these different levels of analysis, and between different geographic domains, helps improve the accuracy of our analysis of Arctic geopolitics.

States participate in multiple geographical security domains, at a smaller scale than 'North America', 'Pacific', 'Europe', or 'Arctic'. Therefore, these security dynamics must be analysed and understood in order to improve our understanding of state behaviour in international relations, as well as the limitations posed by geography, and how states practice their security dynamics in such domains. In essence, what the examination here of the two Arctic SRSCs showcases is the importance of studying the subregional: specific issues and developments that take place a level below the regional (Arctic) and the global, but that constitute more than the 'national'.

Geopolitics in Practice

The focus on security issues and practices amongst states in a regional, bounded setting has increased in the last few decades. The view of global development as a monodirectional path towards globalisation and a 'borderless' society has been replaced with increased attention on territory, national interests, and the immediate region. Decoupling and protectionism are today more in vogue than openness and the removal of barriers. In these trends, attention has shifted towards security developments in regional settings where 'geopolitics' is apparent, as states engage in rivalry and even conflict, such as, for example, in the South China Sea, the Gulf of Aden, and Eastern Europe. The Arctic is also one such geographic space that has increasingly been attracting attention due to its geopolitics.

Do we have our scope properly adjusted and are we looking at the right levels of analysis? Or are we even trying to distinguish between the different security dynamics that take place at different levels and over different sets of issues? Arguably, when we examine these geographic regions, we tend to lean on already preconceived notions of where the boundaries of that region are, and in our analyses reproduce these boundaries that are either (1) a result of region-building efforts, or (2) due to the relevance of these boundaries in other issue areas such as trade, migration, or climate issues, or (3) due to both.

This book fundamentally agrees with the premise that as 'world security actually does fall in "chunks" for mostly geographical and partly cultural and historical reasons, the level of regional security complexes can stabilize as an organizing center from which one ties domestic, interregional and global security together around the regional focus'

(Jacobsen, Wæver, and Pram Gad 2024c: 14). However, the argument presented here is that if we want to understand what happens and what states do in these specific regions, or *geopolitics in practice*, we must scale down our level of analysis.

When examining the geopolitics, regions, and political processes of regionalisation, particularly in the context of security, it is crucial to understand the role and limitations imposed by geography. By leaning on both political geographers' conceptualisation of geopolitics, and international relations approaches to regions and security complexes, we can conceive a more fine-grained framework for analyses of regional affairs and state security. A framework that argues for a restricted or scaled-down approach, where security practices and interactions in, for example, the Eastern Mediterranean do not impact the same practices and interactions in the Baltic region, despite both being part of 'Europe'.

This immediate area for states is arguably smaller than what scholars and analysts usually designate as a 'region', which is a concept that can be vague and overly broad. Most states are neither great powers nor superpowers; they typically do not engage in security interactions across multiple global regions. Instead, their security concerns are often confined to their immediate surroundings. Neither Norway nor Vietnam engage in the Beringia security subregional complex, but Japan, for example, can do by extension of its geography. By considering prevailing geopolitical perceptions and theories such as the 'revenge of geography' and the regional security complex theory, it is clear that we need to focus more on 'subregions'. These are smaller, more defined areas involving fewer states that share borders and/or maritime spaces, where the dynamics of security are intensely localised, and the basic questions of 'survival' still impact state actions.

Moreover, developments in these subregions must be seen in conjunction, but not equated to, the various, and at times contradictory, security dynamics that all occur under the broader region label (in this instance the Arctic), but that in reality relate to different levels of security analysis. Unpacking these dynamics is done by the distinction presented here: systemic or global rivalry and competition (in the Arctic, why the United States started focusing on China in 2018–2019), regional interaction best understood as 'regionalism' (why Arctic states have up until 2022 continued to meet to sign new agreements hailing the cooperative spirit of the North), and national security practices (why, for example, some Arctic states, and not others, have invested heavily in their northern defence posture).

That is not to say that the concepts of regions or regional security complexes are without relevance; this book merely urges caution in slapping on the 'region label' when examining security dynamics between states, without clear definitions or parameters. Moreover, attention must still be paid to geography when focusing on the security of states. Which states border each other, the distances between states, geographical features such as mountain ranges, and enclosed maritime domains surrounded by states are all components of the immediate geographic surroundings that influence and – to an extent – determine security practices and thus also dynamics amongst the same states.

Others have noted the importance of the subregional level in order to understand military concerns or state security practices (in the Arctic specifically, see Åtland 2007; Friis 2021; Greaves 2019). However, as argued throughout this book, we must also connect these subregions to the wider regional and global security dynamics *and* examine the national approaches to the same region and subregion in order to understand why, for example, China engages in the Arctic space and which part of it. In turn, what these nuances imply is that simplistic, one-liner descriptions of security or geopolitics in a regional context such as the Arctic must be taken with a pinch of salt. This should inspire further studies of security politics in regional settings or – more accurately – security subregions around the world that are highly complex security structures.

This matters because, fundamentally, we need to provide accurate analyses of security development and state practices around the world. In the case of the Arctic, perhaps it is indeed a security region, if one is looking through the lens of human or environmental security and examining pan-Arctic (regional) developments since the start of the twenty-first century. Such a characterisation might also link to a critical idea of security regions, or regionalism, where the construction of regional order and institutions is seen as an end goal in itself to solve certain problems. Maybe environmental concerns and corresponding region-building efforts to deal with those are less spatially bound and 'travel' more easily across the Arctic spatial domain.

However, as this book has emphasised, seeing everything as one big bundle of security issues without hierarchy or priority does not help unpack state practices concerning security. Wide-ranging theories or claims such as those offered by either popular geopoliticians concerning the inevitability of geography, or regionalist scholars arguing that regions are inherently just political constructs, fail to capture the essence of state security practices within a geographically confined domain. States do prioritise, and security is still inherently tied up into notions of survival, threat perceptions, and fear of the 'other'.

In this book, the focus has been on the more traditional or conventional forms of security and geopolitics in order to unpack them in a specific regional context and answer the fundamental question in international affairs of whether or not there are certain inherent traits in that regional context that makes it prone to conflict. In other words, by adjusting our scope when examining questions of conflict and escalation between states, it is possible to get to a more fine-grained understanding of these issues at hand. That means we can avoid getting trapped in simple causal statements, such as equating the presence of resources with interstate conflict or arguing that the strategic importance of certain islands is bound to cause tension between states.

That does not mean that we can, and we should not, stop examining the relevance of other forms of security, or the effects of seeing them as bundled, or categorised, for both state practices and region-building. Had this book, for example, prioritised local or human security concerns in the Arctic, or solely the perspectives of the Indigenous Peoples, a very different outlook would have emerged (Greaves 2016). The same goes if the priority had been on global climate change and the role of the Arctic region therein (Palosaari and Tynkkynen 2015; Miner et al. 2023).

This book has also cautioned against attempting to define *one* region, its boundaries, or the *primary* geographic domain that a given state participates in or belongs to. Arguably, such boundaries and regional definitions are inherently subjective to the analyst and dependent on that individual's regional knowledge and which factors are given priority in defining the framework that in turn produces the regional. Instead, this book has been about studying and pinpointing the role of geographic regions, or specific spatial domains, in influencing state security practices. The goal of an analyst should thus be to do two things at once: keep an eye on both various levels of security dynamics of relevance to the subregion in question, while also studying geography and state practices in that specific subregion. This in turn can help avoid geopolitical simplifications and provide more accuracy to our understanding of international relations.

The Arctic Illusion

The conclusions presented here lead to the question of whether or not we should discount the whole notion, and use of, 'Arctic security' or 'Arctic geopolitics' given that they probably hold as much meaning as saying 'European geopolitics' or 'Asian security'. Perhaps the title of this book is inherently a misguided lure in order to play off all the attention given to 'Arctic geopolitics'. However, contrary to what you might think after reading some of these pages, there is arguably still value in keeping the 'Arctic' label.

First, although in terms of security dynamics and state practices the Arctic is too broad in scope, the political idea of the Arctic, as well as general conceptions of security, geopolitics, and actors' roles and interests in the Arctic, influences states' practices. Over the last few years, several non-Arctic states such as Turkey and the United Arab Emirates have shown an interest in Arctic political affairs with a geopolitical flair. For example, Turkey signed the Svalbard Treaty, with Turkish experts highlighting that it might lead to a more active Turkish position in the Arctic in general (Sarıoğlu 2023; Ayaydin 2023). Or, in terms of actual military footprints: as the UK has been eager to showcase its Arctic military credentials post Brexit, British troops sail and train more, and British politicians talk more about their North Atlantic and Barents Sea security capacities and roles (Depledge, Kennedy-Pipe, and Roger 2019; Depledge and Østhagen 2021).

Despite being inaccurate in terms of denoting a specific spatial domain, states (their politicians and bureaucrats) *are* spurred into certain actions due to the increased interest in the Arctic, whether linked to climate change, economics, military posturing or status-seeking. This might in turn have concrete implications for security dynamics in the various subregions. We then must also study and conceptualise the effects of these interests in and ideas of Arctic geopolitics, and how they in turn might influence security practices in the various parts of the Arctic. We cannot ignore the effects of 'Arctic' geopolitics and related narratives and notions that impact state practices, despite the inaccuracy of that terminology in and of itself.

However, studies of the Arctic would benefit from bridging views from geographically oriented security experts and regionally oriented Arctic governance experts. Building on reflections after fifteen years in the 'Arctic bubble',[2] it is interesting to note the emergence of two distinct groups of analysts examining Arctic-related political issues.

The first group comprises members of the Arctic 'epistemic community', as both Beate Steinveg (2023) and Rasmus Bertelsen (2019) have described, where barriers between policymakers, bureaucrats, and scholars are minimal. At Arctic conferences and seminars, discussions often centre on enhancing the role of the Arctic Council, advancing Arctic Ocean governance, and promoting sustainable development for northern communities. Security analyses are present in these forums, but military security is often sidelined, seen as too narrow in scope. Although there are exceptions, one could label these analysts 'doves'. This group, simplified, is the one actively working to shelter the Arctic Council structure (or at least parts of it) so that cooperation can, at some level, continue.

The other group, however, seldom attends Arctic seminars and conferences. In fact, they seldom make use of the term the Arctic to describe their work, since it – as shown in this book – denotes a vast geographical area and often draws attention to issues such as climate change or governance, which are less relevant to their focus on military security and defence in northern countries. This group could be labelled 'hawks', who assert that cooperation with Russia on softer issues through, for example, the Arctic Council holds little relevance for Russian defence doctrines in the North. This group, simplified, tends to argue that maintaining any form of multilateral or bilateral cooperation with Russia is not only redundant, but could also be a liability.

This author has experienced being labelled a dove – or, more bluntly, naïve – when emphasising the importance of governance structures such as the Law of the Sea, or the self-interest of Russian fishers in maintaining relative stability around Svalbard and preventing further escalation over ongoing disputes. Conversely, at Arctic conferences and seminars, particularly before 2014 but also afterwards, this author has been called a hawk or, more bluntly, a warmonger for arguing that

[2] As explored by Beate Steinveg (2023) in *Arctic Governance through Conferencing*.

military security concerns *do* exist in the Arctic, or that escalation over fisheries issues could develop into a serious state-to-state security concern. Which metaphor holds true clearly depends on the context.

This observation points to a more fundamental issue in how we approach regional studies related to security, politics, and state practices. Security and defence analysts have sometimes overlooked the broader political and situational contexts in which states operate, beyond the immediate demands of their defence doctrines and military capabilities. Conversely, Arctic politics and governance scholars have occasionally disregarded military developments at a subregional level and the continued primacy of security and survival for states in the twenty-first century. One potential solution for fostering cross-learning is to bring more Arctic analysts and military analysts into the same room – even if 'the Arctic' is at times an artificial political construct. It could help prevent simplistic statements and misguided policy actions concerning a region that comprises 4 per cent of the globe and is made up of multiple domains and spaces.

Second, there might also be other benefits from combining levels of analysis, types of security, and geographical domains, namely by upholding the Arctic 'illusion'. In the 1960s, Stanley Hoffman (1966: 882, 901) coined the distinction between 'high' and 'low' politics as a way of separating the different policy spheres of European integration. Economic policy constitutes 'low politics', whereas the survival of the state – security policy – constitutes 'high politics'.[3] In the field of low politics, there is in theory room for dialogue between states even as tensions mount, for example, through so-called track two diplomacy.[4]

[3] Scholars have made use of Hoffmann's distinction to highlight how certain areas of low politics – such as environmental policy or trade – have allowed states to cooperate in spite of rivalry in the domain of high politics (Mirovitskaya, Clark, and Purver 1993: 44–45; Young 2010: 173; Østreng 1999: 76). Homans (2011), for example, argues that track two diplomacy has become a core element of international diplomacy. In particular, relations between Asian countries and those in Southeast Asia have been the focal point of much of the related literature. Constructivist scholars have argued that we can draw on cognitive work and ideas of bounded rationality to understand how leaders and policymakers are influenced by the act of arguing and discursive persuasion (Risse 2000; A. I. Johnston 2001).

[4] Some contributed the term to the article 'Foreign Policy According to Freud' by William D. Davidson and Joseph V. Montville (1982), which examined psychological dimensions to conflict resolution.

Military security concerns were dominant in the two highlighted security subregions – the Beringia and Barents domains – both during the Cold War and from 2014 onwards. During the period in between, from the 1990s to the first decade of the 2000s, security dynamics never disappeared, but emphasis was placed on region-building efforts and political (and scientific and social) cooperation, both in a pan-Arctic context and in a subregional context. This can be understood as efforts from certain Arctic states that narrated the actor constellation as something broader, or porous, in order to avoid being trapped by the inherent subregional security dynamics.

Norway, for example, benefited from the circumpolar or pan-Arctic constellation because it could then collaborate with Russia in a setting other than the bilateral one, which has historically been defined by asymmetry and security concerns. Region-building and regional cooperation that expands beyond the SRSCs can enable states to transcend, or at least partly disregard, 'patterns of amity and enmity'. By upscaling to a so-called Arctic level, there was more space for multilateral low politics efforts that otherwise might get 'trapped' by subregional security dynamics or hampered by asymmetry between the engaged actors. Some speculate this might be why Russia is not interested in leaving the Arctic Council structures despite exiting the Barents Cooperation – it provides for a multilateral arena for dialogue with the USA beyond, for example, the rigidity of the UN Security Council or the sensitivities of bilateral relations (Andreeva and Rottem 2024).

With Russia's full-scale invasion of Ukraine in 2022 and the subsequent responses from the other Arctic states, including Finland and Sweden joining NATO, Arctic-wide cooperation on any level has become almost impossible and security concerns now dominate much of the Arctic focus, both in Arctic states and amongst the global community. Moreover, Russia has deliberately turned away from emphasising Arctic cooperation and engagement with the other Arctic states and is rather seeking third-party involvement from particularly the BRICS grouping of states (Edvardsen 2023a), turning to, for example, China (in military and political matters) and India (in scientific and economic matters).

In terms of hard security dialogue, formalised arenas developed to discuss security matters – such as the Arctic Security Forces Roundtable and the Northern Chiefs of Defense Forum – have been cancelled, postponed, or held without Russian participation. At the same time, organising Arctic security roundtables without Russia defeats one of the main purposes why such venues were created. As stated by the US European Command director of strategy and policy before the war in Ukraine: 'not to single one nation out, but having Russia at the table matters' (Vandiver 2012).

The Arctic states consequently find themselves in a catch-22. Reinstating the military dialogue encounters the problem of upholding a strong response to Russian actions in Ukraine. The Arctic states cannot cherry-pick what military-to-military contact to preserve for fear of distorting the effect of a coherent response.

However, the effects of the Russian war in Ukraine have not led to a complete breakdown of the various cooperative mechanisms and hotlines that exist in the Arctic. Relations between states, including in the Arctic, are not binary. Interestingly enough, it is often military practitioners (not analysts) who seem to be the most pragmatic concerning contact with Russia, exemplified by how both the Norwegian and Finnish Armed Forces emphasise the need to maintain a hotline with Russia in case of emergencies (Edvardsen and Molid Martinussen 2024a; Edvardsen 2024).[5]

At the same time, military-to-military structures are also the ones that are cut first in response to deterioration in relations. This, however, does not concern notions of region-building or contact between states in order to achieve wider political or economic goals. This concerns avoiding escalation of incidents and developing joint codes of conduct and operational procedures, and – as put forth in this

[5] Spokesperson for the Norwegian Joint Headquarters, Corporal Jonny Karlsen, highlights that:

> 'Weekly system checks occur on Skype between the NJHQ and the Northern Fleet. In addition, there has been sporadic correspondence via letters or e-mail between the headquarters at Reitan, Northern Norway, and in Severomorsk, North-West Russia. As known, military cooperation with Russia has been stopped. The communications channel between the military headquarters is kept open to ensure the security of our airmen and sailors as they operate in the same area' (Edvardsen and Molid Martinussen 2024a).

Major General Sami Nurmi, the Deputy Chief of Staff for Strategy in the Finnish Defence Forces, outlines similar pragmatic measures: 'Our hotline to the Russian Armed Forces is tested weekly, and the Russian side answers. The conversation is very short. The Finnish Border Guard also maintains regular contacts on appropriate levels with their counterparts in Russia' (Edvardsen 2024).

book – recognising the fundamental interlinkages between geography and state security in a given spatial domain.

Even so, these things cannot be conceived as completely independent of each other. Arctic security debates do not find themselves in a vacuum. They never have, despite ideas of an 'Arctic exceptionalism'. Although what exceptionalism entails is sometimes unspecified, the notion is still relevant to the extent that the Arctic can be a political arena to engage in low-level political dialogue and/or scientific research with Russia. There are still no major internal disputes or drivers for *regional* tension in the Arctic, or in the various subregions of the Arctic (the European Arctic and the North Pacific Arctic) where security interactions compound.

Arctic region-building, political relations, and track two dialogue was never what was going to stop Russia from engaging in rivalry, even conflict, with other states beyond that specific geographic domain. But it might be a place where re-engaging Russia in constructive dialogue, even on practical or low-level security matters, might be possible in the future. If relations between Russia and the other states improve at some future point, region-building efforts and a re-emphasis on low-level politics or positive-sum cooperation might again emerge. Or, at a bare minimum, common procedures for military exercise or incident management could be developed. That in turn can improve security relations in the more tense subregions of the Arctic.

The framework put forth in this book is about recognising the importance of geography and subregional security dynamics that are inherent to subregions (e.g., Norway cannot avoid contact with Russia and vice versa). However, this framework is also about recognising *where* security practices materialise, and *how* these practices are indeed linked to other levels including region-building efforts such as those highlighted here. This makes it possible to pinpoint how some actors have attempted to portray the Arctic as a coherent region, including for security interests. This might in turn offer more clarity surrounding a pathway to creating pan-Arctic arrangements for security discussions that might lean on 'softer' security elements or 'low politics' in order to build trust and confidence amongst Arctic actors.

It is perhaps a paradox that this book, which seemingly relies on traditional conceptions of geography, security, and geopolitics, concludes by emphasising the importance of region-building efforts as a mechanism for managing increasingly tense security practices in the Arctic context. Nonetheless, this is the fundamental point made here when examining state security relations in geographical domains worldwide: the necessity to map and conceptualise state security practices along a broad spectrum that includes geography, regionalism, and various levels of analysis to more effectively address or manage these dynamics. This in turn constitutes the 'geopolitics' of any given spatial domain.

In essence, this book has showcased how geography matters – probably more than we acknowledge in general analyses of international relations, but less than what some are quick to claim.[6] For state practitioners and analysts alike, understanding what the geopolitics of a region entails is crucial, rather than becoming trapped by conceptions and ideas of it.[7] States cannot choose their neighbours. However, they can choose how to deal with their neighbours and avoid getting stuck in traditional rivalries and conflict patterns under the guise of the so-called fate of geopolitics. This is likely what the Russian foreign minister toasted to aboard the icebreaker in Murmansk in 2008 after decades of peace-building efforts in the Arctic, but seems to have since forgotten.

[6] Paraphrasing Dodds (2019: 53).

[7] A fitting example of getting trapped in geographic determinism might be *Foundations of Geopolitics: The Geopolitical Future of Russia* by Aleksandr Dugin (1997).

References

ACGF. 2024. 'The Arctic Coast Guard Forum'. *About the ACGF*. 2024. www.arcticcoastguardforum.com/about-acgf.

Acharya, Amitav. 2001. *Constructing a Security Community in Southeast Asia*. London: Routledge.

Adler, Emanuel. 1997. 'Imagined (Security) Communities: Cognitive Regions in International Relations'. *Journal of International Studies* 26 (2): 249–77.

AFSA. 2021. 'A Balanced Approach to the Arctic – A Conversation with U.S. Coordinator for the Arctic Region James P. DeHart'. American Foreign Service Association. 2021. https://afsa.org/balanced-approach-arctic-conversation-us-coordinator-arctic-region-james-p-dehart.

Agnew, John. 1994. 'The Territorial Trap: The Geographical Assumptions of International Relations Theory'. *Review of International Political Economy* 1 (1): 53–80.

Arctic Council. 2018. *Agreement to Prevent Unregulated High Seas Fisheries in the Central Arctic Ocean*. www.regjeringen.no/en/aktuelt/agreement-on-unregulated-fishing-in-the-arctic-ocean/id2580484.

Al Jazeera. 2022. 'Russia Threatens Retaliation over Norway Arctic Island Goods Ban'. *Al Jazeera*, 29 June 2022. www.aljazeera.com/news/2022/6/29/russia-threatens-retaliation-against-norway-over-artic-access.

Alagappa, Muthiah. 1995. 'Regionalism and Conflict Management: A Framework for Analysis'. *Review of International Studies* 21 (4): 359.

Allen, John, Doreen Massey, and Allan Cochrane. 1998. *Rethinking the Region*. London: Routledge.

Almén, Oscar, and Christopher Weidacher Hsiung. 2022. 'China's Economic Influence in the Arctic Region: The Nordic and Russian Cases'. *Swedish Defence Research Agency*. www.foi.se/rapportsammanfattning?reportNo=FOI-R--5326--SE.

Amable, Dennis Senam. 2022. 'Theorizing the Emergence of Security Regions: An Adaptation for the Regional Security Complex Theory'. *Global Studies Quarterly* 2 (4).

Anderson, David H. 2009. 'The Status under International Law of the Maritime Areas Around Svalbard'. *Ocean Development & International Law* 40 (4): 373–84.

Anderson, Jon, and Kimberley Peters. 2014. *Water Worlds: Human Geographies of the Ocean*. Farnham, UK: Ashgate Publishing Ltd.

Andersson, Patrik, Jesper Willaing Zeuthen, and Per Kalvig. 2018. 'Chinese Mining in Greenland: Arctic Access or Access to Minerals?' *Arctic Yearbook 2018*. https://arcticyearbook.com/images/yearbook/2018/China-and-the-Arctic/7_AY2018_Andersson.pdf.

Andreeva, Serafima, and Svein Vigeland Rottem. 2024. 'How and Why the Arctic Council Survived until Now – An Analysis of the Transition in Chairship between Russia and Norway'. *The Polar Journal* 13 (2): 1–18.

Anstey, Chris. 2022. 'Arctic Rivalry May Shape a New Cold War over Resources'. *Bloomberg*, 3 September 2022. www.bloomberg.com/news/newsletters/2022-09-03/arctic-standoff-may-shape-a-new-cold-war-over-resources-new-economy-saturday.

The Antarctic Treaty, signed 1 December 1959, 402 UNTS 71 (entered into force 23 June 1961). https://documents.ats.aq/keydocs/vol_1/vol1_2_at_antarctic_treaty_e.pdf.

Arctic Council. 1996. 'Ottawa Declaration (1996)'. *Ottawa Declaration (1996)*. 1996. https://oaarchive.arctic-council.org/handle/11374/85.

2011a. 'Agreement on Cooperation on Aeronautical and Maritime Search and Rescue in the Arctic'. Arctic Council. https://oaarchive.arctic-council.org/items/9c343a3f-cc4b-4e75-bfd3-4b318137f8a2.

2011b. 'Nuuk Declaration'. *Arctic Council*. www.arctic-council.org/index.php/en/about/documents/category/5-declarations#.

2013. 'Agreement on Cooperation on Marine Oil Pollution, Preparedness and Response in the Arctic'. Kiruna: Emergency Prevention Preparedness and Response Working Group. https://oaarchive.arctic-council.org/items/ee4c9907-7270-41f6-b681-f797fc81659f.

2015. 'Enhanced Black Carbon and Methane Emissions Reductions: An Arctic Council Framework for Action'. Ministerial Meetings. 2015. https://oaarchive.arctic-council.org/items/8ad2f52f-0ec7-4f52-92e3-6fdfb2ab15db.

2024. 'Observers'. About. 2024. https://arctic-council.org/about/observers/.

Arctic Council Secretariat. 2024. 'Arctic Council Advances Resumption of Project-Level Work'. News. 2024. https://arctic-council.org/news/arctic-council-advances-resumption-of-project-level-work/.

Arctic Futures Symposium. 2024. 'Arctic Futures Symposium'. Past Editions. 2024. www.arcticfutures.org/past-editions.

Arctic Ocean Conference. 2008. 'The Ilulissat Declaration'. *Arctic Ocean Conference*. 2008. www.regjeringen.no/globalassets/upload/ud/080525_arctic_ocean_conference-_outcome.pdf.

Arlov, Thor B. 2003. *Svalbards Historie*. 2nd ed. Trondheim: Tapir Akademiske Forlag.

Armbruster, Ben. 2023. 'How US Media Builds Public Support for Confrontation with China'. *Responsible Statecraft*, August 2023.

Armstrong, Kathryn. 2024. 'Greenland again Tells Trump It Is Not for Sale'. *BBC*, 23 December 2024. www.bbc.com/news/articles/c791xy4pllqo.

Aspinwall, Nick. 2020. 'Taiwan President Asserts Sovereignty over Disputed Islands Claimed by Japan and China'. *The Diplomat*, 26 June 2020. https://thediplomat.com/2020/06/taiwan-president-asserts-sovereignty-over-disputed-islands-claimed-by-japan-and-china/.

The Associated Press. 2020a. 'Over 50 Warships Were Involved in Russian Navy Exercises that Surprised Alaska Trawlers'. *Alaska Public Media*, August 2020.

2020b. 'Over 50 Warships Were Involved in Russian Navy Exercises that Surprised Alaska Trawlers'. *Alaska Public Media*, 30 August 2020. https://alaskapublic.org/2020/08/30/over-50-warships-were-involved-in-russian-navy-exercises-that-surprised-alaskan-trawlers/.

2022. 'Arrival of Final 2 F-35s Completes Complement at Alaska Base'. *Alaska's News Source*, 18 April 2022. www.alaskasnewssource.com/2022/04/18/arrival-final-2-f-35s-completes-complement-alaska-base/.

Åtland, Kristian. 2007. 'The European Arctic after the Cold War: How Can We Analyze It in Terms of Security?' *Kjeller*. https://ffi-publikasjoner.archive.knowledgearc.net/bitstream/handle/20.500.12242/2006/07-00344.pdf.

2008. 'Mikhail Gorbachev, the Murmansk Initiative, and the Desecuritization of Interstate Relations in the Arctic'. *Cooperation and Conflict* 43 (3): 289–311.

2014. 'Interstate Relations in the Arctic: An Emerging Security Dilemma'. *Comparative Strategy* 33 (2): 145–66.

Åtland, Kristian, and Kristin Ven Bruusgaard. 2009. 'When Security Speech Acts Misfire: Russia and the Elektron Incident'. *Security Dialogue* 40 (3): 333–53.

Åtland, Kristian, Thomas Nilsen, and Torbjørn Pedersen. 2022. 'Military Muscle-Flexing as Interstate Communication: Russian NOTAM Warnings off the Coast of Norway, 2015–2021'. *Scandinavian Journal of Military Studies* 5 (1): 63–78.

2024. 'Bolstering the Bastion: The Changing Pattern of Russia's Military Exercises in the High North'. *Scandinavian Journal of Military Studies* 7 (1): 145–60.

Åtland, Kristian, and Torbjørn Pedersen. 2009. 'The Svalbard Archipelago in Russian Security Policy: Overcoming the Legacy of Fear – Or Reproducing It?' *European Security* 17 (2–3): 227–51.

Attinà, Fulvio. 2004. 'The Building of Regional Security Partnership and the Security-Culture Divide in the Mediterranean Region'. *UC Berkeley: Institute of European Studies*. https://escholarship.org/uc/item/7821m737.

Auerswald, David. 2022. '*All Security Is Local: Arctic Defense Policies and Domain Awareness*'. Washington, DC.

Ayaydin, Eda. 2022. '(Research): Evolution of Arctic Exploration from National Interest to Multinational Investment'. In *Building Common Interests in the Arctic Ocean with Global Inclusion*, edited by Paul Arthur Berkman, Alexander N. Vylegzhani, Oran R. Young, David A. Balton, and Ole Rasmus Øvretveit, 189–202. Cham: Springer.

2023. 'Interpretation of the Svalbard Treaty in Turkey'. *High North News*, 10 October 2023. www.highnorthnews.com/en/interpretation-svalbard-treaty-turkey.

Babikian, David, and Julia Nesheiwat. 2024. 'NATO Needs a Strategy to Address Russia's Arctic Expansion'. *Atlantic Council*, 9 July 2024. www.atlanticcouncil.org/blogs/new-atlanticist/nato-needs-a-strategy-to-address-russias-arctic-expansion/.

Baev, Pavel K. 2018. 'Russia's Ambivalent Status-Quo/Revisionist Policies in the Arctic'. *Arctic Review on Law and Politics* 9 (2018): 408–424.

Baker, James S., and Michael Byers. 2012. 'Crossed Lines: The Curious Case of the Beaufort Sea Maritime Boundary Dispute'. *Ocean Development & International Law* 43 (March 2010): 70–95.

Baker, Mike. 2020. '"Are We Getting Invaded?" U.S. Boats Faced Russian Aggression Near Alaska'. *The New York Times*, 20 November 2020. www.nytimes.com/2020/11/12/us/russia-military-alaska-arctic-fishing.html.

Baldacchino, Godfrey, and Anders Wivel, eds. 2020. *Handbook on the Politics of Small States*. Cheltenham, UK: Edward Elgar.

Baldwin, David A. 1997. 'The Concept of Security'. *Review of International Studies* 23 (1–2): 5–26.

Balton, David. 2021. 'The Arctic Fisheries Agreement Enters into Force'. *Polar Points*, 25 June. www.wilsoncenter.org/blog-post/no-9-arctic-fisheries-agreement-enters-force.

2022. 'What Will the BBNJ Agreement Mean for the Arctic Fisheries Agreement?' *Marine Policy 142 (August 2022)*. www.sciencedirect.com/science/article/abs/pii/S0308597X1930449X.

Barents Euro-Arctic Council. 1993. 'Kirkenes Declaration'. *Barentsinfo*. 1993. www.barentsinfo.fi/beac/docs/459_doc_kirkenesdeclaration.pdf.

Bartenstein, Kristin. 2015. 'The "Common Arctic": Legal Analysis of Arctic & Non-Arctic Political Discourses'. *Arctic Yearbook* 2015.

Bases, Daniel. 2008. 'Iceland Sees More Russian Flights in N. Atlantic'. *Reuters*, 14 March 2008. www.reuters.com/article/world/iceland-sees-more-russian-flights-in-n-atlantic-idUSN14619600/.

Baudu, Pauline. 2023. 'Minding the Archipelago: What Svalbard Means to NATO'. *Arctic Review on Law and Politics* 14: 76–82.

BBC News. 2007. 'Russia Plants Flag under N Pole'. *World*, 2 August 2007. http://news.bbc.co.uk/1/hi/world/europe/6927395.stm.

2023. 'The Inquiry: Is Peace in the Arctic Melting?' *The Inquiry*, 26 October 2023. www.bbc.co.uk/programmes/w3ct4wdk.

Beckman, Robert, Ian Townsend-Gault, Clive Schofield, Tara Davenport, and Leonardo Bernard, eds. 2013. *Beyond Territorial Disputes in the South China Sea: Legal Frameworks for the Joint Development of Hydrocarbon Resources*. Cheltenham, UK: Edward Elgar.

Bekkevold, Jo Inge, and Paal Sigurd Hilde. 2024. 'The New U.S. Arctic Strategy Is Wrong to Focus on China'. *Foreign Policy*, November. https://foreignpolicy.com/2024/11/11/us-arctic-strategy-china-russia-alaska-geopolitics/.

Bekkevold, Jo Inge, and Kristine Offerdal. 2014. 'Norway's High North Policy and New Asian Stakeholders'. *Strategic Analysis* 38 (6): 825–40.

Bennett, Mia. 2023. 'The Death of Arctic Exceptionalism'. *Cryopolitics*, 11 October 2023. www.cryopolitics.com/2023/10/11/death-of-arctic-exceptionalism/.

Berg, Roald. 1995. *Norsk Utenrikspolitikks Historie: 1905–1920. Norge På Egen Hånd*. Oslo: Universitetsforlaget.

Bergmann, Eiríkur. 2014. 'Iceland: A Postimperial Sovereignty Project'. *Cooperation and Conflict* 49 (1): 33–54.

Berkman, Paul Arthur, and Oran R. Young. 2009. 'Governance and Environmental Change in the Arctic Ocean'. *Science* 324 (17 April 2009).

Bertelsen, Rasmus Gjedssø. 2019. 'The Role of Transnational Knowledge Networks and Epistemic Communities in Arctic Shipping Governance'. In *Emerging Legal Orders in the Arctic*, edited by Akiho Shibata, Leilei Zou, Nikolas Sellheim, and Marzia Scopelliti. Routledge.

Bertelsen, Rasmus Gjedssø, and Vincent Gallucci. 2016. 'The Return of China, Post-Cold War Russia, and the Arctic: Changes on Land and at Sea'. *Marine Policy* 72 (October): 240–45.

Berthiaume, Lee. 2015. 'Arctic Council Meeting Sets up Awkward Russian Visit to Canada'. World: Ottawa Citizen. 2015. http://ottawacitizen.com/news/politics/arctic-council-meeting-sets-up-awkward-russian-visit-to-canada.

Bethlehem, Daniel. 2014. 'The End of Geography: The Changing Nature of the International System and the Challenge to International Law'. *European Journal of International Law* 25 (1): 9–24.

Bicchi, Federica. 2006. 'Our Size Fits All: Normative Power Europe and the Mediterranean'. *'Our Size Fits All': Normative Power Europe and the Mediterranean* 13 (2): 286–303.

Bjur, Ingeborg. 2021. 'Norsk Sikkerhetspolitikk Er En "balanse Mellom Avskrekking Og Beroligelse". Men Vet Noen Hva Det Betyr?' *Aftenposten*, 28 February 2021. www.aftenposten.no/meninger/kronikk/i/6zXbR3/norsk-sikkerhetspolitikk-er-en-balanse-mellom-avskrekking-og-beroligelse-men-vet-noen-hva-det-betyr.

Bjur, Ingeborg, Paal S. Hilde, and Karen-Anna Eggen. 2020. 'Mellom Varmebølge Og Isfront: Nordområdene, Sikkerhet Og Storpolitikk'. *Internasjonal Politikk* 78 (4): 478–89.

Black, Jeremy. 2015. *Geopolitics and the Quest for Dominance.* Bloomington: Indiana University Press.

Bladet Vesterålen. 2011. 'EU Utfordrer Norsk Svalbard-Politikk', October 2011.

Blanchfield, Mike. 2010. 'Clinton Rebukes Canada on Arctic Meeting'. *The Globe and Mail*, 29 March 2010. www.theglobeandmail.com/news/politics/clinton-rebukes-canada-on-arctic-meeting/article1210187/.

Bloom, Evan T. 1999. 'Establishment of the Arctic Council'. *The American Journal of International Law* 93 (3): 712–22.

Bolongaro, Kait. 2017. 'Oil Lurks beneath EU-Norway Snow Crab Clash'. *Politico*, 18 June 2017. www.politico.eu/article/of-crustaceans-and-oil-the-case-of-the-snow-crab-on-svalbard/.

Bolt, Paul J., and Sharyl N. Cross. 2018. *China, Russia and Twenty-First Century Global Geopolitics*. Oxford: Oxford University Press.

Borger, Julian. 2014. 'Barack Obama: Russia Is a Regional Power Showing Weakness over Ukraine'. *The Guardian*, 25 March 2014.

Borgerson, Scott G. 2008. 'Arctic Meltdown: The Economic and Security Implications of Global Warming'. *Foreign Affairs* 87 (2): 63–77.

2009. 'The Great Game Moves North'. *Foreign Affairs 25 March*. www.foreignaffairs.com/articles/global-commons/2009-03-25/great-game-moves-north.

2013. 'The Coming Arctic Boom: As the Ice Melts, the Region Heats Up'. *Foreign Affairs* 92 (4): 76–89.

Børresen, Jacob, Gullow Gjeseth, and Rolf Tamnes. 2004. *Allianseforsvar i Endring 1970–2000, Vol. 5 in Norsk Forsvarshistorie*. Bergen: Eide forlag.

Börzel, Tanja A., and Thomas Risse, eds. 2016. *The Oxford Handbook of Comparative Regionalism*. Oxford: Oxford University Press.

Boswell, Randy. 2008. 'Astonishing' Data Boost Arctic Claim. *Ottawa Citizen*, 12 November 2008.

Boulègue, Mathieu. 2019. 'Russia's Military Posture in the Arctic: Managing Hard Power in a "Low Tension" Environment'. *London*. https://reader.chathamhouse.org/russia-s-military-posture-arctic-managing-hard-power-low-tension-environment.

2022. 'The Militarization of Russian Polar Politics'. *London*. www.chathamhouse.org/2022/06/militarization-russian-polar-politics.

Bowermaster, Jon. 1993. 'The Last Front of the Cold War'. *The Atlantic, no*. November 1993. www.theatlantic.com/magazine/archive/1993/11/the-last-front-of-the-cold-war/668806/.

Brady, Anne-Marie. 2017. *China as a Polar Great Power*. Cambridge: Cambridge University Press.

Breemer, Jan S. 1989. 'The Soviet Navy's SSBN Bastions: Why Explanations Matter'. *The RUSI Journal* 134 (4): 33–39.

Bremer, Stuart A. 1992. 'Dangerous Dyads: Conditions Affecting the Likelihood of Interstate War, 1816–1965'. *The Journal of Conflict Resolution* 36 (2): 309–41.

Brennan, David. 2025. 'Trump Says US Will "go as Far as We Have to" to Get Control of Greenland'. *ABC News*, March 2025.

Brewster, Murray. 2022. 'NATO Chief Tours Arctic Defences as Canada Comes under Pressure to Guard the Far North'. *CBC News*, 25 August 2022. www.cbc.ca/news/politics/stoltenberg-arctic-trudeau-russia-1.6561950.

Brugeron, Marine. 2022. 'Sécurité Alimentaire : En Norvège, Le Bunker de l'apocalypse'. *Le Parisien*, 3 December 2022. www.leparisien.fr/environnement/securite-alimentaire-en-norvege-le-bunker-de-lapocalypse-03-12-2022-M5VKCYP33FA2TOVBE5LPS7DS5Q.php.

Brunstad, Bjørn, Eivind Magnus, Philip Swanson, Geir Hønneland, and Indra Øverland. 2004. *Big Oil Playground, Russian Bear Preserve Or European Periphery?* Delft: Eburon Academic.

Bruusgaard, Kristin Ven. 2006. 'Fiskerikonflikter i Barentshavet – Potensial for Eskalering? (Fisheries Conflicts in the Barents Sea – Potential for Escalation?)'. Oslo. https://brage.bibsys.no/xmlui/handle/11250/2444396.

Bryant, Miranda. 2023. 'Barentsburg: The Norwegian Town Feeling the Chill of the Ukraine War'. *The Guardian*, 10 October 2023. www.theguardian.com/world/2023/oct/10/barentsburg-the-norwegian-town-feeling-the-chill-of-the-ukraine-war.

Buchanan, Allen, and Margaret Moore, eds. 2003. *States, Nations and Borders: The Ethics of Making Boundaries*. Cambrigde: Cambridge University Press.

Buchanan, Elizabeth. 2023a. *Red Arctic: Russian Strategy under Putin*. Washington, DC: Brookings Institution Press.

2023b. 'Russia's Gain in the Great Arctic Race'. *War on the Rocks*, 4 May 2023. https://warontherocks.com/2023/05/russias-gains-in-the-great-arctic-race/.

Bueger, Christian, Elizabeth Mendenhall, and Rebecca Strating. 2025. 'Introduction'. In *The Politics of Global Ocean Regions*, edited by Christian Bueger, Elizabeth Mendenhall, and Rebecca Strating. Singapore: Palgave Macmillan.

Burgoon, Brian, and Peter Trubowitz. 2023. *Geopolitics and Democracy – The Western Liberal Order from Foundation to Fracture*. Oxford: Oxford University Press.

Burke, Danita Catherine. 2017. 'Romantic Notions about the Arctic Must Include Indigenous Rights'. *The Conversation*. 2017.

Buzan, Barry. 1983. *People, States and Fear: An Agenda for International Security Studies in the Post-Cold War Era*. London: Wheatsheaf Books.

2000. 'The Logic of Regional Security in the Post-Cold War World'. In *The New Regionalism and the Future of Security and Development*, edited by Björn Hettne, András Inotai, and Osvaldo Sunkel, 1–25. London: Palgave Macmillan.

Buzan, Barry, and Ole Wæver. 2003. *Regions and Powers: The Structure of International Security*. Cambridge: Cambridge University Press.

Bye, Hilde-Gunn. 2021. 'Norge Inngår Ny Avtale Om Forsvarssamarbeid Med USA'. *High North News*, April 16, 2021.

2022. 'Russia with Military Exercises in Ocean Vis-à-Vis Alaska'. *High North News*, 20 September 2022. www.highnorthnews.com/en/russia-military-exercises-ocean-vis-vis-alaska.

Byers, Michael. 2012. 'You Can't Replace Real Icebreakers'. *The Globe and Mail*, 27 March 2012. www.theglobeandmail.com/globe-debate/you-cant-replace-real-icebreakers/article534351/.

2013. *International Law and the Arctic*. New York: Cambridge University Press.

2014a. 'Why Canada's Search for an Icebreaker Is an Arctic Embarrassment'. *The Globe and Mail, 2014*. www.theglobeandmail.com/news/national/the-north/why-canadas-search-for-an-icebreaker-is-an-arctic-embarrassment/article16425755/.

2014b. 'The North Pole Is a Distraction'. *The Globe and Mail*. 20 August 2014. www.theglobeandmail.com/globe-debate/the-north-pole-is-a-distraction/article20126915/.

2016. 'Arctic Cruises: Fun for Tourists, Bad for the Environment'. *The Globe and Mail*, 18 April 2016. www.theglobeandmail.com/opinion/arctic-cruises-great-for-tourists-bad-for-the-environment/article29648307/.

2017. 'Crises and International Cooperation: An Arctic Case Study'. *International Relations* 31 (4): 375–402.

2019. 'Cold, Dark, and Dangerous: International Cooperation in the Arctic and Space'. *Polar Record* 55 (1): 32–47.

Byers, Michael, and Suzanne Lalonde. 2009. 'Who Controls the Northwest Passage?' *Vanderbilt Journal of Transnational Law* 42 (2).

Byers, Michael, and Andreas Østhagen. 2017. 'Why Does Canada Have So Many Unresolved Maritime Boundary Disputes?' *Canadian Yearbook of International Law* 54 (October): 1–62.

Bykova, Alina. 2019. 'The Changing Nature of Russia's Arctic Presence: A Case Study of Pyramiden'. *The Arctic Institute*, 9 December 2019. www.thearcticinstitute.org/changing-nature-russia-arctic-presence-case-study-pyramiden/.

2023. 'Svalbard's Extractive Economy: Past, Present, Future'. *The Arctic Institute*. Vol. June 2023. Washington, DC. www.thearcticinstitute.org/wp-content/uploads/2023/06/Svalbards-Extractive-Economy-Report-2023.pdf.

Canada-USA. 1989. 'CANADA—UNITED STATES: AGREEMENT ON ARCTIC COOPERATION AND EXCHANGE OF NOTES CONCERNING TRANSIT OF NORTHWEST PASSAGE'. *International Legal Materials* 28 (1): 141–45.

Canada–United States. 1988. *Agreement between the Government of Canada and the Government of the United States of America on Arctic Cooperation*. www.jstor.org/stable/20693289?seq=1#page_scan_tab_contents.

Canrong, Jin. 2016. 'How America's Relationship with China Changed under Obama'. World Economic Forum. 2016. www.weforum.org/stories/2016/12/america-china-relationship/.

Carroll, Francis M. 2010. 'Diplomats and the Civil War at Sea'. *Canadian Review of American Studies* 40 (1): 117–30.

Carter, David B. 2010. 'The Strategy of Territorial Conflict'. *American Journal of Political Science* 54 (4): 969–87.

Caswell, John Edwards. 1951. 'Materials for the History of Arctic America'. *Pacific Historical Review* 20 (3): 219–26.

CBC News. 2014. 'Canadian Fighter Jets Intercept Russian Bombers in Arctic'. *CBC News Canada*. 2014. www.cbc.ca/news/canada/canadian-fighter-jets-intercept-russian-bombers-in-arctic-1.2772440.

CCTV. 2015. '日美加紧部署隐形战机 美舰或逼近中国南海岛礁? (Japan and the United States Are Stepping up the Deployment of Stealth Fighters. Will US Warships Approach China's South China Sea Islands and Reefs?)', 8 September 2015. https://tv.cctv.com/2015/09/08/VIDE1441723316238486.shtml.

2024a. '[新闻直播间]中俄海警舰艇编队近日抵达北冰洋 专家分析 中国海警首次进入北冰洋意味着什么？ ([Live News Room] A Fleet of Chinese and Russian Coast Guard Ships Recently Arrived in the Arctic Ocean. Experts Analyzed What It Means for the Chinese Coast Guard to Enter the Arctic Ocean for the First *Time?*)' CCTV. 2024. https://tv.cctv.com/2024/10/05/VIDEH418xH0pbC5YWmTiOdmV241005.shtml.

2024b. '《新闻联播》（《News Broadcast》）20240921 19:00'. *News Broadcast*. 2024. https://tv.cctv.com/2024/09/21/VIDEXSNMbGmvA1hGrZMXSD85240921.shtml.

Chan, Ryan. 2025. 'Map Shows Where Chinese Ships Have Been Detected off Alaska'. *Newsweek*, August 2025.

Charron, Andrea, and James Fergusson. 2022. *NORAD: In Perpetuity and Beyond*. Montreal: McGill-Queen's University Press.

Checkel, Jeffrey T. 2005. 'International Institutions and Socialization in Europe: Introduction and Framework'. *International Organization* 59 (04): 801–26.

Chen, Chuan. 2023. 'China-Russia Arctic Cooperation in the Context of a Divided Arctic'. *The Arctic Institute*, 4 April 2023. www.thearcticinstitute.org/china-russia-arctic-cooperation-context-divided-arctic/.

China.com. 2023. '中俄两国海警合兵一处， 联合巡航选在美国家门口 (China and Russia Join Forces to Conduct Joint Patrols at the Doorstep of the United States)'. China.Com. 2023. https://military.china.com/news/13004177/20230504/44932127_all.html#page_2.

Choi, Jong Kun, and Yong-Soo Eun. 2017. 'What Does International Relations Theory Tell Us about Territorial Disputes and Their Resolution?' *International Politics* 55 (141–59).

Choi, Seong Hyeon. 2024. 'What Does China's First Arctic Coastguard Patrol with Russia Reveal about Its Ambitions?' *South China Morning Post*, 6 October 2024. www.scmp.com/news/china/military/article/3281228/what-does-chinas-first-arctic-coastguard-patrol-russia-reveal-about-its-ambitions.

Chotiner, Isaac. 2023. 'What's behind the Chinese Spy Balloon'. *The New Yorker*, 18 February 2023. www.newyorker.com/news/q-and-a/whats-behind-the-chinese-spy-balloon.

Christensen, Miyase. 2013. 'Arctic Climate Change and the Media: The News Story That Was'. In *Media and the Politics of Arctic Climate Change: When the Ice Breaks*, edited by Miyase Christensen, Annika E. Nilsson, and Nina Wormbs, 26–51. Basingstoke: Palgrave Macmillan.

Churchill, Robin R. 2001. 'Claims to Maritime Zones in the Arctic – Law of the Sea Normality or Polar Peculiarity?' *In The Law of the Sea and Polar Maritime Delimitation and Jurisdiction*, 105–24. The Hague: Martinus Nijhoff.

Churchill, Robin R., and Geir Ulfstein. 1992. *Marine Management in Disputed Areas: The Case of the Barents Sea*. London: Routledge.

2010. 'The Disputed Maritime Zones around Svalbard'. In *Changes in the Arctic Environment and the Law of the Sea*, edited by Myron H. Nordquist, John Norton Moore, and Thomas H. Heidar, 551–93. Leiden: Martinus Nijhoff Publishers.

2020. 'The Application of the Svalbard Treaty Offshore'. In *Svalbardtraktaten 100 År*, edited by Irene Dahl and Øystein Jensen, 265–94. Bergen: Fagbokforlaget.

Claes, Dag H., and Arild Moe. 2018. 'Arctic Offshore Petroleum: Resources and Political Fundamentals'. In *Arctic Governance: Energy, Living Marine Resources and Shipping*, edited by Svein Vigeland Rottem, Ida Folkestad Soltvedt, and Geir Hønneland, 9-26f. London: I. B. Tauris.

Clark, Joseph. 2024. 'Pentagon Arctic Policy Official Underscores Critical Role of Alliances in Evolving Region'. *DOD News*, 15 April 2024. www.defense.gov/News/News-Stories/Article/Article/3741920/pentagon-arctic-policy-official-underscores-critical-role-of-alliances-in-evolv/#:~:text=Pentagon Arctic Policy Official Underscores Critical Role of Alliances in Evolving Region,-April 15%25.

Coates, Ken S., Heather Exner-Pirot, Pauline Frost, Peter Johnston, Derek Joyce, Whitne P. Lackenbauer, and Jennifer Spence. 2024. 'Report of the Yukon Arctic Security Advisory Council'. *Whitehorse*. https://yukon.ca/en/arctic-security-advisory-council-report.

Coates, Ken S., P. Whitney Lackenbauer, William R. Morrion, and Greg Poelzer. 2008. *Arctic Front: Defending Canada in the Far North*. Markham: Thomas Allen.

Conley, Heather A., and Sophie Arts. 2023. 'NATO's Policy and Posture in the Arctic: Revisiting Allied Capabilities and Command Plans'. *GMF US*, 5 July 2023. www.gmfus.org/news/natos-policy-and-posture-arctic-revisiting-allied-capabilities-and-command-plans.

Conley, Heather A., Sophie Arts, Bonnie S. Glaser, Kristine Berzina, and Jaine Archambeau. 2024. 'A New Era of Arctic Geopolitics: Russia-PRC Strategic Alignment Is Driving Unprecedented Region al Collaboration'. Washington, DC. www.gmfus.org/sites/default/files/2024-07/A New Era of Arctic Geopolitics.pdf.

Conley, Heather A., Matthew Melino, Nikos Tsafos, and Ian Williams. 2020. 'America's Arctic Moment: Great Power Competition in the Arctic to 2050'. Washington, DC. https://csis-website-prod.s3.amazonaws.com/s3fs-public/publication/Conley_ArcticMoment_layout_WEB FINAL.pdf.

Conley, Heather A., Cyrus Newlin, Colin Wall, Andrew Lohsen, Nikos Tsafos, and Ben Cahill. 2021. 'Russia's Climate Gamble: The Pursuit and Contradiction of Its Arctic Ambitions'. Washington, DC. https://csis-website-prod.s3.amazonaws.com/s3fs-public/publication/210924_Conley_Russia%27s_Climate_Gamble.pdf?PehRoHxLxSI_CzhQ8sI.ssO0AwQeXAXF.

Conley, Heather A., and Caroline Rohloff. 2015. *The New Ice Curtain*. Washington, DC: Center for Strategic & International Studies (CSIS).

Conley, Heather A., Terry Toland, Jamie Kraut, and Andreas Østhagen. 2012. 'A New Security Architecture for the Arctic: An American Perspective'. *CSIS Report*. Vol. January 20. Washington, DC: Center for Strategic & International Studies (CSIS).

Contenta, Sandro. 2010. 'Canadians Have Their Own Oil Worries'. *Global Post: Canada*. www.globalpost.com/dispatch/global-green/100602/canadians-worry-about-oil-spilling-beaufort-sea?page=0,1.

Cook, P. J., and C. M. Carleton, eds. 2000. *Continental Shelf Limits: The Scientific and Legal Interface*. Oxford: Oxford University Press.

Council on Foreign Relations. 2024. 'Trump's Foreign Policy Moments'. 2017–2021: Trump's Foreign Policy Moments. 2024. www.cfr.org/timeline/trumps-foreign-policy-moments.

Cox, James, and Sean Speer. 2016. 'From a Mandate for Change to a Plan to Govern: A New National Defence Policy for A Dangerous World'. *MLI Commentary* 6 (January).

CSIS. 2024. 'China-Russia Joint Military Exercises'. ChinaPower. 2024. https://chinapower.csis.org/data/china-russia-joint-military-exercises/.

Danilov, Peter Bakkemo. 2020. 'Researcher Argues Norwegian Participation in Barents Sea Military Exercise Was Unfortunate'. *High North News*, 22 September 2020. www.highnorthnews.com/en/researcher-argues-norwegian-participation-barents-sea-military-exercise-was-unfortunate.

Davidson, William D., and Joseph V. Montville. 1982. 'Foreign Policy According to Freud'. *Foreign Policy* 45: 145–57.

Dean, Ryan, and Whitney P. Lackenbauer. 2017. 'Conceiving and Executing Operation Gauntlet: The Canadian-Led Raid on Spitzbergen, 1941'. *Canadian Military History* 26 (2).

Depledge, Duncan. 2020. 'Train Where You Expect to Fight: Why Military Exercises Have Increased in the High North'. *Scandinavian Journal of Military Studies* 3 (1): 288–301.

Depledge, Duncan, Caroline Kennedy-Pipe, and James Roger. 2019. 'The UK and the Arctic: Forward Defence'. *Arctic Yearbook*. https://arcticyearbook.com/images/yearbook/2019/Scholarly-Papers/18_AY2019_Depledge.pdf.

Depledge, Duncan, and Whitney P. Lackenbauer, eds. 2021. *On Thin Ice: Persepctives on Arctic Security*. Peterborough: North American and Arctic Defence and Security Network (NAADSN). www.naadsn.ca/wp-content/uploads/2021/03/Depledge-Lackenbauer-On-Thin-Ice-final-upload.pdf.

Depledge, Duncan, and Andreas Østhagen. 2021. 'Scotland: A Touchstone for Security in the High North?' *RUSI Journal* 166 (6–7): 46–63. www.tandfonline.com/doi/full/10.1080/03071847.2022.2036229.

Deutsch, Karl W. 1981. 'On Nationalism, World Regions, and the Nature of the West'. In *Mobilization, Center-Periphery Structures and Nation-Building: Volume in Commemoration of Stein Rokkan*, edited by Per Torsvik. Bergen: Universitetsforlaget.

Deutsch, Karl W., Sidney A. Burrell, Robert A. Kann, and Maurice Lee Jr. 1957. *Political Community and the North Atlantic Area*. Princeton, NJ: Princeton University Press.

Deutsche Welle. 2022. 'The Race for the Arctic Is Ramping up. Here's Why'. *DW Planet*. 2022. www.youtube.com/watch?v=hvRzWzQW2go&ab_channel=DWPlanetA.

Devyatkin, Pavel. 2019. 'Russian Strategic Intentions in the Arctic'. *The Arctic Institute*, 21 May 2019. www.thearcticinstitute.org/russian-strategic-intentions-arctic/.

2023. 'Arctic Exceptionalism: A Narrative of Cooperation and Conflict from Gorbachev to Medvedev and Putin'. *Polar Journal* 13 (2): 336–57.

Diesen, Sverre. 2013. 'Towards an Affordable European Defence and Security Policy? The Case for Extensive European Force Integration'. In *NATO's European Allies: Military Capability and Political Will*, edited by Magnus Petersson and Janne Haaland Matlary, 57–70. Basingstoke: Palgrave Macmillan.

Dodds, Klaus. 2010. 'Flag Planting and Finger Pointing: The Law of the Sea, the Arctic and the Political Geographies of the Outer Continental Shelf'. *Political Geography* 29 (2): 63–73.

2011. 'We Are a Northern Country: Stephen Harper and the Canadian Arctic'. *Polar Record* 47 (4): 371–74.

2019. *Geopolitics: A Very Short Introduction*. Oxford: Oxford University Press.

Dodds, Klaus, Merje Kuus, and Joanne P. Sharp. 2013. 'Introduction: Geopolitics and Its Critics'. In *The Ashgate Research Companion to Critical Geopolitics*, edited by Klaus Dodds, Merje Kuus, and Joanne Sharp, 1–14. Farnham: Ashgate.

Doshi, Rush, Alexis Dale-Huang, and Zhang Gaoqi. 2021. '*Northern Expedition: China's Arctic Activities and Ambitions*'. Washington, DC.

Duffy, Cian. 2013. *'The Region of Beauty and Delight': Reimagining the Polar Sublime. The Landscapes of the Sublime, 1700–1830*. London: Palgrave Macmillan.

Dugin, Aleksandr. 1997. *Foundations of Geopolitics: The Geopolitical Future of Russia*. Moscow: Arktogeja Publication.

Dutton, Peter A. 2009. *Scouting, Signaling, and Gatekeeping: Chinese Naval Operations in Japanese Waters and the International Law Implications*. Newport, RI: China Maritime Studies Institute, U.S Naval War College. https://digital-commons.usnwc.edu/cmsi-red-books/8/.

Dyner, Anna Maria. 2024. 'No 138: Ocean 2024 Exercises Demonstrate Russian Naval Capabilities, Strategic Signalling'. *Warsaw*. https://pism.pl/publications/ocean-2024-exercises-demonstrate-russian-naval-capabilities-strategic-signalling.

Ebel, Francesca. 2022. 'Putin and Xi Highlight Russia-China Cooperation against Backdrop of War'. *The Washington Post*, 30 December 2022. www.washingtonpost.com/world/2022/12/30/putin-xi-call-cooperation-ukraine/.

Eckstein, Megan. 2018. 'Truman Carrier Strike Group Operating North of Arctic Circle; First Time for US Navy since 1991'. *USNI News*, 19 October 2018. https://news.usni.org/2018/10/19/truman-carrier-strike-group-operating-north-arctic-circle-first-time-us-navy-since-1991#:~:text=Truman's excursion north of the,%2C' Truman Commanding Officer Capt.

ECON. 2005. '2025 Ringer i Vannet'. Oslo. www.aksjonsprogrammet.no/vedlegg/ECON_ringer06.pdf.

The Economist. 2015. 'Not So Cool'. *The Economist*, 14 February 2015. www.economist.com/news/international/21641240-hype-over-arctic-recedes-along-summer-ice-not-so-cool.

The Economist. 2024. 'The World Faces Its Worst Trade Wars since the 1930s'. *The Economist*, 7 November 2024. www.economist.com/international/2024/11/07/the-world-faces-its-worst-trade-wars-since-the-1930s.

Edstrøm, Anders Christoffer, Guðbjörg Ríkey Th. Hauksdóttir, Whitney P. Lackenbauer, and Tessa Varvares. 2025. 'Cutting through Narratives on Chinese Arctic Investments'. Cambridge, MA. www.belfercenter.org/research-analysis/china-arctic-investments.

Edstrøm, Anders Christoffer, Iselin Stensdal, and Gørild Heggelund. 2020. 'Den "nye Supermakten": Hva Vil Kina i Arktis?' *Internasjonal Politikk* 78 (4): 523–34.

2023. 'The "New Superpower": What Are China's Intentions in the Arctic'. In *Norway's Arctic Policy: Geopolitics, Security, and Identity in the High North*, edited by Andreas Østhagen. Cheltenham: Edward Elgar.

Edvardsen, Astri. 2023a. 'Russia Wants to Cooperate with BRICS Countries on Research on Svalbard'. *High North News*, 14 April 2023. www.highnorthnews.com/en/russia-wants-cooperate-brics-countries-research-svalbard.

2023b. 'Russia on Svalbard: Stepping Down on Coal, Investing More in Tourism'. *High North News*, 21 April 2023. www.highnorthnews.com/en/russia-svalbard-stepping-down-coal-investing-more-tourism.

2024. 'De-Escalation Is Best Achieved with Credible Deterrence and Defense, Says Finnish Major General'. *High North News*, 28 September 2024. www.highnorthnews.com/en/de-escalation-best-achieved-credible-deterrence-and-defense-says-finnish-major-general.

Edvardsen, Astri, and Birgitte Annie Molid Martinussen. 2024a. 'Norway's Navy Chief: "High North, Low Tension" Depends on Our Own Capability at Sea'. *High North News*, 19 April 2024. www.highnorthnews.com/en/norways-navy-chief-high-north-low-tension-depends-our-own-capability-sea.

2024b. 'Canada and the US Move Forward with Negotiations on the Beaufort Sea Boundary'. *High North News*, 2 October 2024. www.highnorthnews.com/en/canada-and-us-move-forward-negotiations-beaufort-sea-boundary.

Edwards, Bernard. 2002. *The Road to Russia: Arctic Convoys*. Barnsley: Leo Cooper.

Eilperin, Juliet. 2015. 'The Fascinating (and Scenic) History of Presidential Visits to Alaska'. *The Washington Post*, 25 August 2015. www.washingtonpost.com/news/the-fix/wp/2015/08/25/a-scenic-visual-history-of-presidential-visits-to-alaska/.

Elden, Stuart. 2013. *The Birth of Territory*. Chicago: University of Chicago Press.

Emmerson, Charles. 2010. *The Future History of the Arctic*. New York: PublicAffairs.

Etteretningstjenesten. 2016. 'Fokus 2016'. Oslo. www.etterretningstjenesten.no/publikasjoner/fokus/fokus-norsk/Fokus 2016.pdf.

2017. 'Fokus 2017'. Oslo. www.etterretningstjenesten.no/publikasjoner/fokus/fokus-norsk/Fokus 2017.pdf.

2018. 'Fokus 2018'. Oslo. www.etterretningstjenesten.no/publikasjoner/fokus/fokus-norsk/Fokus 2018.pdf.

2020. 'Fokus 2020'. Oslo. www.etterretningstjenesten.no/publikasjoner/fokus/fokus-norsk/Fokus 2020.pdf.

2021. 'Fokus 2021'. Oslo. www.etterretningstjenesten.no/publikasjoner/fokus/fokus-norsk/Fokus2021-highres.pdf.

European Commission. 2008. 'The European Union and the Arctic Region'. Edited by D G Relex. *Communcation from the Commission to the European Parliament and the Council*. Brussels: European Commission. http://ec.europa.eu/external_relations/arctic_region/docs/com_08_763_en.pdf.

2021. 'JOINT COMMUNICATION TO THE EUROPEAN PARLIAMENT, THE COUNCIL, THE EUROPEAN ECONOMIC AND SOCIAL COMMITTEE AND THE COMMITTEE OF THE REGIONS: A Stronger EU Engagement for a Peaceful, Sustainable and Prosperous Arctic'. Brussels. www.eeas.europa.eu/sites/default/files/documents/Joint Communication on a stronger EU engagement for a peaceful%2C sustainable and prosperous Arctic.pdf.pdf.

'European Parliament Resolution of 9 October 2008 on Arctic Governance'. 2008. *European Parliament: Texts Adopted*. Brussels: European Parliament. www.europarl.europa.eu/sides/getDoc.do?type=TA&language=EN&reference=P6-TA-2008-0474.

Evans, Grace, and Andreas Østhagen. 2023. 'In Hot Water: Arctic Fisheries as a Proxy for Geopolitical Tensions'. *RUSI Commentary* 11 July. https://rusi.org/explore-our-research/publications/commentary/hot-water-arctic-fisheries-proxy-geopolitical-tensions.

Exner-Pirot, Heather. 2012. 'New Directions for Governance in the Arctic Region'. *Arctic Yearbook*.

2013. 'What Is the Arctic a Case of? The Arctic as a Regional Environmental Security Complex and the Implications for Policy'. *The Polar Journal* 3 (1): 120–35.

2015. 'Arctic Council: The Evolving Role of Regions in Arctic Governance'. *Alaska Dispatch*, 2015. www.adn.com/article/20150109/arctic-council-evolving-role-regions-arctic-governance.

2016. 'Canada's Arctic Council Chairmanship (2013–2015): A Post-Mortem'. *Canadian Foreign Policy Journal* 22 (1): 84–96.

2018. 'How to Write an Arctic Story in 5 Easy Steps'. *Arctic Today*, 4 December 2018. www.arctictoday.com/write-arctic-story-5-easy-steps/.

Exner-Pirot, Heather, and Robert W. Murray. 2017. 'Regional Order in the Arctic: Negotiated Exceptionalism'. *Politik* 20 (3).

Falkenek, Clara. 2024. 'Who's at 2 Percent? Look How NATO Allies Have Increased Their Defense Spending since Russia's Invasion of Ukraine'. Atlantic Council. 2024. www.atlanticcouncil.org/blogs/econographics/whos-at-2-percent-look-how-nato-allies-have-increased-their-defense-spending-since-russias-invasion-of-ukraine/#:~:text=The war in Ukraine has,of their GDP on defense.

Faulconbridge, Guy. 2010. 'Russian Navy Boss Warns of China's Race for Arctic'. *Reuters*, 4 October 2010. www.reuters.com/article/legal/government/russian-navy-boss-warns-of-chinas-race-for-arctic-idUSLDE6931GL/.

Fearon, James D. 1995. 'Rationalist Explanations for War'. *International Organization* 49 (3): 379–414.

1998. 'Bargaining, Enforcement, and International Cooperation'. *International Organization* 52 (2): 269–305.

Fermann, Gunnar, and Tor Håkon Inderberg. 2015. 'Norway and the 2005 Elektron Affair: Conflict of Competencies and Competent Realpolitik'. In *War: An Introduction to Theories and Research on Collective Violence*, edited by Tor Georg Jakobsen, 2nd ed., 373–402. New York: Nova Science.

Fernandez, Georgina. 2023. 'Alaska Military Projects Receive $168M from 2024 National Defense Authorization Act'. *Alaska's News Source*, 24 August 2023. www.alaskasnewssource.com/2023/08/24/alaska-military-projects-receive-168m-2024-national-defense-authorization-act/.

Fife, Rolf Einar. 2021. 'Hvordan Skal Man Tolke Svalbardtraktaten Til Havs? "Territorialfarvannet": Etterforskningskritt i Fem Episoder'. In *Historiker, Strateg Og Brobygger: Festskrift Til Rolf Tamnes 70 År*, edited by Helge Ø. Pharo, Kjell Inge Bjerga, Sunniva Engh, Gunnar D. Hatlehol, and Kristine Offerdal, 46–76. Oslo: Pax Forlag.

Finger, Matthias, and Gunnar Rekvig, eds. 2022. *Global Arctic: An Introduction to the Multifaceted Dynamics of the Arctic*. Cham: Springer.

Finnemore, Martha, and Kathryn Sikkink. 1998. 'International Norm Dynamics and Political Change'. *International Organization* 52 (4): 887–917.

Finnset, Knut Anders, Aurora Berg, and Ida Louise Rostad. 2022. 'Russland Anklager Norge for Økt Militær Tilstedeværelse På Svalbard (Russia Accuses Norway of Increased Military Presence on Svalbard)'. *NRK*, 7 October 2022.

Fjærtoft, Daniel, Moe Arild, Natalia Smirnova, and Alexey Cherepovitsyn. 2018. 'Unitization of Petroleum Fields in the Barents Sea: Towards a Common Understanding?' *Arctic Review on Law and Politics* 9: 72–96.

Forbes, Vivian Louis. 1995. *The Maritime Boundaries of the Indian Ocean Region*. Singapore: Singapore University Press.

Forsberg, Tuomas. 1996. 'Explaining Territorial Disputes: From Power Politics to Normative Reasons'. *Journal of Peace Research* 33 (4): 433–49.

Forsvarskommisjonen av 2021. 'NOU 2023: 14. Forsvarskommisjonen Av 2021 – Forsvar for Fred Og Frihet'. Oslo.

France24. 2024. 'For Sale: Unique Piece of Land in Strategic Arctic Archipelago'. *France24*, 17 May 2024. www.france24.com/en/live-news/20240517-for-sale-unique-piece-of-land-in-strategic-arctic-archipelago.

Fravel, M. Taylor, Kathryn Lavelle, and Liselotte Odgaard. 2022. 'China Engages the Arctic: A Great Power in a Regime Complex'. *Asian Security* 18 (2): 138–58.

Fredriksen, Benjamin, Beth Mørch Pettersen, Gyda Katrine Hesla, Inghild Eriksen, and Håvard Gulldahl. 2022. 'Kabelmysteriene'. NRK. 26 June 2022. www.nrk.no/nordland/xl/russiske-tralere-krysset-kabler-i-vesteralen-og-svalbard-for-brudd-1.16007084#intro-authors--expand.

French Ministry of Armed Forces. 2019. 'France and the New Strategic Challenges in the Arctic'. *Paris*. www.defense.gouv.fr/english/layout/set/print/content/download/565142/9742558/version/3/file/France+and+the+New+Strategic+Challenges+in+the+Arctic+-+DGRIS_2019.pdf.

Fridtjof Nansen Institute. 2024a. 'Arctic Security Conference 2024 (Day 2)'. YouTube. www.youtube.com/live/5Oit8vTurc4?t=155s.

2024b. 'Recording: Great Power Competition in the Arctic'. News. www.fni.no/news/recording-great-power-competition-in-the-arctic.

Friedman, Thomas L. 2005. *The World Is Flat: A Brief History of the Twenty-First Century*. New York: Farrar, Straus and Giroux.

Friis, Karsten. 2021. 'Analyzing Security Subregions: Forces of Push, Pull, and Resistance in Nordic Defense Cooperation'. *Journal of Global Security Studies* 6 (4).

Fukuyama, Francis. 1989. 'The End of History?' *The National Interest* No. 16 (Summer 1989): 3–18.

2006. *The End of History and the Last Man*. New York: Free Press.

2011. *The Origins of Political Order: From Prehuman Times to the French Revolution*. New York: Farrar, Straus and Giroux.

Funston, Bernard. 2014. '*Emergency Preparedness in Canada's North: An Examination of Community Capactiy*'. Toronto, ON.

Furgal, Christopher, and Jacinthe Seguin. 2006. 'Climate Change, Health, and Vulnerability in Canadian Northern Aboriginal Communities'. *Environmental Health Perspectives* 114 (12): 1964–70.

Gaasland, Stine B. 2024. 'Bygger Opp i Norfolk'. *Forsvaret*. 2024. www.forsvaret.no/aktuelt-og-presse/aktuelt/bygger-opp-i-norfolk.

Gade, Jo, and Paal S. Hilde. 2015. 'Nordområdenes Sikkerhetspolitiske Betydning for NATO'. In *Norge Og Russland: Sikkerhetspolitiske Utfordringer i Nordområdene*, edited by Tormod Heier and Anders Kjølberg, 96–110. Oslo: Universitetsforlaget.

Gady, Franz-Stefan. 2015. 'China and Russia to Stage Amphibious Assault Exercise in Sea of Japan'. *The Diplomat*, 20 July 2015. https://thediplomat.com/2015/07/china-and-russia-to-stage-amphibious-assault-exercise-in-sea-of-japan/.

Gahr Støre, Jonas. 2008. *Å Gjøre En Forskjell*. Oslo: Cappelen Damm.

2010a. 'Norge, Russland Og Delelinjen (Norway, Russia and the Delimitation)'. Utenriksdepartementet (Norwegian Ministry of Foreign Affairs). 2010. www.regjeringen.no/no/dokumentarkiv/stoltenberg-ii/ud/taler-og-artikler/2010/blogg_delelinje/id604680/.

2010b. 'Speech by the Norwegian Foreign Minister at the China Institute of International Studies'. Regjeringen. No. 2010.

2011. 'Arctic State'. *The Parliament Magazine*, April 2011.

2012. 'The High North and the Arctic: The Norwegian Perspective'. *The Arctic Herald* 2 (June 2012): 8–15. www.regjeringen.no/no/aktuelt/nord_arktis/id685072/.

2019. 'Arktis Smelter. Geografien Forandres. Politikk i Nord Må Handle Om Mer Enn Forholdet Til Russland'. *Aftenposten*, 23 July 2019. www.aftenposten.no/meninger/kronikk/i/P9Eone/arktis-smelter-geografien-forandres-politikk-i-nord-maa-handle-om-mer-enn-forholdet-til-russland-jonas-gahr-stoere.

2023a. 'Norway and the Oceans. Speech given by the Prime Minister Jonas Gahr Støre'. *Norwegian Government, Speeches and Statements*. 2023. www.regjeringen.no/en/aktuelt/norway-and-the-oceans.-speech-given-by-prime-minister-jonas-gahr-store/id2971768/.

2023b. 'Opening Remarks by the Prime Minister at the High North Dialogue Conference in Bodø'. Norwegian Government, Speeches and Statements. 2023. www.regjeringen.no/en/aktuelt/the-high-north-dialogue-conference-at-the-high-north-centre/id2975403/.

Garamone, Jim. 2022. 'DOD Establishes Arctic Strategy and Global Resilience Office'. *U.S. Indo-Pacific Command*, 27 September 2022. www.pacom.mil/Media/News/News-Article-View/Article/3172196/dod-establishes-arctic-strategy-and-global-resilience-office/.

Gåsemyr, Hans Jørgen, and Bjørnar Sverdrup-Thygeson. 2017. 'Chinese Investments in Norway: A Typical Case Despite Special Circumstances'. In *Chinese Investment in Europe: A Country-Level Approach*, edited by John Seaman, Mikko Huotari, and Miguel Otero-Iglesias, 101–9. Oslo: IFRI.

Gautam, P. K. 2011. 'The Arctic as a Global Common'. *New Dehli*. www.files.ethz.ch/isn/135416/IB_TheArcticasaGlobalCommon.pdf.

The General and the Ambassador. 2021. 'Power and Presence: The US & the Arctic with Admiral James Foggo and Ambassador (and SECNAV) Kenneth Braithwaite'. *Podcast Episode*, 25 March 2021. https://generalandambassador.libsyn.com/power-and-presence-the-us-the-arctic-with-admiral-james-foggo-and-ambassador-and-secnav-kenneth-braithwaite.

Geri, Maurizio. 2024. 'War with Russia Will Start in the Arctic, Unless Leaders Move to Stop Climate *Change'*. Friends of Europe. www.friendsofeurope.org/insights/critical-thinking-war-with-russia-will-start-in-the-arctic-unless-leaders-at-the-munich-security-conference-move-to-stop-climate-change/.

Gilpin, Robert. 1981. *War and Change in World Politics*. Cambridge: Cambridge University Press History. Vol. 18. http://ebooks.cambridge.org/ref/id/CBO9780511664267.

Gjelsten, Roald. 2013. 'Fremmede Ubåter i Norske Fjorder (Foreign Submarines in Norwegian Fjords)'. *Oslo Files on Defence and Security*. Vol. 1. Oslo.

Goertz, Gary, and Paul F. Diehl. 1992. *Territorial Changes and International Conflict*. 1st ed. New York: Routledge.

Goodsite, Michael, and Niklas Swanström, eds. 2023. *Towards a Sustainable Arctic: International Security, Climate Change and Green Shipping*. Singapore: World Scientific Publishing.

Google Trends. 2024. '"What Is Geopolitics' Search Term'. *Interest over Time*. 2024. https://trends.google.com/trends/explore?date=all&q=what is geopolitics.

Gorbachev, Mikhail. 1987. 'The Speech in Murmansk at the Ceremonial Meeting on the Occasion of the Presentation of the Order of Lenin and the Gold Star Medal to the City of Murmansk'. *Novosti Press Agency*. Moscow. www.barentsinfo.fi/docs/Gorbachev_speech.pdf.

Gosnell, Rachael, and Lars Saunes. 2024. 'Report No.2: Integrated Naval Deterrence in the Arctic Region–Strategic Options for Enhancing Regional Naval Cooperation'. Newport, RI. https://digital-commons.usnwc.edu/nasi/2/.

Government of Canada. 2010. 'Speech from the Throne', *March 3, 2010*. www.speech.gc.ca/eng/media.asp?id=1388.

2014. 'The Canada-U.S. Defence Relationship'. News. 2014. www.canada.ca/en/news/archive/2014/04/canada-defence-relationship.html.

Government of India. 2022. 'India's Arctic Policy: Building a Partnership for Sustainable Development'. *New Dehli*. www.moes.gov.in/sites/default/files/2022-03/compressed-SINGLE-PAGE-ENGLISH.pdf.

Government of the Russian Federation. 2007. 'Speech and the Following Discussion at the Munich Conference on Security Policy'. Transcripts. 2007. http://en.kremlin.ru/events/president/transcripts/copy/24034.

Graczyk, Piotr, and Svein Vigeland Rottem. 2020. 'The Arctic Council: Soft Actions, Hard Effects?' In *Routledge Handbook of Arctic Security*, edited by Gunhild Hoogensen Gjørv, Marc Lanteigne, and Horatio Sam-Aggrey, 221–34. Abingdon, UK: Routledge.

Gramer, Robbie. 2017. 'Here's What Russia's Military Build-Up in the Arctic Looks Like'. *Foreign Policy*, 25 January 2017. https://foreignpolicy.com/2017/01/25/heres-what-russias-military-build-up-in-the-arctic-looks-like-trump-oil-military-high-north-infographic-map/.

Great Britain–Russia. 1825. 'Great Britain–Russia: Limits of Their Respective Possessions on the North-West Coast of America and the Navigation of the Pacific Ocean'. 75 CTS 95: 16 February 1825.

Greaves, Wilfrid. 2016. 'Thinking Critically about Security and the Arctic in the Anthropocene'. *The Arctic Institute*, 22 March 2016. www.thearcticinstitute.org/thinking-critically-security-arctic-anthropocene/.

2019. 'Arctic Break up: Climate Change, Geopolitics, and the Fragmenting Arctic Security Region'. *Arctic Yearbook* 2019: 1–17. https://arcticyearbook.com/images/yearbook/2019/Scholarly-Papers/9_AY2019_Greaves.pdf.

Greaves, Wilfrid, and Whitney Lackenbauer, eds. 2021. *Breaking Through: Understanding Sovereignty and Security in the Circumpolar Arctic*. Toronto, ON: University of Toronto Press.

Greenpeace. 2012. 'Environmental Resistance to Oil and Gas in the Arctic'. Edited by Andreas Østhagen. *Greenpeace U.S.*

Grieco, Jospeh, John Ikenberry, and Michael Mastanduno. 2022. *Introduction to International Relations: Perspectives, Connections and Enduring Questions*. Third. London: Bloomsbury Academic.

Griffiths, Franklyn. 2011a. 'Arctic Security: The Indirect Approach'. In *Arctic Security in an Age of Climate Change*, edited by James Kraska, 3–19. New York: Cambridge University Press.

2011b. 'Towards a Canadian Arctic Strategy'. In *Canada and the Changing Arctic: Sovereignty, Security and Stewardship*, edited by Franklyn Griffiths, Rob Huebert, and Whitney P. Lackenbauer, 181–225. Waterloo, ON: Wilfrid Laurier University Press.

Griffiths, Sian. 2010. 'US-Canada Arctic Border Dispute Key to Maritime Riches'. *BBC News*, 2 August 2010. www.bbc.com/news/world-us-canada-10834006.

Grindheim, Astrid. 2009. '*The Scramble for the Arctic? A Discourse Analysis of Norway and the EU's Strategies Towards the European Arctic*'. Oslo: Fridtjof Nansen Institute.

Grønning, Ragnhild. 2016. 'Why Military Security Should Be Kept out of the Arctic Council'. *High North News*, 7 June 2016. www.highnorthnews.com/op-ed-why-military-security-should-be-kept-out-of-the-arctic-council/.

Grønningsæter, Fredrik. 2018. 'USAs Forsvarsminister: – Norge Gjør Mer Enn Forventet'. *Dagen*, 14 July 2018.

Grossman, Vladimir. 1946. 'Scandinavia's Strategic Position'. *International Journal* 1 (1): 71–75.

The Guardian. 2019. 'US Warns Beijing's Arctic Activity Risks Creating "New South China Sea"'. *The Guardian*, 6 May 2019. www.theguardian.com/world/2019/may/06/pompeo-arctic-activity-nsew-south-china-sea.

Gunnarsson, Björn, and Arild Moe. 2022. 'International Shipping and the Northern Sea Route'. In *Global Development in the Arctic: International Cooperation for the Future*, edited by Andrey Mineev, Anatoli Bourmistrov, and Frode Mellemvik, 216–31. London & New York: Routledge.

Guo, Ling, and Steven Lloyd Wilson. 2020. 'China, Russia, and Arctic Geopolitics'. *The Diplomat*, 29 March 2020. https://thediplomat.com/2020/03/china-russia-and-arctic-geopolitics/.

Guo, Peiqing. 2012. 'An Analysis of New Criteria for Permanent Observer Status on the Arctic Council and the Road of Non-Arctic States to Arctic'. *KMI International Journal of Maritime Affairs and Fisheries* 4 (2): 21–38.

Guy, Jack. 2024. 'Got a Spare $324 Million? Buy a Slice of This Strategic Arctic Archipelago'. *CNN*, 14 May 2024. https://edition.cnn.com/2024/05/14/travel/svalbard-land-sale-scli-intl/index.html#:~:text=The last piece of privately,listing from agent Knight Frank.

Guzzini, Stefano, ed. 2012. *The Return of Geopolitics in Europe? Social Mechanisms and Foreign Policy Identity Crises*. Cambridge: Cambridge University Press.

Haas, Peter M. 1989. 'Do Regimes Matter? Epistemic Communities and Mediterranean Pollution Control'. *International Organization* 43 (3): 377–403.

Håkenstad, Magnus, and Olav Bogen. 2015. *Balansegang: Forsvarets Omstilling Etter Den Kalde Krigen*. Oslo: Dreyer.

Hanger, Mari Rian. 2018. 'Ny Forskningsstrategi: Vil Markere at Svalbard Er Norsk'. *Forskning.No*, 16 January 2018. www.forskning.no/om-forskning-politikk-svalbard/ny-forskningsstrategi-vil-markere-at-svalbard-er-norsk/1280783.

2019. '– Overraskende at Kina Går Så Hardt Ut Mot Norges Svalbard-Politikk'. *Universitetsavisa*, 24 January 2019.

Haraldstad, Marie. 2014. 'Embetsverkets Rolle i Utformingen Av Norsk Sikkerhetspolitikk: Nærområdeinitiativet [The Role of the Bureaucracy in the Shaping of Norway's Security Policy: The Close Area Initiative]'. *Internasjonal Politikk* 72 (4): 431–51.

Harsem, Øistein, Arne Eide, and Knut Heen. 2011. 'Factors Influencing Future Oil and Gas Prospects in the Arctic'. *Energy Policy* 39 (12): 8037–45.

Haugli, Børre, and Einar Lohne Bjøru. 2023a. 'Russerne Har Økt Aktiviteten På Svalbard'. *Dagsavisen*, 5 May 2023. www.dagsavisen.no/nyheter/2023/05/05/russerne-har-okt-aktiviteten-pa-svalbard/.

2023b. 'Frykter Tap Av Kontroll På Svalbard'. *Dagsavisen*, 29 May 2023. www.dagsavisen.no/nyheter/2023/05/29/frykter-tap-av-kontroll-pa-svalbard/.

Haugseth, Peter. 2017. 'Local Border Traffic (LBT) – A New Era in the Norwegian-Russian Arctic Schengen Borderland'. *Bratislava*. https://borderua.sk/data/PP_Haughseth_cbc01018.pdf.

Heier, Tormod. 2018. 'Avoiding War: How Should Northern Europe Respond to the US-Russian Rivalry?' *Arctic Review on Law and Politics* 9: 267–86.

2021. *En Randstat På Avveie? Norges Vei Inn i Den Nye Kalde Krigen, 2014–2021*. Oslo: Cappelen Damm.

Heininen, Lassi. 2012. 'Arctic Strategies and Policies – Inventory and Comparative Study'. *Northern Research Forum*, no. April: 1–97.

2013. 'Arctic Security – Global Dimensions and Challenges, and National Policy Responses'. In *The Yearbook of Polar Law Volume 5*, edited by Gudmundur Alfredsson, Timo Koivurova, and Adam Stepien, 93–115. Leiden & Boston: Martinus Nijhoff Publishers.

Heininen, Lassi, Karen Everett, Barbora Padrtová, and Anni Reissell. 2020. 'Arctic Policies and Strategies – Analysis, Synthesis, and Trends'. Laxenburg, Austria. http://pure.iiasa.ac.at/id/eprint/16175/1/ArticReport_WEB_new.pdf.

Heininen, Lassi, Heather Exner-Pirot, and Justin Barnes, eds. 2019. *Arctic Yearbook 2019: Redefining Arctic Security*. Akureyri, Iceland: Arctic Portal. www.arcticyearbook.com/.

Henriksen, Tore, and Geir Ulfstein. 2011. 'Maritime Delimitation in the Arctic: The Barents Sea Treaty'. *Ocean Development & International Law* 42 (1–2): 1–21.

Henrikson, Alan K. 2002. 'Distance and Foreign Policy: A Political Geography Approach'. *International Political Science Review* 23 (4): 437–66.

Hensel, Paul R. 1999. 'Charting a Course to Conflict: Territorial Issues and Interstate Conflict, 1816–1992'. In *The Road Map to War*, edited by Paul F. Diehl, 115–46. Nashville, TN: Vanderbilt University Press.

2017. 'Territory, Contentious Issues, and International Relations'. *Oxford Research Encyclopedias*. https://doi.org/10.1093/acrefore/9780190228637.013.565.

Hensel, Paul R., S. McLaughlin Mitchell, Thomas E. Sowers II, and Clayton L. Thyne. 2008. 'Bones of Contention: Comparing Territorial, Maritime and River Issues'. *Journal of Conflict Resolution* 52 (1): 117–43.

Herrmann, Victoria. 2019. 'WANTED: America as a Sustained, Engaged, and Committed Arctic Nation'. *Georgetown Journal of International Affairs* 12 April. https://gjia.georgetown.edu/2019/04/12/wanted-america-as-committed-arctic-nation/.

Herrmann, Victoria, and Lillian Hussong. 2021. 'No UNCLOS, No Icebreakers, No Clue? U.S. Arctic Policy through the Eyes of Congress'. In *Handbook on Geopolitics and Security in the Arctic: The High North between Cooperation and Confrontation*, edited by Joachim Weber, 23–40. Cham: Springer.

Hettne, Björn. 1999. 'Globalization and the New Regionalism: The Second Great Transformation'. In *Globalism and the New Regionalism*, edited by Björn Hettne, András Inotai, and Osvaldo Sunkel, 1–24. London: Palgrave Macmillan.

2007. 'Interregionalism and World Order : The Diverging EU and US Models'. *European Union and New Regionalism – Regional Actors and Global Governance in a Post-Hegemonic Era* 127: 107–23.

2009. *Thinking about Development: Development Matters*. London: Zed Books.

High North News. 2016. 'UD Avviser at Norges Posisjon På Svalbard Er Truet'. *High North News*, 11 November 2016. www.highnorthnews.com/nb/ud-avviser-norges-posisjon-pa-svalbard-er-truet.

Higraff, Marit, Ståle Hansen, Øyvind Bye Skille, and Eirik Veum. 2015. 'Stiller Forsvarsultimatum'. *NRK*, September 2015.

Hilde, Paal S. 2014. 'Armed Forces and Security Challenges in the Arctic'. In *Geopolitics and Security in the Arctic: Regional Dynamics in a Global World*, edited by Rolf Tamnes and Kristine Offerdal, 147–65. London: Routledge.

2019. 'Forsvar Vår Dyd, Men Kom Oss Ikke for Nær. Norge Og Det Militære Samarbeidet i NATO [Defend Our Virtue, but Do Not Get Too Close. Norway and the Military Cooperation in NATO]'. *Internasjonal Politikk* 77 (1): 60–70.

Hilde, Paal S., and Helene F. Widerberg. 2010. 'NATOs Nye Strategiske Konsept Og Norge [NATO's New Strategic Concept and Norway]'. *Norsk Militært Tidsskrift, no*. 4: 10–20.

Hoel, Alf Håkon. 2009. 'Do We Need a New Legal Regime for the Arctic Ocean?' *The International Journal of Marine and Coastal Law* 24 (2): 443–56.

2014. 'The Legal-Political Regime in the Arctic'. In *Geopolitics and Security in the Arctic*, 49–72. New York: Routledge.

2023. 'Kina – Fisk Og Geopolitikk (China – Fish and Geopolitics)'. *Framsenteret*, 4 September 2023. https://framsenteret.no/kina-fisk-og-geopolitikk/.

Hoffman, Stanley. 1966. 'Obstinate or Obsolete: The Fate of the Nation State and the Case of Western Europe'. *Daedalus, 95:3*, 862–915.

Holm, Arne O. 2020. 'The Arctic and Its Good Helpers'. *High North News*, 28 January 2020. www.highnorthnews.com/en/arctic-and-its-good-helpers.

2025. 'Norway's High North Strategy: The Same Slogans, but an Entirely Different and Dramatic Situation'. *High North News*, 27 August 2025. www.highnorthnews.com/en/norways-high-north-strategy-same-slogans-entirely-different-and-dramatic-situation.

Holsbø, Ida. 2011. '*Hvordan Var Det Mulig å Komme Frem Til En Endelig Løsning i Delelinjespørsmålet Mellom Norge Og Russland i Barentshavet i 2010, Og Ikke i 1978? (How Was It Possible to Find a Final Solution to the Boundary Dispute between Norway and Russia?)*'. University of Tromsø.

Holst, Johan Jørgen. 1966. 'Norsk Sikkerhetspolitikk i Strategisk Perspektiv'. *Internasjonal Politikk* 24 (5): 463–90.

Holsti, Kalevi J. 1991. *Peace and War: Armed Conflicts and International Order, 1648–1989*. Cambridge: Cambridge University Press.

Holtsmark, Sven G. 1993. 'A Soviet Grab for the High North. USSR. Svalbard and Northern Norway 1920–1953'. *Forsvarsstudier* 7 (1993).

ed. 1995. *Norge Og Sovjetunionen 1917–1955. En Utenrikspolitisk Dokumentasjon*. Oslo: Cappelen.

ed. 2015. *Naboer i Frykt Og Forventning: Norge Og Russland 1917–2014 (Neighbours in Fear and Anticipation: Norway and Russia 1917–2014)*. Oslo: Pax Forlag.

Homans, Charles. 2011. 'Track II Diplomacy: A Short History'. *Foreign Policy*, 20 June 2011. http://foreignpolicy.com/2011/06/20/track-ii-diplomacy-a-short-history/.

Homolar, Alexandra. 2011. 'How to Last Alone at the Top: US Strategic Planning for the Unipolar Era'. *Journal of Strategic Studies* 34 (2): 189–217.

Hønneland, Geir. 1998. 'Identity Formation in the Barents Euro-Arctic Region'. *Cooperation and Conflict* 33: 277–97.

1999. 'Co-Management and Communities in the Barents Sea Fisheries'. *Human Organization* 58 (4): 397–404.

2012a. *Making Fishery Agreements Work: Post-Agreement Bargaining in the Barents Sea*. Cheltenham, UK: Edward Elgar.

2012b. 'Norsk-Russisk Miljø- Og Ressursforvaltning i Nordområdene [Norwegian-Russian Environmental and Resource Management in the High North]'. *Nordlit 29*. http://septentrio.uit.no/index.php/nordlit/article/view/2303/2134.

2014. 'Norway and Russia: Bargaining Precautionary Fisheries Management in the Barents Sea'. *Arctic Review on Law and Politics* 5 (1): 75–99.

2016. *Russia and the Arctic: Environment, Identity and Foreign Policy*. London: I. B. Tauris.

2024. 'Kvoteavtale Med Russland På Plass'. *Nordlys*, October 2024.

Hønneland, Geir, and Anne-Kristin Jørgensen. 2015. 'Kompromisskulturen i Barentshavet (the Culture of Compromise in the Barents Sea)'. In *Norge Og Russland: Sikkerhetspolitiske Utfordringer i Nordområdene (Norway and Russia: Security Challenges in the High North)*, edited by Tormod Heier and Anders Kjølberg, 57–68. Oslo: Universitetsforlaget.

2025. 'Urovekkende i Barentshavet – Har Aldri Vært Nærmere Stupet'. *Dagens Næringsliv*, September 2025.

Hønneland, Geir, and Lars Rowe. 2010. *'Nordområdene – Hva Nå?' – The High North – What Now?* Trondheim: Tapir Akademiske Forlag.

Hoogensen Gjørv, Gunhild, Dawn R. Bazely, Marina Goloviznina, and Andrew J. Tanentzap, eds. 2014. *Environmental and Human Security in the Arctic*. Abingdon, UK: Routledge.

Hoogensen Gjørv, Gunhild, and Marina Goloviznina. 2014. 'Introduction: Can We Broaden Our Understanding of Security in the Arctic?' In *Environmental and Human Security in the Arctic*, edited by Gunnhild Hoogensen Gjørv, Dawn R. Bazely, Marina Goloviznina, and Andrew J. Tanentzap, 1–14. Abingdon, UK: Routledge.

Hoogensen Gjørv, Gunhild, and Kara K. Hodgson. 2019. '"Arctic Exceptionalism" or "Comprehensive Security"? Understanding Security in the Arctic'. *Arctic Yearbook*.

Hoogensen Gjørv, Gunhild, Marc Lanteigne, and Horatio Sam-Aggrey, eds. 2020. *Routledge Handbook of Arctic Security*. Abingdon: Routledge.

Hoogensen, Gunhild. 2005. 'Bottoms Up! A Toast to Regional Security?' *International Studies Review* 7 (2): 269–74.

Hopf, Ted. 2002. *Social Construction of Foreign Policy: Identities and Foreign Policies, Moscow, 1955 and 1999*. Ithaca, NY: Cornell University Press.

Hopper, Tristin. 2014. 'A New Cold War: Denmark Gets Aggressive, Stakes Huge Claim in Race for the Arctic'. National Post. December 2014.

Houck, Olivia Wynne. 2022. 'Infrastructure in the Arctic: The Arctic Institute Infrastructure Series'. *The Arctic Institute*, 22 March 2022. www.thearcticinstitute.org/infrastructure-arctic-the-arctic-institute-infrastructure-series/.

Hovelsrud, Grete K., Bjørn P. Kaltenborn, and Julia Olsen. 2020. 'Svalbard in Transition: Adaptation to Cross-Scale Changes in Longyearbyen'. *The Polar Journal* 10 (2): 420–42.

Hsiung, Christopher Weidacher. 2016. 'China and Arctic Energy: Drivers and Limitations'. *The Polar Journal* 6 (2): 243–58.

2022. 'China's Perspective on Russia: Assessing How Beijing Views and Values Its Relationship with Moscow Now and Later'. *Stockholm*. www.foi.se/rapportsammanfattning?reportNo=FOI-R--5267--SE.

Huebert, Rob. 1995. 'Polar Vision or Tunnel Vision the Making of Canadian Arctic Waters Policy. The Making of Canadian Arctic Waters Policy'. *Marine Policy* 19 (4): 343–63.

2003. 'The Shipping News Part II: How Canada's Arctic Sovereignty Is on Thinning Ice'. *International Journal* 58 (3): 295–308.

2009. 'United States Arctic Policy: The Reluctant Arctic Power'. *SPP Briefing Papers* 2 (2). https://doi.org/10.2307/j.ctv6gqr43.13.

2010. 'The Newly Emerging Arctic Security Environment'. *Canadian Defence & Foreign Affairs Institute* March 2010.

2011. 'Canadian Arctic Sovereignty and Security in a Transforming Circumpolar World'. In *Canada and the Changing Arctic: Sovereignty, Security and Stewardship*, edited by Franklyn Griffiths, Rob Huebert, and Whitney P. Lackenbauer, 13–68. Waterloo, ON: Wilfrid Laurier University Press.

2013a. 'Submarines, Oil Tankers and Icebreakers'. *International Journal* 66 (4): 809–24. www.jstor.org/stable/23104394.

2013b. 'U.S. Arctic Policy: The Reluctant Arctic Power'. In *The Fast-Changing Arctic: Rethinking Arctic Security for a Warmer World*, edited by Barry Scott Zellen, 189–226. Calgary: Calgary University Press.

Humpert, Malte. 2014. 'Arctic Shipping: An Analysis of the 2013 Northern Sea Route Season'. Edited by The Arctic Institute. Vol. October 20. Washington, DC: The Arctic Institute. www.thearcticinstitute.org/2014/10/NSR-Shipping-Report.html.

2020. 'U.S. Warns of Russian Arctic Military Buildup: "Who Puts Missiles on Icebreakers?"' *High North News*, 25 May 2020. www.highnorthnews.com/en/us-warns-russian-arctic-military-buildup-who-puts-missiles-icebreakers.

2024a. 'U.S. Polar Icebreaker Faces New Delays and Cost Overruns Report Says'. *High North News*, 30 May 2024. www.highnorthnews.com/en/us-polar-icebreaker-faces-new-delays-and-cost-overruns-report-says.

2024b. 'China Deploys Three Icebreakers to Arctic as U.S. Presence Suffers after "Healy" Fire'. *GCaptain*, 20 August 2024. https://gcaptain.com/china-dispatches-three-icebreakers-to-arctic-us-suffers-healy-fire/.

2025. 'U.S. Coast Guard to Receive 11 New Icebreakers under Landmark Finland Deal'. *High North News*, 16 October 2025. www.highnorthnews.com/en/us-coast-guard-receive-11-new-icebreakers-under-landmark-finland-deal.

Huntington, Samuel P. 1993. 'The Clash of Civilizations?' *Foreign Affairs* 73 (3): 22–49.

2003. *The Clash of Civilizations and the Remaking of World Order*. New York: Simon & Schuster Paperbacks.

Huth, Paul K. 1998. *Standing Your Ground: Territorial Disputes and International Conflict*. Ann Arbor, MI: University of Michigan Press.

Hvinden, Ida Soltvedt, and Andreas Østhagen. 2024. 'Mineralvirksomhet På Norsk Sokkel: Mulige Konsekvenser i Lys Av Svalbardtraktaten Og Norsk Forvaltning Av Havområdene Rundt Svalbard (Seabed Mining on the Norwegian Continental Shelf: Possible Consequences in Light of the Svalbard Treaty and Norwegian Man'. *Internasjonal Politikk* 82 (1): 83–101.

Icelandic Coast Guard. 2016. 'Icelandic Coast Guard – Always Prepared'. *About Us*. 2016. www.lhg.is/english/icg/about-us/.

Icelandic Ministry of Foreign Affairs. 2006. 'Joint Understanding'. *Goverment of Iceland*. 2006. www.government.is/library/01-Ministries/Ministry-for-Foreign-Affairs/Joint_Understanding.pdf.

Ikenberry, G. John. 2014. 'The Illusion of Geopolitics: The Enduring Power of the Liberal Order'. *Foreign Affairs* 93 (3): 80–90.

The Ilulissat Declaration. 2008. 'Arctic Ocean Conference'. In. Ilulissat.

Immerwahr, Daniel. 2022. 'Are We Really Prisoners of Geography?' *The Guardian*, 10 November 2022. www.theguardian.com/world/2022/nov/10/are-we-really-prisoners-of-geography-maps-geopolitics.

Inderberg, Tor Håkon Jackson. 2007. 'Den Utenrikspolitiske Håndteringen Av Elektronsaken: Kompetent Realpolitikk Eller Kompetansestrid?' Oslo.

Indian Ministry of Defence. 2023. 'EXERCISE YUDH ABHYAS-23' *SET TO COMMENCE IN FORT WAINWRIGHT, ALASKA, USA – INDIAN CONTINGENT DEPARTS'*. Press Release. 2023. https://pib.gov.in/PressReleasePage.aspx?PRID=1960137.

Ingimundarson, Valur. 2011. 'Territorial Discourses and Identity Politics'. In *Arctic Security in an Age of Climate Change*, edited by James Kraska, 174–89. New York: Cambridge University Press.

Ivanova, Maria. 2011. 'Oil Spill Emergency Preparedness in the Russian Arctic : A Study of the Murmansk Region'. *Polar Research* 30 (7285). www.polarresearch.net/index.php/polar/article/view/7285.

Jacobsen, Marc. 2015. 'The Power and Paradoxes of Collective Identity Narration: Greenland's Way to a More Autonomous Foreign Policy (2009–2014)'. *Arctic Yearbook*. https://arcticyearbook.com/arctic-yearbook/2015/2015-scholarly-papers/127-the-power-of-collective-identity-narration-greenland-s-way-to-a-more-autonomous-foreign-policy.

2018. 'Ilulissat Declaration's 10-Year Anniversary'. *High North News*, 23 May 2018. www.highnorthnews.com/en/ilulissat-declarations-10-year-anniversary.

Jacobsen, Marc, and Signe Lyngholm Lindbjerg. 2024. 'Denmark's Security Perspectives on USA, China, and Russia in Greenland: How Great Power Threats Made Danish Politicians Talk about Independent Greenlandic Agency'. In *Greenland in Arctic Security: (De)Securitization Dynamics under Climatic Thaw and Geopolitical Freeze*, edited by Marc Jacobsen, Ole Wæver, and Ulrik Pram Gad, 196–226. Ann Arbor, MI: University of Michigan Press.

Jacobsen, Marc, Ole Wæver, and Ulrik Pram Gad. 2024a. 'Conclusion'. In *Greenland in Arctic Security: (De)Securitization Dynamics under Climatic Thaw and Geopolitical Freeze*, edited by Marc Jacobsen, Ole Wæver, and Ulrik Pram Gad. Ann Arbor, MI: University of Michigan Press.

eds. 2024b. *Greenland in Arctic Security: (De)Securitization Dynamics under Climatic Thaw and Geopolitical Freeze*. Ann Arbor, MI: University of Michigan Press.

2024c. 'Introduction: Analyzing Greenland in Arctic Security'. In *Greenland in Arctic Security: (De)Securitization Dynamics under Climatic Thaw and Geopolitical Freeze*, edited by Marc Jacobsen, Ole Wæver, and Ulrik Pram Gad, 1–39. Ann Arbor, MI: University of Michigan Press.

Jakobson, Linda. 2010. 'China Prepares for an Ice-Free Arctic'. *Arctic, no.* 2010: 1–16. http://books.sipri.org/files/insight/SIPRIInsight1002.pdf.

Jensen, Leif Christian. 2014. 'The Times They Are A-Changin': Norsk Sikkerhet Og Usikkerhet i Nordområdene'. *Internasjonal Politikk* 72 (1): 7–29.

2017. 'An Arctic "Marriage of Inconvenience": Norway and the Othering of Russia'. *Polar Geography* 40 (2): 121–43.

Jensen, Leif Christian, and Geir Hønneland. 2011. 'Framing the High North: Public Discourses in Norway after 2000'. *Acta Borealia* 28 (1): 37–54.

Jensen, Leif Christian, Øystein Jensen, and Svein Vigeland Rottem. 2011. 'Norwegian Foreign Policy in the High North: Energy, International Law and Security'. *Atlantisch Perspectief* 35 (3): 1–7.

Jensen, Øystein. 2014a. *Norge Og Havets Folkerett*. Trondheim: Akademia Forlag.

2014b. 'The Commission on the Limits of the Continental Shelf: An Administrative, Scientific, or Judicial Institution?' *Ocean Development & International Law* 45 (2): 171–85. www.tandfonline.com/doi/abs/10.1080/00908320.2014.898921.

2016. 'The International Code for Ships Operating in Polar Waters: Finalization, Adoption and Law of the Sea Implications'. *Arctic Review on Law and Politics* 7 (1): 60–82. https://arcticreview.no/index.php/arctic/article/view/236/793.

2020. 'Svalbardtraktaten Og Norsk Rett'. In *Svalbardtraktaten 100 År*, edited by Irene Dahl and Øystein Jensen, 19–50. Bergen: Fagbokforlaget.

2022. 'End of the Common Arctic Seabed: Recent State Practice in the Establishment of Continental Shelf Limits beyond 200 Nm'. *The Polar Journal* 12 (1): 108–21.

Jensen, Øystein, and Svein Vigeland Rottem. 2010. 'The Politics of Security and International Law in Norways Arctic Waters'. *Polar Record* 46 (236): 75–83.

Jervis, Robert. 1976. *Perception and Misperception in International Politics*. Princeton, NJ: Princeton University Press.

Jian, Yang. n.d. *Arctic Governance and China's Engagement*. Shanghai: Shanghai Institutes for International Studies.

Job, Brian. 1992. *The Insecurity Dilemma: National Security of Third World States*. Boulder, CO: Lynne Rienner.

Jockel, Joseph T., and Joel J. Sokolsky. 2009. 'Canada and NATO: Keeping Ottawa in, Expenses down, Criticism out… and the Country Secure'. *International Journal* 64 (2): 315–36.

2012. 'Continental Defence: "Like Farmers Whose Lands Have a Common Concession Line"'. In *Canada's National Security in the Post-9/11 World*, edited by David McDonough, 114–36. Toronto, ON: University of Toronto Press.

Johnsen, Alf Bjarne, and Terje Bringedal. 2021. 'Solberg Til NATO-Toppmøtet Med Ny Russland-Avtale Om å Unngå Militære Sammenstøt'. *VG*, 14 June 2021. www.vg.no/nyheter/utenriks/i/OQwja1/solberg-til-nato-toppmoetet-med-ny-russland-avtale-om-aa-unngaa-militaere-sammenstoet.

Johnsen, Gunnar. 2024. 'Norge Sover i Timen. Kinas Aktivitet På Svalbard Truer Vår Sikkerhet, Mener Professorer'. *Aftenposten*, November 15, 2024. www.aftenposten.no/verden/i/AvxM3n/kinas-aktivitet-paa-svalbard-truer-vaar-sikkerhet-mener-professorer-norge-sover-i-timen.

Johnston, A I. 2001. 'Treating, International Institutions as Social Environments'. *International Studies Quarterly* 45 (4): 487–515.

Johnston, Douglas M. 1988. *The Theory and History of Ocean Boundary-Making*. Montreal: McGill-Queen's University Press.

2002. 'The Northwest Passage Revisited'. *Ocean Development and International Law*.

Jonassen, Trine. 2022. 'The Arctic Council Nominated for the Nobel Peace Prize'. *High North News*, 2 February 2022. www.highnorthnews.com/en/arctic-council-nominated-nobel-peace-prize.

2023. 'Professor: – Kan Komme Til å Se Kinesisk Fiskeflåte i Arktis'. *High North News*, 4 October 2023. www.highnorthnews.com/nb/professor-kan-komme-til-se-kinesisk-fiskeflate-i-arktis.

Jonsson, Gauti. 2023. 'NATO Needs a Unified Arctic Command'. *Georgetown Security Studies Review*, 10 April. https://georgetownsecuritystudiesreview.org/2023/04/10/nato-needs-a-unified-arctic-command/.

Jørgensen, Anne-Kristin. 2022. 'Stock-Shifts and Regime Resilience in the Barents Sea'. In *Marine Resources, Climate Change and International Management Regimes*, edited by Olav Schram Stokke, Andreas Østhagen, and Andreas Raspotnik, 153–78. London: Bloomsbury.

Jørgensen, Anne-Kristin, and Arild Moe. 2023. 'Russisk Nærværspolitikk På Svalbard'. Oslo.

Jørgensen, Anne-Kristin, and Andreas Østhagen. 2020. 'Norges Vern Av Suverene Rettigheter Rundt Svalbard: Russiske Persepsjoner Og Reaksjoner'. *Internasjonal Politikk* 78 (2): 167–92.

Jørgensen, Jørgen Holten. 2010. *Russisk Svalbardspolitikk (Russian Svalbard Policy)*. Trondheim: Tapir Akademiske Forlag.

2013. 'Hvor Normalt Kan Svalbard Bli? Et Lite Stykke Russland i Møte Med Norsk Forvaltning (How Normal Can Svalbard Be? A Small Part of Russia in Encounter with Norwegian Governance)'. *Nordisk Østforum* 27 (4): 327–51.

Jouan, Nicolas. 2024. 'How to Take Advantage of NATO Enlargement in the Arctic'. *RAND Research and Commentary*, 24 June 2024. www.rand.org/pubs/commentary/2024/06/how-to-take-advantage-of-nato-enlargement-in-the-arctic.html.

Juul Stensrud, Cecilie, and Andreas Østhagen. 2024. 'Hybrid Warfare at Sea? Russia, Svalbard and the Arctic'. *Scandinavian Journal of Military Studies* 7 (1): 111–30.

Kaczynski, Vlad M. 2007. 'US-Russian Bering Sea Marine Border Dispute: Conflict over Strategic Assets, Fisheries and Energy Resources'. *Russian Analytical Digest* 20 (1 May 2007): 2–6.

Kaplan, Robert D. 2009. 'The Revenge of Geography'. *Foreign Policy* 172 (May/June): 96–105.

2012. *The Revenge of Geography: What the Map Tells Us about Coming Conflicts and the Battle against Fate*. New York: Random House.

Käpylä, Juha, and Harri Mikkola. 2019. 'Contemporary Arctic Meets World Politics: Rethinking Arctic Exceptionalism in the Age of Uncertainty'. In *The GlobalArctic Handbook*, edited by Matthias Finger and Lassi Heininen, 153–69. Cham.

Karayiğit, Hakkı Ozan. 2021. 'Maps That Map the Mind: Abstraction of Geography by IR Discipline'. *Journal of Research and Didactics in Geography* 1.

Katzenstein, Peter J. 2005. *A World of Regions: Asia and Europe in the American Imperium*. Ithaca, NY: Cornell University Press.

Kee, Randy 'Church'. 2021. '"Securing the Northern Flank": Reflections toward Establishing the Department of Defense's Newest Regional Center: The Ted Stevens Center for Arctic Security Studies'. *Journal of Indo-Pacific Affairs*, 20 October. https://media.defense.gov/2021/Oct/24/2002878833/-1/-1/1/JIPA – KEE – ARCTIC.PDF.

Keil, Kathrin. 2013. 'Cooperation and Conflict in the Arctic: The Cases of Energy, Shipping and Fishing'. PhD Thesis: Berlin Graduate School for Transnational Studies. Freie Universität Berlin.

2014. 'The Arctic – A New Region of Conflict? The Case of Oil and Gas'. *Cooperation and Conflict* 49 (2): 162–90.

Keil, Kathrin, and Andreas Raspotnik. 2014. 'The European Union's Gateways to the Arctic'. *European Foreign Affairs Review* 19 (1): 101–20.

Kelly, Robert E. 2007. 'Security Theory in the "New Regionalism"'. *International Studies Review* 9 (2): 197–229.

Kennan, George F. 1947. 'The Sources of Soviet Conduct'. *Foreign Affairs* 25 (4): 566–82.

Keohane, Robert O. 1984. *After Hegemony: Cooperation and Discord in the World Political Economy*. Princeton, NJ: Princeton University Press.

Keohane, Robert O., and Joseph S. Nye. 1977. *Power and Interdependence*. 4th-Kindle ed. Boston, MA: Longman.

Keskitalo, E. Carina H. 2004. *Negotiating the Arctic: The Construction of an International Region*. New York: Routledge.

2007. 'International Region-Building: Development of the Arctic as an International Region'. *Cooperation and Conflict* 42 (2): 187–205.

Khorrami, Nima, and Jason C. Moyer. 2025. 'Sweden's Arctic Priorities: Balancing Domestic, European Union, and NATO Expectations'. In *The Nordic States and Arctic Security: The Future of Arctic Geopolitics between the European Union and NATO*, edited by Andreas Raspotnik and Andreas Østhagen. Cheltenham, UK: Edward Elgar.

Kienko, E. V. 2018. 'China's Stance towards the Applicability of the Common Heritage of Mankind Governance Regime to the Arctic'. *Journal of Law and Administration* 2 (13–21).

Kimura, Yasuhiro. 2022. 'ロシア「ボストーク2022演習」 ウクライナ戦争の劣勢を否定も、西側諸国には逆効果 (Russia's Vostok 2022 Drills Deny Ukrainian War Weakness, but Have a Negative Effect on Western Countries)'. JNF. 2022. https://forum.j-n.co.jp/narrative/4320/.

Kinnunen, Terhi. 2012. 'Finland, Sweden to Help NATO in Iceland Air Policing'. *Reuters*, 30 October 2012. http://uk.reuters.com/article/2012/10/30/uk-finland-iceland-idUKBRE89T0OU20121030.

Kirby, Paul. 2025. 'US Tells Denmark to "Calm Down" over Alleged Greenland Influence Operation'. *BBC*, August 2025.

Kjellén, Jonas. 2022. 'The Russian Northern Fleet and the (Re)Militarisation of the Arctic'. *Arctic Review on Law and Politics* 13 (2022): 34–52.

Kjellén, Rudolf. 1924. *Der Staat Als Lebensform*. 4th ed. Berlin-Grunewald: Vowinckel.

Klausen, Dan Henrik, Hanne Wilhelms, and Martin Fjørtoft. 2022. 'Mener Norge Kan Utestenge Alle Russiske Fiskebåter'. *NRK*, 7 October 2022. www.nrk.no/tromsogfinnmark/eksperter-mener-norge-ikke-har-rettslig-forpliktelse-til-gi-russiske-bater-havneadgang-1.16127752.

Klimenko, Ekaterina, Annika Nilsson, and Miyase Christensen. 2019. 'Narratives in the Russian Media of Conflict and Cooperation in the Arctic'. *SIPRI Insights on Peace and Security* 2019/5: 1–32.

Knutsen, Bjørn Olav, and Marius Nyquist Pedersen. 2024. 'How to Understand Climate Change as a Threat Multiplier in the Arctic'. *Arctic Review on Law and Politics* 15: 153–76.

Koivurova, Timo. 2010. 'Limits and Possibilities of the Arctic Council in a Rapidly Changing Scene of Arctic Governance'. *Polar Record* 42 (6): 146–56.

2019. 'Is This the End of the Arctic Council and Arctic Governance as We Know It?' *The Polar Connection*, 11 December 2019. https://polarconnection.org/arctic-council-governance-timo-koivurova/.

Koivurova, Timo, and Sanna Kopra, eds. 2020. *Chinese Policy and Presence in the Arctic*. Leiden, NLD: Brill Nijhoff.

Koivurova, Timo, and Akiho Shibata. 2023. 'After Russia's Invasion of Ukraine in 2022: Can We Still Cooperate with Russia in the Arctic?' *Polar Record 59*. https://doi.org/10.1017/S0032247423000049.

Konyshev, Valery, and Alexander Sergunin. 2014. 'Is Russia a Revisionist Military Power in the Arctic?' *Defense and Security Analysis* 30 (4): 323–35.

Kopra, Sanna, and Harri Mikkola. 2025. 'Finland in the Arctic: From Détente to Defence'. In *The Nordic States and Arctic Security: The Future of Arctic Geopolitics between the European Union and NATO*, edited by Andreas Raspotnik and Andreas Østhagen. Cheltenham, UK: Edward Elgar.

Koremenos, Barbara, Charles Lipson, Duncan Snidal, Barbara Koremenos, Charles Lipson, and Duncan Snidal. 2001. 'International Organization Foundation the Rational Design of International Institutions'. *International Organization* 55 (4): 761–99.

Koschut, Simon. 2016. *Normative Change and Security Community Disintegration*. London: Palgrave Macmillan.

Kotani, Tetsuo. 2019. 'China and Russia in the Western Pacific: Implications for Japan and the United States'. *Maritime Awareness Project*, 13 May 2019.

Kraska, James. 2007. 'The Law of the Sea Convention and the Northwest Passage'. *The International Journal of Marine and Coastal Law* 22 (2): 257–82.

ed. 2011. *Arctic Security in an Age of Climate Change*. New York: Cambridge University Press.

Krasner, Stephen D. 1982. 'Structural Causes and Regime Consequences: Regimes as Intervening Variables'. *International Organization* 36 (02): 185–205.

1999. *Sovereignty: Organized Hypocrisy*. Princeton, NJ: Princeton University Press.

Kratochwil, Friedrich. 1986. 'Of Systems, Boundaries, and Territoriality: An Inquiry into the Formation of the State System'. *World Politics* 39 (1): 27–52.

Krever, Mick. 2015. 'Norway: "We Are Faced with a Different Russia"'. *CNN*. 2015. https://edition.cnn.com/2015/02/25/world/amanpour-norway-ine-eriksen-soreide/index.html.

Kristensen, Kristian Søby, and Jon Rahbek-Clemmensen. 2017. 'Greenlandic Sovereignty in Practice: Uranium, Independence, and Foreign Relations in Greenland between Three Logics of Security'. In *Greenland and the International Politics of a Changing Arctic*, edited by Kristian Søby Kristensen and Jon Rahbek-Clemmensen. London: Routledge.

KSAT. 2021. 'KSAT Has Installed Antenna Number 100 at Svalbard Ground Station'. *News Archive*. 2021. www.ksat.no/news/news-archive/2021/ksat-has-installed-antenna-number-100-at-svalbard-ground-station/.

Kuersten, Andreas. 2016. 'The Arctic Five Versus the Arctic Council'. *Arctic Yearbook* 2016.

Kuus, Merje. 2013. 'Part III: Agents – Introduction: Human Agency in Geopolitics'. In *The Ashgate Research Companion to Critical Geopolitics*, edited by Klaus Dodds, Merje Kuus, and Joanne Sharp, 383–86. Farnham: Ashgate.

2014. *Geopolitics and Expertise: Knowledge and Authority in European Diplomacy*. Chichester: Wiley Blackwell.

Kvam, Ina Holst-Pedersen. 2021. 'Bastionforsvaret Og Russlands Militærmakt, Et Utdatert Trusselbilde?' Oslo.

Kystvakten. 2024. 'Kystvakten Var Vertskap for Arctic Coast Guard Forum'. *Forsvaret*. 2024. www.forsvaret.no/om-forsvaret/organisasjon/sjoforsvaret/kystvakten/kystvakten-var-vertskap-for-arctic-coast-guard-forum.

Lackenbauer, Whitney P. 2010. 'Mirror Images? Canada, Russia, and the Circumpolar World'. *International Journal* 65 (4): 879–97.

2011a. 'From Polar Race to Polar Saga: An Integrated Strategy for Canada and the Circumpolar World'. In *Canada and the Changing Arctic: Sovereignty, Security and Stewardship*, edited by Franklyn Griffiths, Rob Huebert, and Whitney P. Lackenbauer, 69–180. Waterloo, ON: Wilfrid Laurier University Press.

2011b. 'Polar Race or Polar Saga? Canada and the Circumpolar World'. In *Arctic Security in an Age of Climate Change*, edited by James Kraska, 218–43. New York: Cambridge University Press.

2013a. *The Canadian Rangers: A Living History*. Vancouver, BC: UBC Press.

2013b. 'Harper's Arctic Evolution'. *The Globe and Mail*, 2013. www.theglobeandmail.com/globe-debate/harpers-arctic-evolution/article13852195/.

2021. 'Threats Through, To, and In the Arctic: A Canadian Perspective'. In *On Thin Ice: Perspectives on Arctic Security*, edited by Duncan Depledge and Whitney P. Lackenbauer, 35–47. Peterborough: North American and Arctic Defence and Security Network (NAADSN). www.naadsn.ca/wp-content/uploads/2021/03/Depledge-Lackenbauer-On-Thin-Ice-final-upload.pdf.

Lackenbauer, Whitney P., and Suzanne Lalonde. 2019. 'Breaking the Ice Curtain? Russia, Canada, and Arctic Security in a Changing Circumpolar World'. *Ottawa*.

Lähteenmäki, Kaisa, and Jyrki Käkönen. 1999. 'Regionalization and Its Impact on the Theory of International Relations'. In *Globalism and the New Regionalism*, edited by Björn Hettne, András Inotai, and Osvaldo Sunkel, 203–27. London: Palgrave Macmillan UK. http://dx.doi.org/10.1007/978-1-349-27268-6_8.

Lajeunesse, Adam. 2008. 'The Northwest Passage in Canadian Policy: An Approach for the 21st Century'. *International Journal* 63 (4): 1037–52.

Lajeunesse, Adam, and Timothy Choi. 2022. 'Here There Be Dragons? Chinese Submarine Options in the Arctic'. *Journal of Strategic Studies* 45 (6–7): 1044–70.

Lake, David A. 1997. 'Regional Security Complexes: A Systems Approach'. In *Regional Orders: Building Security in a New World*, edited by David A. Lake and Patrick Morgan. University Park: Pennsylvania State University Press.

Lalonde, Suzanne, and Frédéric Lasserre. 2013. 'The Position of the United States on the Northwest Passage: Is the Fear of Creating a Precedent Warranted?' *Ocean Development & International Law* 44 (1): 28–72.

Lamazhapov, Erdem. 2024. 'Polar Contradictions: China's Dialectical Thinking about the Arctic'. *Geopolitics*, 1–36. https://doi.org/10.1080/14650045.2024.2408601.

Landriault, Mathieu. 2024. 'Russian Media Coverage of Arctic Issues: Changes since the Invasion of Ukraine'. *Arctic Yearbook 2024*. https://arcticyearbook.com/arctic-yearbook/2024/2024-briefing-notes/532-russian-media-coverage-of-arctic-issues-changes-since-the-invasion-of-ukraine.

Lanteigne, Marc. 2016. 'Ties that Bind: The Emerging Regional Security Complex in the Arctic'. *NUPI Policy Brief* 4: 1–4.

Laruelle, Marlene. 2014. *Russia's Arctic Strategies and the Future of the Far North*. New York: Sharpe.

Lasserre, Frédéric, and Pierre-louis Têtu. 2016. 'Russian Air Patrols in the Arctic: Are Long-Range Bomber Patrols a Challenge to Canadian Security and Sovereignty ?' *Arctic Yearbook 2016*, 2016.

Lavrov, Sergei, and Jonas Gahr Støre. 2010. 'Canada, Take Note: Here's How to Resolve Maritime Disputes'. *The Globe and Mail*, 21 September 2010. www.theglobeandmail.com/opinion/canada-take-note-heres-how-to-resolve-maritime-disputes/article4326372/.

Leland, Sigve R., Alf Håkon Hoel, Pami Aalto, Helge Blakkisrud, and Hanna Smith. 2008. 'Learning by Doing: The Barents Cooperation and Development of Regional Collaboration in the North'. In *The New Northern Dimension of the European Neighbourhood*, 36–54. Brussels: Centre for European Policy Studies.

Lemke, Douglas. 2002. *Regions of War and Peace*. Cambridge, UK: Cambridge University Press.

2005. 'Regions and Powers: The Structure of International Security'. *Perspectives on Politics* 3 (1): 197–98.

Lepperød, Trond. 2005. 'Russiske Myndigheter Sier Tirsdag at de Vil Bringe Tråleren "Elektron" Til Havn i Murmansk. (Russian Authorities a Say Tuesday that They Will Bring the Trawler "Elektron" to Port in Murmansk)'. *Nettavisen*, 18 October 2005.

Levy, Marc A., Oran R. Young, and Michael Zürn. 1995. 'The Study of International Regimes'. *European Journal of International Relations* 1 (3): 267–330.

Lim, Kong Soon. 2018. 'China's Arctic Policy and the Polar Silk Road Vision'. *Arctic Yearbook* 2018: 420–32. https://papers.ssrn.com/sol3/papers.cfm?abstract_id=3603710.

Little, Alex. 2024. 'What Will NATO Do with Its Giant Arctic Footprint? What Will NATO Do with Its Giant Arctic Footprint?' *Responsible Statecraft*, 17 April 2024. https://responsiblestatecraft.org/arctic-nato-russia/.

Liu, Nengye. 2021. 'China and One Hundred Years of the Svalbard Treaty: Past, Present and Future'. *Marine Policy* 124 (February 2021).

Liu, Xuanzun. 2022. 'China, Russia Hold 2nd Joint Naval Patrol in Pacific Ocean'. *Global Times*, 16 September 2022. www.globaltimes.cn/page/202209/1275377.shtml.

Liu, Xuanzun, and Yuandan Guo. 2022. 'For 1st Time China Sends Ground, Naval, Air Forces to Join Russian Vostok Drills'. *Global Times*, 25 August 2022. www.globaltimes.cn/page/202208/1273883.shtml.

2024. 'China, Russia Commence Joint Drills, Patrol'. *Global Times*, 11 September 2024. www.globaltimes.cn/page/202409/1319659.shtml.

Lohne, Lone, Synne Eggum Myrvang, and Gisle Oddstad. 2019. 'Erna Solberg i Toppmøte Med Putin: – Vi Har Ingen Tid å Miste'. *VG*, 9 April 2019. www.vg.no/nyheter/utenriks/i/kJ0nEB/erna-solberg-i-toppmoete-med-putin-vi-har-ingen-tid-aa-miste.

Luhn, Alec. 2020. 'Freezing Cold War: Militaries Move in as Arctic Ice Retreats'. *The Guardian*, 16 October 2020.

Łuszczuk, Michal. 2016. 'Military Cooperation and Enhanced Arctic Security in the Context of Climate Change and Growing Global Interest in the Arctic'. In *Future Security of the Global Arctic: State Policy, Economic Security and Climate*, edited by Lassi Heininen, 35–54. London: Palgave Macmillan.

Luttwak, Edward N. 1990. 'From Geopolitics to Geo-Economics: Logic of Conflict, Grammar of Commerce'. *The National Interest* 20 (Summer 1990): 17–23.

MacAskill, Ewen. 2007. 'Canada Uses Military Might in Arctic Scramble'. *The Guardian*, 11 August 2007. www.theguardian.com/world/2007/aug/11/oil.arctic.

Machiavelli, Niccolò. 1950. *The Prince, and the Discourses*. New York: Random House.

Mackinder, Halford John. 1904. 'The Geographical Pivot of History'. *The Geographical Journal* 23 (4): 421–37.

Macnab, Ron. 2004a. 'The Case for Transparency in the Delimitation of the Outer Continental Shelf in Accordance with LOSC Article 76'. *Ocean Development & International Law* 35: 1–17.

2004b. 'The Outer Limit of the Continental Shelf in the Arctic Ocean'. In *Legal and Scientific Aspects of Continental Shelf Limits*, edited by M. H. Nordquist, J. N. Moore, and T. H. Heidar. Leiden: Martinus Nijhoff Publishers.

Mahadzir, Dzirhan. 2023. 'Russian, Chinese Warships in East China Sea after Sailing Near Alaska'. *USNI News*, 17 August 2023. https://news.usni.org/2023/08/17/russian-chinese-warships-in-east-china-sea-after-sailing-near-alaska.

Maher, Kit, and Kevin Liptak. 2025. 'Vance Previously Slammed Zelensky for Lack of Gratitude. This Time, It Placed Him at Center of Tense Oval Office Exchange'. *CNN*, March 2025.

Mandraud, Isabelle. 2014. 'Russia Prepares for Ice-Cold War with Show of Military Force in the Arctic'. *The Guardian*. Europe section. 2014. www.theguardian.com/world/2014/oct/21/russia-arctic-military-oil-gas-putin.

Manicom, James. 2011. 'Maritime Boundary Disputes in East Asia: Lessons for the Arctic'. *International Studies Perspectives* 12 (3): 327–40.

Marines. 2020. 'New Marine Rotational Force Deploys to Norway for Winter Training Alongside Norwegian Allies'. *News Display*, 28 October 2020. www.marines.mil/News/News-Display/Article/2397771/new-marine-rotational-force-deploys-to-norway-for-winter-training-alongside-nor/.

Maritime Executive. 2019. 'U.S. 2nd Fleet Reactivates to Counter Russian Navy'. *Maritime Executive*, 31 December 2019. https://maritime-executive.com/article/u-s-2nd-fleet-reactivated-to-counter-russian-navy.

Marshall, Tim. 2015. *Prisoners of Geography: Ten Maps that Tell You Everything You Need to Know about Global Politics*. London: Elliot and Thompson.

2019. *Prisoners of Geography*. 2019 ed. London: Elliot and Thompson.

2021. *The Power of Geography: Ten Maps that Reveal the Future of Our World*. London: Elliott & Thompson.

Martinez, Michael. 2015. 'Russian Aggressions Prompt "a New Cold War" in Rhetoric Only'. *CNN World*, 30 October 2015. www.cnn.com/2015/10/29/world/new-cold-war-russia-west/.

McCarthy, Simone. 2024. 'China's Coast Guard Claims to Have Entered the Arctic Ocean for the First Time as It Ramps Up Security Ties with Russia'. *CNN*, 3 October 2024. https://edition.cnn.com/2024/10/03/china/china-russia-coast-guard-arctic-ocean-intl-hnk.

McDorman, Ted L. 2002. 'The Role of the Commision on the Limits of Continental Shelf: A Technical Body in a Political World'. *International Journal of Marine and Coastal Law* 17: 301.

2009. *Salt Water Neighbors: International Ocean Law Relations between the United States and Canada*. New York: Oxford University Press.

McGwin, Kevin. 2020. 'An Arctic Treaty Has Been Rejected by the Region's Leaders. Again'. *Arctic Today*, 12 February 2020.

Mead, Walter Russell. 2014. 'The Return of Geopolitics: The Revenge of the Revisionist Powers'. *Foreign Affairs May/June*. www.foreignaffairs.com/articles/china/2014-04-17/return-geopolitics.

Mearsheimer, John J. 2001. *The Tragedy of Great Power Politics*. New York: W. W. Norton & Company.

2010. 'The Gathering Storm'. *The Chinese Journal of International Politics* 3 (4): 381–96.

Medby, Ingrid A. 2014. 'Arctic State, Arctic Nation? Arctic National Identity among the Post-Cold War Generation in Norway'. *Polar Geography* 37 (3): 252–69.

Melia, Michael. 2015. 'Arctic Coast Guards Pledge Co-Operation at U.S. Meeting'. *CBC News: North*, 30 October 2015. www.cbc.ca/news/canada/north/arctic-coast-guards-pledge-co-operation-at-u-s-meeting-1.3296921.

Mendenhall, Elizabeth. 2018. 'Fluid Foundations: Ocean Transparency, Submarine Opacity, and Strategic Nuclear Stability'. *Journal of Military and Strategic Studies* 19 (1).

2019. 'The Ocean Governance Regime: International Conventions and Institutions'. In *Climate Change and Ocean Governance*, edited by Paul G. Harris, 27–42. Cambridge: Cambridge University Press.

Midford, Paul. 2015. 'Sino-Japanese Conflict and Reconciliation in the East China Sea'. In *The China-Japan Border Dispute: Islands of Contention in Multidisciplinary Perspective*, edited by Tim F. Liao, Krista E. Wiegand, and Kimie Hara. Farnham: Ashgate.

Midford, Paul, and Andreas Østhagen. 2024. 'The East China Sea: A Case of Ocean Geopolitics and Maritime Conflict'. *East Asia* 41 (1).

Mineev, Andrey, Anatoli Bourmistrov, and Frode Mellemvik, eds. 2022. *Global Development in the Arctic: International Cooperation for the Future*. London: Routledge.

Miner, Kimberley, Leslie Canavera, Jared Gonet, Kelly Luis, Marisol Maddox, Paul McCarney, Gwen Bridge, David Schimel, and James Rattlingleaf. 2023. 'The Co-Production of Knowledge for Climate Science'. *Nature Climate Change* 13 (2023): 307–8.

Ministry of Foreign Affairs of the Russian Federation. 2020. 'Press Release on Foreign Minister Sergey Lavrov's Message to Norwegian Foreign Minister Ine Eriksen Soreide on the Occasion of the 100th Anniversary of the Spitsbergen Treaty'. News. www.mid.ru/en/web/guest/maps/no/-/asset_publisher/f4MKo6byouc4/content/id/4019093.

Mirovitskaya, Natalia S., Margaret Clark, and Ronald G. Purver. 1993. 'North Pacific Fur Seals: Regime Formation as a Means of Resolving Conflict'. In *Polar Politics: Creating International Environmental Regimes*, edited by Oran Young and Gail Osherenko, 22–55. New York: Cornell University Press.

Mishra, Rahul. 2017. 'Code of Conduct in the South China Sea: More Discord than Accord'. *Maritime Affairs* 13 (2): 62–75.

Mitchell, James R. 2013. 'The Canadian Coast Guard in Perspective: A Paper Prepared for Action Canada'. Vol. August. Ottawa: Action *Canada.* www.actioncanada.ca/en/wp-content/uploads/2013/08/Canadian-Coast-Guard-In-Perspective_EN.pdf.

Mock, William B.T. 1992. 'Game Theory, Signalling and International Legal Relations'. *George Washington Journal of International Law & Economics* 26: 33–60.

Moe, Arild. 2014. 'The Northern Sea Route: Smooth Sailing Ahead?' *Strategic Analysis* 38 (6): 784–802. www.tandfonline.com/doi/abs/10.1080/09700161.2014.952940.

2020. 'Forskningspolitikk På Svalbard'. In *Svalbardtraktaten 100 År*, edited by Irene Dahl and Øystein Jensen, 109–34. Bergen: Fagbokforlaget.

2021. 'En Kinesisk Jernbane Til Arktis? Historien Om Belkomur – Så Langt ("A Chinese Railway to the Arctic? The Story of Belkomur – so Far")'. *Nordisk Østforum* 35 (2021): 126–43.

Moe, Arild, Daniel Fjærtoft, and Indra Øverland. 2011. 'Space and Timing: Why Was the Barents Sea Delimitation Dispute Resolved in 2010?' *Polar Geography* 34 (3): 145–62.

Moe, Arild, and Øystein Jensen. 2020. 'Svalbard Og Havområdene – Nye Utenrikspolitiske Utfordringer for Norge? (Svalbard and the Maritime Areas – Challenges for Norway?)'. *Internasjonal Politikk* 78 (4): 511–22.

Moe, Arild, Erdem Lamazhapov, and Oleg Anisimov. 2022. 'Russia's Expanding Adaptation Agenda and Its Limitations'. *Climate Policy* 23 (2): 184–98.

Moe, Arild, Andreas Østhagen, and Svein Vigeland Rottem. 2022. 'Aftenposten Bommer Om Konfliktfaren i Arktis'. *Aftenposten*, 19 May 2022. www.aftenposten.no/meninger/debatt/i/G3E5Lq/aftenposten-bommer-om-konfliktfaren-i-arktis.

Moe, Arild, and Olav Schram Stokke. 2019. 'Asian Countries and Arctic Shipping: Policies, Interests and Footprints on Governance'. *Arctic Review on Law and Politics* 10 (2019): 24–52.

Moens, Barbara, and Gabriela Galindo. 2021. 'EU Faces Arctic Cod War with Oslo over Post-Brexit Rights'. *Politico*, 9 August 2021.

Mohan, C Raja. 2024. 'India's New Alignment: "Non-West" but Not "Anti-West"'. *ISAS Briefs*, 19 February 2024. www.isas.nus.edu.sg/papers/indias-new-alignment-non-west-but-not-anti-west/.

Molenaar, Erik J. 2012. 'Fisheries Regulation in the Maritime Zones of Svalbard'. *The International Journal of Marine and Coastal Law* 27 (1): 3–58.

2022. 'A Response to "The Politics of Research Presence in Svalbard" by Torbjørn Pedersen'. *The Polar Journal* 11 (2): 427–32.

Moon, Bruce. 1998. 'Regionalism Is Back! Now What?' *International Studies Review* 42: 338–42.

Moravcsik, Andrew. 1997. 'Taking Preferences Seriously: A Liberal Theory of International Politics'. *International Organization* 51 (04): 229.

Morgenthau, Hans J. 1954. *Politics among Nations: The Struggle for Power and Peace*. Second. New York: Alfred A. Knopf.

Morris, Ian. 2023. *Geography Is Destiny: Britain and the World, a 10,000 Year History*. London: Profile Books Ltd.

Mortensen, Martin, and Arne Egil Tønset. 2021. 'Forsvarsministeren Med Kritikk Til Tromsø Kommune: – Har Sabotert Beredskapsplanen'. *NRK*, May 2021.

The Moscow Times. 2024. 'Russia, China Conduct Strategic Bomber Patrol Near Alaska'. *Reuters*, 25 July 2024. www.themoscowtimes.com/2024/07/25/russia-china-conduct-strategic-bomber-patrol-near-alaska-a85822.

Mottola, Lee. 2023. 'NATO's Arctic Command: A Case for the Expansion of NATO's Mission in the High North'. *The Arctic Institute*, 17 January 2023. www.thearcticinstitute.org/nato-arctic-command-case-expansion-nato-mission-high-north/.

Müller, Harald. 2006. 'A Theory of Decay of Security Communities with an Application to the Present State of the Atlantic Alliance'. *Institute of European Studies UC Berkeley 04-04-2006*. http://escholarship.org/uc/item/95n4b4sp.

Müller, Martin. 2013. 'Text, Discourse, Affect and Things'. In *The Ashgate Research Companion to Critical Geopolitics*, edited by Klaus Dodds, Merje Kuus, and Joanne Sharp, 49–68. Farnham: Ashgate.

Murkowski, Lisa. 2023a. 'Murkowski, Sullivan Statements on Chinese and Russian Vessels in U.S. Waters off Coast of Aleutians'. *Press Releases*. 2023. www.murkowski.senate.gov/press/release/murkowski-sullivan-statements-on-chinese-and-russian-vessels-in-us-waters-off-coast-of-aleutians.

2023b. 'Murkowski, Sullivan Statements on Chinese and Russian Vessels in U.S. Waters off Coast of Aleutians'. *Press Releases*. 2023.

2023c. 'Murkowski Reintroduces Resolution Calling on U.S. Senate to Ratify Law of the Sea'. *Press Releases*. 2023. www.murkowski.senate.gov/press/release/murkowski-reintroduces-resolution-calling-on-us-senate-to-ratify-law-of-the-sea.

Nahmen, Alexandra von. 2023. 'Ukraine War Raises Tensions in Svalbard Archipelago'. *DW*, 9 June 2023. www.dw.com/en/ukraine-war-raises-tensions-in-svalbard-archipelago/video-66740399.

The National Archives of Norway. 2020. 'Svalbarði Fundinn'. *Nyere Historie (1814–)*. 2020. www.arkivverket.no/utforsk-arkivene/nyere-historie-1814-/svalbardi-fundinn.

National Defence. 2024. 'Operation NANOOK'. *Curren Operations List.* www.canada.ca/en/department-national-defence/services/operations/military-operations/current-operations/operation-nanook.html.

National Park Service. 2023. 'About Beringia'. About. www.nps.gov/subjects/beringia/about.htm.

National Science Foundation. 1984. 'Arctic Research and Policy Act of 1984'. Washington, DC.

National Security Council. 1971. 'National Security Decision Memorandum 144'. Washington, DC.

NATO-Russia Council. 2013. 'Russia and Norway Military Cooperation'. News.

NATO. 2023a. 'Speech by NATO Secretary General Jens Stoltenberg at the Annual Conference of the Confederation of Norwegian Enterprises'. Speeches & Transcripts.

2023b. 'The Future of the High North'. *Allied Command Transformation.* 2023. www.act.nato.int/article/the-future-of-the-high-north/.

NATO SHAPE. 2023. 'ICELAND'S "PEACETIME PREPAREDNESS NEEDS"'. *Operations and Missions.* https://shape.nato.int/icelands-peacetime-preparedness-needs.

NBC News. 2007. 'Russia Defends North Pole Flag-Planting'. *World*, 8 August 2007.

Neffenger, Peter V. 2014. 'Testimony of Vice Admiral Peter V. Neffenger Vice Commandant on Implementing U.S. Policy in the Arctic'. *House Coast Guard and Maritime Transportation Subcommittee.* U.S. Department of Homeland Security.

Németh Winther, Iselin. 2024. 'Situating the Local Level in Arctic Politics: How Can It Engage in Arctic Governance, and How Is It Affected by Arctic Geopolitics?' *Strategic Analysis* 48 (6): 588–603.

2025. 'The Arctic Mayors' Forum: The Emergence of a New Circumpolar Inter-Subnational Governance Mechanism'. *The Polar Journal* 15 (2).

Nemeth, Stephen C., Sara McLaughlin Mitchell, Elizabeth A. Nyman, and Paul R. Hensel. 2014. 'Ruling the Sea: Managing Maritime Conflicts through UNCLOS and Exclusive Economic Zones'. *International Interactions* 40 (5): 711–36.

Neuman, Scott. 2019. 'No Joke: Trump Really Does Want to Buy Greenland'. *NPR*, 19 August 2019.

Neumann, Iver B. 1994. 'A Region-Building Approach to Northern Europe'. *Review of International Studies* 20 (1): 53–74.

2001. 'Regionalization and Democratic Consolidation'. In *Democratic Consolidation in Eastern Europe*, edited by Zielonka J. and Pravda A., 58–77. Oxford: Oxford University Press.

2008. 'Russia as a Great Power, 1815–2007'. *Journal of International Relations and Development* 11: 128–51.

Nichols, Christopher McKnight, and Eda Ayaydin. 2025. 'Deeper U.S. Cooperation with Russia Is Coming in the Arctic'. *Foreign Policy* 4 August.

Nilsen, Thomas. 2017. 'Kommersant: Russia Lists Norway's Svalbard Policy as Potential Risk of War'. *The Independent Barents Observer*, 4 October 2017. https://thebarentsobserver.com/en/security/2017/10/kommersant-russia-lists-norways-svalbard-policy-potential-risk-war.

2021a. 'U.S. B-1 Bomber Makes First Landing Inside Norway's Arctic Circle'. *The Barents Observer*, 9 March 2021. www.thebarentsobserver.com/security/us-b1-bomber-makes-first-landing-inside-norways-arctic-circle/123806.

2021b. 'Heavily-Armed Frigate Sails Norwegian Sea to "Ensure Safety of Russian Fishermen"'. *The Barents Observer*, 11 April 2021. www.thebarentsobserver.com/security/heavilyarmed-frigate-sails-norwegian-sea-to-ensure-safety-of-russian-fishermen/159681.

2021c. 'Russia Launches New Borei-A Class Ballistic Missile Sub'. *The Barents Observer*, 25 December 2021. www.thebarentsobserver.com/security/russia-launches-new-boreia-class-ballistic-missile-sub/159564.

2022. 'Russia Steps Up Military Posturing in the Arctic Ahead of NATO's Nuclear Drill'. *Barents Observer*, 16 October 2022. https://thebarentsobserver.com/en/security/2022/10/russia-escalates-arctic-military-poseur-ahead-natos-nuclear-drill.

2023a. 'Russia's Coast Guard Cooperation with China Is a Big Step, Arctic Security Expert Says'. *Barents Observer*, 28 April 2023. https://thebarentsobserver.com/en/security/2023/04/russias-arctic-coast-guard-cooperation-china-big-step-expert.

2023b. 'Svalbard Governor Orders War-Glorifying Cross Demolished. Russian Official Protests on Behalf of the Entire Orthodox World'. *Barents Observer*, 17 October 2023. www.thebarentsobserver.com/arctic/svalbard-governor-orders-warglorifying-cross-demolished-russian-official-protests-on-behalf-of-the-entire-orthodox-world/126145.

2023c. 'Isolated Russia Invites Faraway Countries to Upcoming Svalbard Science Center in Pyramiden'. *Barents Observer*, 30 October 2023. https://thebarentsobserver.com/en/arctic/2023/10/ghost-town-pyramiden-will-be-home-russias-planned-international-svalbard-science.

2024a. 'Small Norwegian Fishing Boat Confronted Russian Warship during Live-Shooting Exercise'. *The Barents Observer*, 27 September 2024. www.thebarentsobserver.com/security/small-norwegian-fishing-boat-confronted-russian-warship-during-liveshooting-exercise/231529.

2024b. 'Norwegian Fishermen Snagged U.S. Nuclear-Powered Submarine'. *The Barents Observer*, 13 November 2024. www.thebarentsobserver.com/security/norwegian-fishermen-snagged-us-nuclearpowered-submarine/420357.

Nilsson, Annika E. 2018. 'The United States and the Making of an Arctic Nation'. *Polar Record* 54 (2): 95–107.

Nilsson, Annika E., and Miyase Christensen. 2019. *Arctic Geopolitics, Media and Power*. Abingdon, UK: Routledge.

NOAA. 2024a. 'How Many Oceans Are There?' Ocean Facts. 2024. https://oceanservice.noaa.gov/facts/howmanyoceans.html#:~:text=Historically%2C there are four named,the 'newest' named ocean.

2024b. 'Report of the 28th Annual Conference of the Parties to the Convention on the Conservation and Management of Pollock Resources in the Central Bering Sea'. www.fisheries.noaa.gov/s3/2024-07/report-28th-annual-conference-central-being-sea-pollock.pdf.

Norway–Russian Federation. 2010. 'Treaty between Norway and the Russian Federation Concerning Maritime Delimitation and Cooperation in the Barents Sea and Arctic Ocean, 15 September 2010'.

Norway–Soviet Union. 1957. 'Agreement between Norway and the Soviet Union Concerning the Sea Frontier in the Varanger Fjord'. 27 February 1957. https://lovdata.no/dokument/TRAKTAT/traktat/1957-02-15-1.

Norwegian Government. 2010. 'Joint Statement on Maritime Delimitation and Cooperation in the Barents Sea and the Arctic Ocean'. *Press Release*, 27 April 2010. www.regjeringen.no/globalassets/upload/ud/vedlegg/folkerett/030427_english_4.pdf.

2018. 'Strategi for Forskning Og Høyere Utdanning På Svalbard'. Oslo. www.uio.no/forskning/tverrfak/nordomradene/aktuelt/2018/strategi-for-forskning-og-hoyere-utdanning-pa-svalbard.pdf.

2021. 'Hurdalsplattformen: For En Regjering Utgått Fra Arbeiderpartiet Og Senterpartiet'. *Hurdal*. www.regjeringen.no/no/dokumenter/hurdalsplattformen/id2877252/.

Norwegian Intelligence Service. 2023. 'Fokus 2023'. Oslo. www.etterretningstjenesten.no/publikasjoner/fokus/fokus-norsk/Fokus2023 – NO – Weboppslag v3.pdf/_/attachment/inline/c1a9a458-aa1d-4bf6-a558-9cec57acde8f:9b2050d897a2b2db1bddc8e505db7b666e608b98/Fokus2023 – NO – Weboppslag v3.pdf.

Norwegian Ministry of Defence. 2015. 'Unified Effort: Expert Commission on Norwegian Security and Defence Policy'. Oslo: Norwegian Ministry of Defence. www.regjeringen.no/globalassets/departementene/fd/dokumenter/rapporter-og-regelverk/unified-effort.pdf.

2023. 'NOU 2023: 14. Norwegian Defence Commission 2021 – English Short Version'. Oslo: Norwegian Ministry of Defence. https://files.nettsteder.regjeringen.no/wpuploads01/sites/495/2023/06/forsvarskommisjonen_kortversjon-1.7_singel.pdf.

Norwegian Ministry of Foreign Affairs. 2005. 'St.Meld. Nr. 30 (2004–2005): Muligheter Og Utfordringer i Nord'. Vol. 30. Oslo. www.regjeringen.no/contentassets/30b734023f6649ee94a10b69d0586afa/no/pdfs/stm200420050030000dddpdfs.pdf.

2006. 'The Norwegian Government's High North Strategy'. Oslo. www.regjeringen.no/globalassets/upload/ud/vedlegg/strategien.pdf.

2009. 'New Building Blocks in the North: The next Step in the Government's High North Strategy'. Oslo. www.regjeringen.no/globalassets/upload/ud/vedlegg/nordomradene/new_building_blocks_in_the_north.pdf.

Norwegian Ministry of Justice and Public Security. 2016. 'Meld. St. 32 (2015–2016) Svalbard' 32 (11 May 2016). www.regjeringen.no/no/dokumenter/meld.-st.-32-20152016/id2499962/.

2022a. 'Endrer Reglene for Valg Til Longyearbyen Lokalstyre'. *Pressemeldinger*. 2022. www.regjeringen.no/no/aktuelt/endrer-reglene-for-valg-til-longyearbyen-lokalstyre/id2919502/.

2022b. 'Meld. St. 9 (2022–2023) Report to the Storting (White Paper): National Control and Cyber Resilience to Safeguard National Security'. Oslo. www.regjeringen.no/en/dokumenter/meld.-st.-9-20222023/id2950130/.

2024. 'Meld. St. 26 (2023–2024), Melding Til Stortinget: Svalbard'. Oslo. www.regjeringen.no/no/dokumenter/meld.-st.-26-20232024/id3041130/.

Norwegian Ministry of Trade Industry and Fisheries. 1976. *Lov Om Norges Økonomiske Sone [Økonomiske Soneloven] (Law on Norway's Economic Zone)*. Norway: https://lovdata.no/dokument/NL/lov/1976-12-17-91.

Norwegian Research Council. 2019. 'Ny-Ålesund Research Station'. Oslo. www.forskningsradet.no/siteassets/publikasjoner/2019/ny-alesund-research-station-research-strategy2.pdf.

Novosti, RIA. 2007. 'Russia Guided By International Law in Its Polar Shelf Probe'. *RIA Novosti*, 3 August 2007.

NTB. 2015. 'Russisk Opptrapping En Utfordring for Forsvaret'. *E24*, 7 April 2015. https://e24.no/norsk-oekonomi/i/Opgmvq/russisk-opptrapping-en-utfordring-for-forsvaret.

2016. 'Solberg Løsnet Latteren Hos Obama'. *Nettavisen*, 13 May 2016. www.nettavisen.no/na24/solberg-losnet-latteren-hos-obama/s/12-95-3423223994.

Nuttall, Mark, Torben R. Christensen, and Martin J. Siegert, eds. 2018. *The Routledge Handbook of the Polar Regions*. Abingdon: Routledge.

O'Brien, Phillips P. 2023. 'There's No Such Thing as a Great Power'. *Foreign Affairs*, 29 June. www.foreignaffairs.com/ukraine/theres-no-such-thing-great-power.

O'Neill, Aaron. 2024. 'Switzerland: Ratio of Military Spending to Gross Domestic Product (GDP) from 2012 to 2022'. Statista. 2024. www.statista.com/statistics/810820/ratio-of-military-expenditure-to-gross-domestic-product-gdp-switzerland/#:~:text=International-,Ratio of military expenditure to,product (GDP) in Switzerland 2022&text=The ratio of military expenditure,ratio than.

Ó Tuathail, Gearóid. 2010. 'Localizing Geopolitics: Disaggregating Violence and Return in Conflict Regions'. *Political Geography* 29 (5): 256–65.

Ó Tuathail, Gearóid, Simon Dalby, and Paul Routledge, eds. 2003. *The Geopolitics Reader*. London & New York: Taylor & Francis e-Library.

Odgaard, Liselotte. 2024. 'NATO Is Unprepared for Russia's Arctic Threats'. *Foreign Policy*, 1 April 2024. https://foreignpolicy.com/2024/04/01/nato-russia-arctic-steadfast-defender-2024/.

Offerdal, Kristine. 2010. 'Arctic Energy in EU Policy: Arbitrary Interest in the Norwegian High North'. *Arctic, 63:1*, 30–42.

2011. 'Norge, Nordområdene Og EU'. #20. Europautredningen.

Office of the under Secretary of Defense for Policy. 2019. 'Report to Congress: Department of Defense Arctic Strategy'. Washington, DC. https://media.defense.gov/2019/Jun/06/2002141657/-1/-1/1/2019-DOD-ARCTIC-STRATEGY.PDF.

Olesen, Mikkel Runge. 2020. 'The End of Arctic Exceptionalism? A Review of the Academic Debates and What the Arctic Prospects Mean for the Kingdom of Denmark'. *Danish Foreign Policy Review* 103 (2020). www.diis.dk/en/research/the-of-arctic-exceptionalism.

Olsen, John Andreas. 2017. 'NATO and the North Atlantic: Revitalising Collective Defence'. *London*. https://rusi.org/publication/whitehall-papers/nato-and-north-atlantic-revitalising-collective-defence.

Oma, Ida Maria. 2021. 'Avskrekking Og Beroligelse – Den "doble Strategi" Som Kom Inn Fra Kulden'. Oslo. www.forsvaret.no/forskning/fou-formidling-ved-forsvarets-hogskole/ifs-insights-avskrekking-og-beroligelse-den-doble-strategi.

Oma Vyvial, Juni. 2023. 'Videreføring Av Samarbeid: Norsk-Russisk Kontakt i En Geopolitisk Anstrengt Tid'. Vol. 8/2023. Oslo. www.fni.no/publikasjoner/videreforing-av-samarbeid-norsk-russisk-kontakt-i-en-geopolitisk-anstrengt-tid.

Ondraskova, Jana, Kristine Berzina, and Sophie Arts. 2024. 'Navigating Murky Waters'. Washington, DC. www.gmfus.org/news/navigating-murky-waters.

Onsager, Jonas Ekkeren, and Kirsti Stuvøy. 2022. 'Barentssamarbeidet Og Geopolitisk Spenning: En Analyse Av Norske Erfaringer Og Legitimeringspraksiser'. *Internasjonal Politikk* 80 (3): 350–74.

Organski, A. F. K. 1968. *World Politics*. 2nd ed. New York: Alfred A. Knopf.

Orheim, Olav, Ingvild Broch, Leif Forsell, Ivan Kristoffersen, Vigdis Stordal, Else G. Broderstad, Rasmus Hansson, et al. 2003. *NOU 2003:32 Mot Nord! Utfordringer Og Muligheter i Nordområdene. Official Norwegian Report*. Vol. 32. Oslo: Norwegian Ministry of Foreign Affaris. www.regjeringen.no/contentassets/28ed358f13704ed2bb3c2a7f13a02be9/no/pdfs/nou200320030032000dddpdfs.pdf.

Oskanian, Kevork. 2013. *Fear, Weakness and Power in the Post-Soviet South Caucasus*. London: Palgave Macmillan.

Østhagen, Andreas. 2012. 'The Arctic and the Need for Greater Differentiation in a Non-Coherent Region'. *The Arctic Institute*, 23 February 2012. www.thearcticinstitute.org/2012/02/arctic-and-need-for-greater.html.

2013. 'The European Union – An Arctic Actor?' *Journal of Military and Strategic Studies* 15 (2): 71–92.

2015. 'Coastguards in Peril: A Study of Arctic Defence Collaboration'. *Defence Studies* 15 (2): 143–60.

2016. 'High North, Low Politics: Maritime Cooperation with Russia in the Arctic'. *Arctic Review on Law and Politics* 7 (1): 83–100.

2018. 'Managing Conflict at Sea: The Case of Norway and Russia in the Svalbard Zone'. *Arctic Review on Law and Politics* 9: 100–23.

2020a. *Coast Guards and Ocean Politics in the Arctic*. Singapore: Palgrave Macmillan.

2020b. 'Maritime Boundary Disputes: What Are They and Why Do They Matter?' *Marine Policy* 120 (October): 104–18.

2021a. 'Norway's Arctic Policy: Still High North, Low Tension?' *The Polar Journal* 11 (1): 75–94.

2021b. 'The Arctic Security Region: Misconceptions and Contradictions'. *Polar Geography* 44 (1): 55–74.

2021c. 'Troubled Seas? The Changing Politics of Maritime Boundary Disputes'. *Ocean & Coastal Management 205 (May 2021): 105535*. https://linkinghub.elsevier.com/retrieve/pii/S096456912100020X.

2022. *Ocean Geopolitics: Marine Resources, Maritime Boundary Disputes and the Law of the Sea*. Cheltenham, UK: Edward Elgar.

2023a. 'Five Misconceptions in Arctic Security and Geopolitics'. *The Arctic Institute June 1*. www.thearcticinstitute.org/five-misconceptions-arctic-security-geopolitics/.

2023b. 'Ine Eriksen Søreide: Nordområdene, USA Og Balansekunst'. *Internasjonal Politikk* 81 (1): 60–76.

ed. 2023c. *Norway's Arctic Policy: Geopolitics, Security and Identity in the High North*. Cheltenham: Edward Elgar.

2023d. 'The Arctic after Russia's Invasion of Ukraine: The Increased Risk of Conflict and Hybrid Threats'. *Finland*. www.hybridcoe.fi/publications/hybrid-coe-paper-18-the-arctic-after-russias-invasion-of-ukraine-the-increased-risk-of-conflict-and-hybrid-threats/.

2024. 'The Myths of Svalbard Geopolitics: An Arctic Case Study'. *Marine Policy* 167 (September 2024).

2025a. 'Norge under Press: En Ny Strategi for Nordområdene Anno 2025'. Lysaker, Norway. www.fni.no/publikasjoner/norge-under-press-en-ny-strategi-for-nordomradene-anno-2025.

2025b. 'Trump's Greenland Flirt Is Clumsy Arctic Geopolitics'. *The Arctic Institute*, 10 January 2025. www.thearcticinstitute.org/trumps-greenland-flirt-clumsy-arctic-geopolitics/.

2025c. 'Trump Has Misunderstood the Importance of Greenland'. *The Arctic Institute*, 29 March 2025. www.thearcticinstitute.org/trump-misunderstood-importance-greenland/.

Østhagen, Andreas, Anne-Kristin Jørgensen, and Arild Moe. 2020. 'Рыбоохранная Зона Шпицбергена: Как Россия и Норвегия Разрешают Арктические Разногласия ("The Svalbard Fisheries Protection Zone: How Russia and Norway Manage an Arctic Dispute")'. Арктика и Север *(Arctic and North)* 40: 183–205. www.arcticandnorth.ru/article_index_years.php?ELEMENT_ID=348131.

Østhagen, Andreas, and Whitney P. Lackenbauer. 2023. 'Security Dynamics In, Through, and Over the Arctic "Region"'. In *Towards a Sustainable Arctic: International Security, Climate Change and Green Shipping*, edited by Michael Goodsite and Niklas Swanström, 1–24. Singapore: World Scientific Publishing.

Østhagen, Andreas, and Andreas Raspotnik. 2018. 'Crab! How a Dispute over Snow Crab Became a Diplomatic Headache between Norway and the EU'. *Marine Policy 98 (December 2018): 58–64*. https://linkinghub.elsevier.com/retrieve/pii/S0308597X18302550.

2019. 'Why Is the European Union Challenging Norway over Snow Crab? Svalbard, Special Interests, and Arctic Governance'. *Ocean Development & International Law* 50 (2–3): 190–208.

Østhagen, Andreas, and Svein Vigeland Rottem. 2023. 'Geopolitics and Increased Tension? Differentiating between Political Dynamics in the

Arctic'. In *Norway's Arctic Policy: Geopolitics, Security, and Identity in the High North*, edited by Andreas Østhagen, 23–37. Cheltenham, UK: Edward Elgar.

Østhagen, Andreas, and Clive H. Schofield. 2021. 'An Ocean Apart? Maritime Boundary Agreements and Disputes in the Arctic Ocean'. *The Polar Journal*. https://doi.org/10.1080/2154896X.2021.1978234.

Østhagen, Andreas, Greg L. Sharp, and Paal S. Hilde. 2018. 'At Opposite Poles: Canada's and Norway's Approaches to Security in the Arctic'. *Polar Journal* 8 (1): 163–81.

Østreng, Willy. 1999. 'Strategic, Legal and Political Implications of International Shipping on the NSR'. In *The 21st Century – Turning Point for the Northern Sea Route?*, edited by Claes L. Ragner, 71–87. Oslo: Kluwer Academic.

Paasi, Anssi, and Eeva Kaisa Prokkola. 2008. 'Territorial Dynamics, Cross-Border Work and Everyday Life in the Finnish-Swedish Border Area'. *Space and Polity* 12 (1): 13–29.

Padrtova, Barbora. 2019. 'Frozen Narratives: How Media Present Security in the Arctic'. *Polar Science* 21 (September): 37–46.

Padrtová, Barbora. 2015. 'The Arctic Regional Security Complex?' In *Panorama of Global Security Environment 2015*, edited by P. Bátor and R. Ondrejcsák. Bratislava.

2017. 'Securitization of the Arctic: U.S. Securitizing Actors and Their Strategies'. Masaryk University. https://is.muni.cz/th/o6uon/Padrtova_PhD_thesis_FINAL.pdf.

Palmer, David, and Rachael Gosnell. 2024. 'An Evolution in Arctic Collective Defense'. *The Arctic Institute*, 7 November 2024. www.thearcticinstitute.org/evolution-arctic-collective-defense/.

Palosaari, Teemu, and Nina Tynkkynen. 2015. 'Arctic Securitization and Climate Change'. In *Handbook of the Politics of the Arctic*, edited by Leif Christian Jensen and Geir Hønneland, 87–104. Cheltenham, UK: Edward Elgar.

Parfitt, Tom. 2007. 'Russia Plants Flag on North Pole Seabed'. *The Guardian*, 2 August 2007.

Parlato, Nicholas, Alexandra Middleton, Susana Hancock, Zia Madani, and Paul Arthur Berkman. 2024. 'The 'Bear' in the Room: How Do We Prepare for the Future with Russia?' *UArctic*. 2024.

Påsche, Eirik Sørenmo. 2021. 'Amerikanske Bombefly Skal Muligens Operere Fra Norge'. *NRK*, January 2021.

Paul, Michael. 2024. 'Back to the Future of the Arctic: The Enduring Relevance of Arms Control'. *Berlin*.

Pedersen, Torbjørn. 2008a. 'Endringer i Internasjonal Svalbard Politikk'. *Internasjonal Politikk* 67 (1): 31–44.

2008b. 'The Constrained Politics of the Svalbard Offshore Area'. *Marine Policy* 32 (6): 913–19.

2017. 'The Politics of Presence: The Longyearbyen Dilemma'. *Arctic Review on Law and Politics* 8: 95–108.

2018. 'Når Nordområdene Lever Sitt Eget Liv'. *Internasjonal Politikk* 76 (3): 140–58.

2021. 'The Politics of Research Presence Revisited: A Response to Molenaar and Ulfstein'. *Polar Journal* 11 (2).

Pedersen, Torbjørn, and Odd Gunnar Skagestad. 2020. 'En Nordområdepolitikk Om Alt – Eller Ingenting?' *Internasjonal Politikk* 78 (4): 490–99.

Pedersen, Torbjørn, and Beate Steinveg. 2024. 'Russia's Clashing Ambitions: Arctic Status Quo and World-Order Revision'. *Politics and Governance* 12: 7311.

Penikett, Tony. 2017. *Hunting the Northern Character*. Vancouver: UBC Press.

Penney, Christopher. 2024. 'The Fiscal Implications of Meeting the NATO Military Spending Target'. *Ottawa*. www.pbo-dpb.ca/en/publications/RP-2425-020-S--fiscal-implications-meeting-nato-military-spending-target--repercussions-financieres-atteinte-cible-depenses-militaires-fixee-otan.

Perreault, Francois. 2011. 'The Arctic Linked to the Emerging Dominant Ideas in Canada's Foreign and Defence Policy'. *The Northern Review* 33 (Spring 2011): 47–67.

Pes, Arianna Muro. 2020. 'Far but Not So Far: Italy's Role and Interests in the Arctic'. *The Arctic Institute*, 10 November 2020. www.thearcticinstitute.org/italy-role-interests-arctic/.

Peters, Kimberley, Phil; Steinberg, and Elaine Stratford, eds. 2018. *Territory Beyond Terra*. London: Rowman & Littlefield International.

Pettersen, Beth Mørch, Inghild Eriksen, Håvard Gulldahl, Benjamin Fredriksen, Erlend Koppergård, Jørn Norstrøm, and Egil Ursin. 2023. 'Spionskipene'. *NRK*, 19 April 2023. www.nrk.no/nordland/xl/fiskebater-og-andre-fartoy-fra-russland-kan-drive-spionasje-og-etterretning-i-norge-1.16371100#intro-authors--expand.

Pettersen, Beth Mørch, Vicky Schaubert, and Inghild Eriksen. 2022. 'Ber Norge Stramme Inn Kontrollen Med Satellittstasjonene På Svalbard (Asks Norway to Better Control Satellite Stations on Svalbard)'. *NRK*, 5 April 2022.

Pettersen, Vibeke Lund. 2010. 'Delelinjeavtalen Undertegnet'. *NRK*, 15 September 2010. www.nrk.no/tromsogfinnmark/delelinjeavtalen-undertegnet-1.7293991.

Pezard, Stephanie, Stephen J. Flanagan, Scott W. Harold, Irina A. Chindea, Benjamin J. Sacks, Abbie Tingstad, Tristan Finazzo, and Soo Kim. 2022. '*China's Strategy and Activities in the Arctic: Implications for North American and Transatlantic Security*'. Santa Monica, CA.

Pharand, Donat. 2007. 'The Arctic Waters and the Northwest Passage: A Final Revisit'. *Ocean Development and International Law* 38 (1–2): 3–69.

Pincus, Rebecca. 2015. 'The Arctic Coast Guard Forum: A Welcome and Important Step'. *Arctic Yearbook 2015*. https://arcticyearbook.com/images/yearbook/2015/Commentaries/COMM_R_Pincus.pdf.

2020. 'Three-Way Power Dynamics in the Arctic'. *Strategic Studies Quarterly* 14 (1): 40–63.

Polar Bear Science. 2018. 'Polar Bears Do Not Outnumber People in Svalbard and the Backlash against Ecotourism over a Justified Defensive Kill'. Polar Bear Science. 2018. https://polarbearscience.com/2018/08/02/polar-bears-do-not-outnumber-people-in-svalbard-and-the-backlash-against-ecotourism-over-a-justified-defensive-kill/.

Power, Elaine M. 2008. 'Conceptualizing Food Security for Aboriginal People in Canada'. *Canadian Journal of Public Health / Revue Canadienne de Santé Publique* 99 (2): 95–97.

Price, Richard. 2008. 'Moral Limit and Possibility in World Politics'. *International Organization* 62 (2): 191–220.

Rachman, Gideon. 2023. *The Age of the Strongman: How the Cult of the Leader Threatens Democracy around the World*. New York: Other Press.

Rack, Ursula. 2018. 'Exploring and Mapping the Antarctic: Histories of Discovery and Knowledge'. In *Routledge Handbook of the Polar Regions*, edited by Mark Nuttall, Torben R. Christensen, and Martin J. Siegert, 34–44. Abingdon, UK: Routledge.

Rahbek-Clemmensen, Jon. 2017. 'The Ukraine Crisis Moves North. Is Arctic Conflict Spill-over Driven by Material Interests?' *Polar Record* 53 (1): 1–15.

Rainwater, Shiloh. 2013. 'Race to the North: China's Arctic Strategy and Its Implications'. *Naval War College Review* 66 (2): Article 7.

Ramutsindela, Maano. 2015. 'Regionalisms and the "Exceptionality" of Security Regions'. *Nordia Geographical Publications* 44 (4): 81–86.

Rantanen, Mika, Alexey Yu. Karpechko, Antti Lipponen, Kalle Nordling, Otto Hyvärinen, Kimmo Ruosteenoja, Timo Vihma, and Ari Laaksonen. 2022. 'The Arctic Has Warmed Nearly Four Times Faster than the Globe since 1979'. *Communications Earth & Environment Volume* 3 (168).

Rappaport, Sarah. 2024. 'For $323 Million, Last Private Land in Arctic Archipelago Can Be Yours'. *Bloomberg*, 10 May 2024. www.bloomberg.com/news/articles/2024-05-10/rare-arctic-land-in-svalbard-is-now-for-sale-for-323-million?embedded-checkout=true.

Raspotnik, Andreas. 2018. *The European Union and the Geopolitics of the Arctic*. Cheltenham, UK: Edward Elgar.

Raspotnik, Andreas, and Andreas Østhagen. 2014. 'From Seal Ban to Svalbard – The European Parliament Engages in Arctic *Matters'*. The Arctic Institute. March 2014.

2019. 'What about the Arctic? The European Union's Geopolitical Quest for Northern Space'. *Geopolitics*.

Raspotnik, Andreas, and Adam Stepien. 2023. 'The European Union's Polar Ambitions: Regional Geo-Policies yet Limited Geo-Strategic Vision'. *Journal of European Integration* 45 (8): 1181–97.

Rayfuse, Rosemary. 2014. 'Coastal State Jurisdiction and the Polar Code: A Test Case for Arctic Oceans Governance?' In *Polar Oceans Governance in an Era of Environmental Change*, edited by Tim Stephens and David L. VanderZwaag, 235–52. Cheltenham and Northampton: Edward Elgar.

Regjeringen. 2025a. 'Norge i Nord'. *Rapporter Og Planer*. 2025. www.regjeringen.no/no/dokumenter/norge-i-nord-nordomradepolitikken-i-en-ny-virkelighet/id3116799/?ch=1.

2025b. 'Norge i Nord'. Rapporter Og Planer. 2025.

Reuters. 2016. 'Ban on New Arctic Drilling Gives Canada Leg Up, U.S. Lawmakers Say'. *Reuters*, 21 December 2016. www.reuters.com/article/us-canada-arctic-drilling-idUSKBN14A261.

Rieker, Pernille. 2016. *External Governance as Security Community Building*. London: Palgrave Macmillan.

Rishikesh, Kumar. 2021. 'Indian Navy Presence in Arctic Could Act as a Strategic Counterweight to China: Analyst'. *Sputnik International*, 11 November 2021. https://sputnikglobe.com/20211111/indian-navy-presence-in-arctic-could-act-as-a-strategic-counterweight-to-china-analyst-1090622956.html.

Risse, Thomas. 2000. '"Let's Argue!": Communicative Action in World Politics'. *International Organization* 54 (1): 1–39.

Riste, Olav. 2005. *Norway's Foreign Relations: A History*. Oslo: Universitetsforlaget.

Roach, J. Ashley, and Robert W. Smith. 2012. *Excessive Maritime Claims*. 3rd ed. Leiden/Boston: Martinus Nijhoff Publishers.

Rosen, Armin. 2015. 'This Chart Shows How Almost Every Country in the World Has a Disputed Border'. *Business Insider*, 4 February 2015. www.businessinsider.com/this-chart-shows-how-almost-every-country-in-the-world-has-a-disputed-border-2015-2.

Rosen, Kenneth R. 2022. 'A Battle for the Arctic Is Underway. And the U.S. Is Already Behind'. *Politico*, 17 December 2022.

Røseth, Tom. 2014. 'Russia's China Policy in the Arctic'. *Strategic Analysis* 38 (6): 841–59.

Rothwell, Donald. 1996. *The Polar Regions and the Development of International Law. Cambridge Studies in International and Comparative Law CN – KZ4110.P65 R68 1996.*

Rottem, Svein Vigeland. 2010. 'Klima Og Sikkerhet i Arktis'. *Internasjonal Politikk* 68 (2): 183–204.

2014. 'The Arctic Council and the Search and Rescue Agreement: The Case of Norway'. *Polar Record* 50 (3): 284–92.

2017. 'The Arctic Council: Challenges and Recommendations'. In *Arctic Governance: Law and Politics. Volume 1*, edited by Svein Vigeland Rottem and Ida Folkestad Soltvedt, 231–51. London: I. B. Tauris.

2020. *The Arctic Council: Between Environmental Protection and Geopolitics*. Singapore: Palgrave Macmillan.

Roucek, Joseph S. 1951. 'The Geopolitics of the Aleutians'. *Journal of Geography* 50 (1): 24–29.

Rowe, Lars. 2015. 'Fra Unntakstilstand Til En Ny Normal ("From State of Emergency to New Normalcy")'. In *Naboer i Frykt Og Forventning: Norge Og Russland 1917–2014 ('Neighbors in Fear and Expectation: Norway and Russia 1917–2014')*, edited by Sven G. Holtsmark, 628–32. Oslo: Pax Forlag.

2018. 'Fornuft Og Følelser: Norge Og Russland Etter Krim (Sense and Sensibility: Norway and Russia after Crimea)'. *Nordisk Østforum* 32: 1–20.

Rowe, Lars, and Geir Hønneland. 2010. 'Hva Er Nordområdepolitikk? Utfordringer Innenrikspolitisk, i Nærområdene Og Globalt'. *Plan* 42 (2): 10–15.

Rozpedowski, Joanna. 2024. 'Resource Wars: How Climate Change Is Fueling Militarization of the Arctic'. *RealClear Defense*. 2024. www.realcleardefense.com/articles/2024/08/07/resource_wars_how_climate_change_is_fueling_militarization_of_the_arctic_1050051.html.

Russia–Norway. 2007. 'Agreement between the Russian Federation and the Kingdom of Norway on the Maritime Delimitation in the Varangerfjord Area (2007)'. *UN Law of the Sea Bulletin* 42, 2007.

Russia–United States. 1867. 'Convention Ceding Alaska between Russia and the United States, 30 March 1867'. Article I, reprinted in C. Parry, ed. Consolidated Treaty Series, Vol. 134 (pp. 331–35).

Sack, Robert. 1986. *Human Territoriality: Its Theory and History*. Cambridge: Cambridge University Press.

Sadurski, Leszek Krzysztof. 2022. 'Regional Security Complex Theory: Why Is This Concept Still Worth Developing?' *Athenaeum. Polskie Studia Politologiczne* 75: 137–53.

2024. 'The Arctic as a Regional Security Complex'. *Polar Geography* 47 (1): 1–15.

Sale, Richard, and Eugene Potapov. 2010. *The Scramble for the Arctic: Ownership, Exploitation and Conflict in the Far North*. London: Frances Lincoln.

Sapsford, Jay. 2024. 'Russia, China, and Arctic Geopolitics'. U.S. Chamber of Commerce. 2024. www.uschamber.com/international/russia-china-and-arctic-geopolitics.

Sarıoğlu, Bülent. 2023. 'Türkiye Joins "Arctic Treaty," Gains Access to North Pole'. *Hürriyet Daily News*, 7 October 2023. www.hurriyetdailynews.com/turkiye-joins-arctic-treaty-gains-access-to-north-pole-186842.

Saxena, Abhishek. 2020. 'The Return of Great Power Competition to the Arctic'. *The Arctic Institute*, 22 October 2020.

Saxi, Håkon Lunde. 2019. 'The Rise, Fall and Resurgence of Nordic Defence Cooperation'. *International Affairs* 95 (3): 659–80.

Schaller, Benjamin. 2015. 'The Arctic Security Community: Proving Ground or Sub-Plot of a Tensed European Security Environment?' *Arctic Yearbook*. https://arcticyearbook.com/arctic-yearbook/2015/2015-scholarly-papers/140-the-arctic-security-community-proving-ground-or-sub-plot-of-a-tensed-european-security-environment.

Schenoni, Luis L., Gary Goertz, Andrew P. Owsiak, and Paul F. Diehl. 2020. 'Settling Resistant Territorial Disputes: The Territorial Boundary Peace in Latin America'. *International Studies Quarterly* 64 (1): 57–70.

Schofield, Clive H. 2015. 'Dividing and Managing Increasingly International Waters: Delimiting the Bering Sea, Strait and Beyond'. In *Science, Technology and New Challenges to Ocean Law*, edited by James Kraska and H. Scheiber, 313–44. Leiden/Boston: Martinus Nijhoff Publishers.

Schofield, Clive, Rashid Sumaila, and William Cheung. 2016. 'Fishing, Not Oil, Is at the Heart of the South China Sea Disput'. *The Conversation*, 15 August 2016.

Science Daily. 2021. 'Alaska Infrastructure at Risk of Earlier Failure'. *Science News*, 24 June 2021. www.sciencedaily.com/releases/2021/06/210624161653.htm.

Senior Arctic Officials. 2011. 'Senior Arctic Officials (SAO) Report to Ministers, Nuuk, Greenland, May 2011'. Nuuk.

Sergunin, Alexander. 2014. 'Four Dangerous Myths about Russia's Plans for the Arctic'. *Russia Direct*, 25 November 2014. www.russia-direct.org/analysis/four-dangerous-myths-about-russias-plans-arctic.

Sergunin, Alexander, and Valery Konyshev. 2014. 'Russia in Search of Its Arctic Strategy: Between Hard and Soft Power?' *Polar Journal* 4 (1): 69–87.

2017. 'Russian Military Strategies in the Arctic: Change or Continuity?' *European Security* 26 (2): 171–89.
Sevunts, Levon. 2016. 'Arctic Nations Deepen Coast Guard Cooperation'. *Eye on the Arctic*, 10 June 2016.
2020. 'Experts Warn of Potentially 'Deadly' Great Power Games in the Arctic'. *The Barents Observer*, September 2020.
Sfraga, Michael. 2017. 'The Arctic: From Romance to Reality'. *The Wilson Quarterly*, Summer 201.
Shadian, Jessica. 2007. 'In Search of an Identity Canada Looks North'. *American Review of Canadian Studies* 37 (3): 323–53.
Sharma, Bipandeep. 2021. 'High Time for India's First Polar Research Vessel'. *IDSA Comment*, 25 November 2021. www.idsa.in/idsacomments/indias-first-polar-research-vessel-bsharma-251121.
2024. 'India'. In *The Big Picture of Arctic Geopolitics: An Actor-Oriented Analysis*, edited by Iselin Nemeth Winther and Andreas Østhagen, 14–17. Lysaker: Fridtjof Nansen Institute. www.fni.no/getfile.php/1318301-1706208108/Filer/Publikasjoner/FNI Report 1 2024.pdf.
Simon, Sheldon W. 2012. 'Conflict and Diplomacy in the South China Sea'. *Asian Survey* 52 (6): 995–1018.
Singer, J. D. 1961. 'The Level-of-Analysis Problem in International Relations'. *World Politics* 14 (01): 77–92. www.journals.cambridge.org/abstract_S0043887100010303.
Singh, Abhijit, and Andreas Østhagen. 2024. 'India's Arctic Imperative'. *The Hindu*, 16 April 2024. www.thehindu.com/opinion/op-ed/indias-arctic-imperative/article68067583.ece.
Sjoberg, Laura. 2008. 'Scaling IR Theory: Geography's Contribution to Where IR Takes Place'. *International Studies Review* 10 (3): 472–500.
Skedsmo, Pål. 2007. 'Demokratisering Og Miljøkamp På Kolahalvøya ("Democratisation and Environmentalism on the Kola Peninsula")'. *Norsk Antropologisk Tidsskrift* 18 (3–4): 241–52.
Skeie, Kari, and Beth Mørch Pettersen. 2016. 'USAs Forsvarsminister Skryter Av Norge Som Sin Beste Venn'. *NRK*, 8 September 2016. www.nrk.no/nordland/usas-forsvarsminister-skryter-av-norge-som-sin-beste-venn-1.13126378.
Skogrand, Kjetil. 2004. *Norsk Forsvarshistorie 1940–1970: Alliert i Krig Og Fred*. Bergen: Eide forlag.
Skogrand, Kjetil, and Rolf Tamnes. 2001. *Fryktens Likevekt. Atombomben, Norge Og Verden 1945–1970*. Oslo: Tiden Norsk Forlag.
Skram, Arild-Inge. 2017. *Alltid Til Stede: Kystvakten 1997–2017 (Always Present: Coast Guard 1997–2017)*. Bergen: Fagbokforlaget.
Sloan, Geoffrey, and Colin S. Gray. 1999. 'Geopolitics, Geography and Strategy'. In *Geopolitics, Geography and Strategy*, 1–14.

Smith, Julianne, Jerry Hendrix, and Robert D. Kaplan. 2017. 'Forgotten Waters – Minding the GIUK Gap'. Washington, DC. www.cnas.org/publications/reports/forgotten-waters.

Smith, Robert W. 1994. 'United States–Russia Maritime Boundary'. In *World Boundaries, Vol.5*, edited by Gerald H. Blake, 91–99. London: Routledge.

Snidal, Duncan. 1985. 'The Game Theory of International Politics'. *World Politics* 38 (1): 25–57. www.theguardian.com/science/2014/jun/20/game-theory-life-qanta-magazine.

Söderbaum, Fredrik. 2012. 'Theories of Regionalism'. In *Routledge Handbook of Asian Regionalism*, edited by Mark Beeson and Richard Stubbs, 11–21. Abingdon, UK: Routledge.

Sokolíčková, Zdenka. 2022. 'The Trouble with Local Community in Longyearbyen, Svalbard: How Big Politics and Lack of Fellesskap Hinder a Not-yet-Decided Future'. *Polar Record* 58 (e36).

Solski, Jan. 2024. 'The US Arctic Gambit: Testing the Limits of UNCLOS'. *The Arctic Institute*, 16 January 2024. www.thearcticinstitute.org/us-arctic-gambit-testing-limits-unclos/.

Solstad, Silje Charlotte. 2012. '*Spillet Om Delelinja: En Tonivåanalyse Av Forhandlingene Om Grense i Barentshavet Og Polhavet (The Game over the Boundary: A Two-Level Analysis of the Negotiations over the Boundary)*'. University of Tromsø.

Solvik, Bjørn Magne. 2021. 'Utenriksministeren Deltok På Kirkeneskonferansen'. *RadioNordkapp*, 20 February 2021.

Søreide, Ine Eriksen. 2016. 'NATO and the North Atlantic: Revitalizing Collective Defense and the Maritime Domain'. *PRISM* 2 (2): 49–57.

2017. 'En Balansert Sikkerhetspolitikk'. Regjeringen. 2017. www.regjeringen.no/no/aktuelt/avskrekkingberoligelse/id2536988/.

2020. 'Mennesker, Muligheter Og Norske Interesser i Nord – Ny Giv for Nordområdepolitikken'. *Internasjonal Politikk* 78 (4): 456–60.

Søreide, Ine Eriksen, and Monica Mæland. 2020. 'Svalbardtraktaten 100 År (Svalbard Treaty 100 Years)'. *Historisk Arkiv*. 2020. www.regjeringen.no/no/dokumentarkiv/regjeringen-solberg/aktuelt-regjeringen-solberg/ud/taler-og-innlegg/imes_taler/2020/traktat_hundre/id2689324/.

Sørensen, Camilla T. N. 2019. 'Intensifying U.S.-China Security Dilemma Dynamics Play out in the Arctic: Implications for China's Arctic Strategy'. *Arctic Yearbook*, no. 2019. https://arcticyearbook.com/arctic-yearbook/2019/2019-scholarly-papers/323-intensifying-u-s-china-security-dilemma-dynamics-play-out-in-the-arctic-implications-for-china-s-arctic-strategy.

Sparke, M. 1998. 'From Geopolitics to Geoeconomics: Transnational State Effects in the Borderlands'. *Geopolitics* 3 (2): 62–98.

Spence, Jennifer. 2017. 'Is a Melting Arctic Making the Arctic Council Too Cool? Exploring the Limits to the Effectiveness of a Boundary Organization'. *Review of Policy Research* 34 (6): 790–811.

Spence, Jennifer, Edward Alexander, Rolf Rødven, and Sara Harriger. 2023. 'What Makes the Arctic Exceptional? Stories of Geopolitics, Environments and Homelands'. *Arctic Yearbook*. https://arcticyearbook.com/arctic-yearbook/2023/2023-scholarly-papers/499-what-makes-the-arctic-exceptional-stories-of-geopolitics-environments-and-homelands.

Spykman, Nicholas J. 1944. *The Geography of the Peace*. New York: Harcourt Brace & Co.

Srivastava, Vanita. 2023. 'Why Indian Scientists Are on a Winter Voyage to the Arctic'. *Nature India*, 29 December 2023. www.nature.com/articles/d44151-023-00203-z.

Staalesen, Atle. 2021. 'Polar Bears Face Extinction in Svalbard and Arctic Russia'. *Barents Observer*, 2 July 2021. https://thebarentsobserver.com/en/climate-crisis/2021/07/polar-bears-face-extinction-svalbard-and-arctic-russia.

2023a. 'Russia Stages Military-Style Propaganda Parade on Norway's Svalbard Archipelago'. *Barents Observer*, 9 May 2023. https://thebarentsobserver.com/en/security/2023/05/russia-stages-military-style-propaganda-parade-norways-svalbard-archipelago.

2023b. 'Moscow Aims for Another Grab: Navy Chief Says Russia Will Expand into Arctic Ocean'. *The Barents Observer*, 7 December 2023. https://thebarentsobserver.com/en/arctic/2023/12/moscow-aims-another-grab-navy-chief-says-russia-will-expand-arctic-ocean.

2024. 'China Commissions New Icebreaker'. *Barents Observer*, 10 July 2024. www.thebarentsobserver.com/arctic/china-commissions-new-icebreaker/119223.

Stabrun, Kristoffer. 2009. 'The Grey Zone Agreement of 1978: Fishery Concerns, Security Challenges and Territorial Interests'. *FNI Report* 13: 1–43.

State Council of the People's Republic of China. 2018. 'China's Arctic Policy'. *Chinese Government*. 2018. http://english.gov.cn/archive/white_paper/2018/01/26/content_281476026660336.htm.

Steinberg, Philip E. 2001. *The Social Construction of the Ocean*. Cambridge: Cambridge University Press.

Steinberg, Philip E., and Klaus Dodds. 2015. 'The Arctic Council after Kiruna'. *Polar Record* 51 (1): 108–10.

Steinveg, Beate. 2017. 'The Backdoor into Arctic Governance?' UiT: The Arctic University of Norway. 2017.

2023. *Arctic Governance through Conferencing. Actors, Agendas and Arenas*. Cham: Springer.

2024. 'Small States in World Politics: Norwegian Interests and Foreign Policy Challenges in the Arctic'. *Arctic Review on Law and Politics* 15: 3–24.

Steinveg, Beate, and Ingrid A. Medby. 2020. 'Nordområdenarrativer Og Identitetsbygging i Nord'. *Internasjonal Politikk* 78 (4): 535–44.

2023. 'Norwegian High North Narratives and Identity Construction in the North'. In *Norway's Arctic Policy: Geopolitics, Security, and Identity in the High North*, edited by Andreas Østhagen. Cheltenham, UK: Edward Elgar.

Stenersen, Øivind, and Ivar Libæk. 2007. *Norges Historie: Fra Istid Til i Dag (History of Norway: From Ice Age until Today)*. Oslo: Dinamo forlag.

Stensdal, Iselin. 2014. 'Chinese Climate-Change Policy, 1988–2013: Moving on Up'. *Asian Perspective* 38 (1): 111–35.

Stensdal, Iselin, and Gørild Heggelund, eds. 2024. *China-Russia Relations in the Arctic: Friends in the Cold?* Singapore: Palgrave Macmillan.

Stephen, Kathrin, and Sebastian Knecht, eds. 2017. *Governing Arctic Change: Global Perspectives*. London: Palgave Macmillan.

Stephens, Tim. 2011. 'The Arctic and Antarctic Regimes and the Limits of Polar Comparativism'. *German Yearbook of International Law* 54: 315–49.

Stokke, Olav Schram. 2006. 'Examing the Consequences of Arctic Institutions'. In *International Cooperation and Arctic Governance: Regime Effectiveness and Northern Region Building*, edited by Olav Schram Stokke and Geir Hønneland, 13–26. New York: Routledge.

2014a. 'Asian States and Arctic Governance'. *Strategic Analysis* 38 (6): 770–83. http://dx.doi.org/10.1080/09700161.2014.952946.

2014b. 'International Environmental Governance and Arctic Security'. In *Geopolitics and Security in the Arctic*, edited by Rolf Tamnes and Kristine Offerdal, 121–46. Abingdon: Routledge.

2017. 'Geopolitics, Governance, and Arctic Fisheries Politics'. In *Global Challenges in the Arctic Region: Sovereignty, Environment and Geopolitical Balance*, edited by E. Conde and S. S. Iglesias, 170–95. London: Routledge.

2022. 'External Shocks, Resilience and Barents Sea Fisher Compliance'. In *Marine Resources, Climate Change and International Management Regimes*, edited by Olav Schram Stokke, Andreas Østhagen, and Andreas Raspotnik, 179–96. London: Bloomsbury.

Stokke, Olav Schram, Andreas Østhagen, and Andreas Raspotnik, eds. 2022. *Marine Resources, Climate Change, and International Management Regimes*. London: Bloomsbury.

Stoltenberg, Jens, Kristin Halvorsen, Åslaug Haga, Hill-Marta Solberg, Øystein Djupedal, Marit Arnstad, Martin Kolberg, Henriette Westhrin, and Magnhild Meltveit Kleppa. 2005. 'Plattform for Regjeringssamarbeidet Mellom Arbeiderpartiet, Sosialistisk Venstreparti Og Senterpartiet'. *Soria Moria*. www.regjeringen.no/globalassets/upload/smk/vedlegg/2005/regjeringsplatform_soriamoria.pdf.

Storey, David. 2012. *Territories: The Claiming of Space*. 2nd ed. Abingdon, UK: Routledge.

Stormark, Kjetil. 2017. 'Russland Trente På Invasjon Av Svalbard'. Aldrimer No. 2017. www.aldrimer.no/russland-trente-pa-invasjon-av-svalbard/.

Suhag, Vikram. 2022. 'India's Arctic Security Dimension: "Thought through" or "a Remedial Process to Address an Insecurity"'. *The Arctic Institute*, 22 February 2022. www.thearcticinstitute.org/india-arctic-security-dimension-thought-through-remedial-process-address-insecurity/.

Sun, Yun. 2020. 'Defining the Chinese Threat in the Arctic'. The Arctic Institute. 2020. www.thearcticinstitute.org/defining-the-chinese-threat-in-the-arctic/.

Supreme Court of Norway. 2023. 'The Svalbard Treaty Does Not Give a Latvian Shipping Company the Right to Catch Snow Crab on the Continental Shelf Outside Svalbard'. *Rulings*. 2023. www.domstol.no/en/supremecourt/rulings/2023/supreme-court-civil-cases/HR-2023-491-P/.

Svalbard Treaty. 1920. 'The Svalbard Treaty'. University of Oslo. www.jus.uio.no/english/services/library/treaties/01/1-11/svalbard-treaty.html.

Sveen, Eirik Hind. 2021. 'Ann Helen Bor Ved En Ny Atomhavn: – Om Man Ikke Har Informasjon, Så Lager Man Sin Egen'. *NRK*, April 2021.

Tamnes, Rolf. 1987. 'Svalbard Mellom Øst Og Vest. Kald Krig Og Lavspenning i Nord 1947–1953'. *Forsvarsstudier*. Vol. 4. Oslo.

1991. 'Svalbard Og Stormaktene. Fra Ingenmannsland Til Kald Krig (1870–1953)'. *Forsvarsstudier*. Vol. 7. Oslo.

1997. *Oljealder 1965–1995: Norsk Utenrikspolitisk Historie*. Oslo: Cappelen Damm.

2011. 'Arctic Security and Norway'. In *Arctic Security in an Age of Climate Change*, edited by James Kraska, 47–64. New York: Cambridge University Press.

2019. 'Småstatsrealisme i 70 År'. *Internasjonal Politikk* 77 (1): 49–59.

2020. 'Sikkerhetspolitiske Utfordringer Og Militær Virksomhet'. In *Svalbardtraktaten 100 År*, edited by Irene Dahl and Øystein Jensen, 135–84. Bergen: Fagbokforlaget.

Tamnes, Rolf, and Kristine Offerdal, eds. 2014. *Geopolitics and Security in the Arctic: Regional Dynamics in a Global World*. Abingdon, UK: Routledge.

Tannenwald, Nina. 2005. 'Ideas and Explanation: Advancing the Theoretical Agenda'. *Journal of Cold War Studies* 7 (2): 13–42.

TASS. 2024. 'Морская Полиция КНР и ФСБ Обсудили Расширение Сотрудничества На Море (PRC Marine Police and FSB Discuss Expanding Maritime Cooperation)'. *TASS*, 24 October 2024. https://tass.ru/politika/22218327.

Tatlow, Didi Kirsten. 2024a. 'China's Expanding Arctic Ambitions Challenge the U.S. and NATO'. *Newsweek*, July 2024.

2024b. 'China's Expanding Arctic Ambitions Challenge the U.S. and NATO'. *Newsweek*, 21 July 2024. www.newsweek.com/2024/08/09/china-russia-us-arctic-north-pole-strategy-svalbard-norway-sea-route-1916641.html.

Thucydides. 1910. *The Peloponnesian War*. London: J. M. Dent.

Till, Geoffrey. 2004. *Seapower: A Guide for the Twenty-First Century*. London: Frank Cass.

Tiller, Rachel. 2009. 'New Resources and Old Regimes: Will the Harvest of Zooplankton Bring Critical Changes to the Svalbard Fisheries Protection Zone?' *Ocean Development & International Law* 40 (4): 309–18.

2010. 'New Resources in Old Waters: The Potential of National and International Conflicts Deriving from the Future Harvest of Calanus Finmarchicus'. *Norges teknisk-naturvitenskapelige universitet*.

Tiller, Rachel, and Elizabeth Nyman. 2015. 'Having the Cake and Eating It Too: To Manage or Own the Svalbard Fisheries Protection Zone'. *Marine Policy* 60: 141–48.

2017. 'The Clear and Present Danger to the Norwegian Sovereignty of the Svalbard Fisheries Protection Zone: Enter the Snow Crab'. *Ocean and Coastal Management* 137: 24–33.

Tilly, Charles. 1990. *Coercion, Capital and European States: AD 990–1990*. First Ed. Cambridge, MA: Basil Blackwell.

Tingstad, Abbie. 2023. 'The US Is Taking an Important, but Imperfect Step in Initiating Extended Continental Shelf Claims – What Are the Implications for the Arctic?' *Wilson Center*, 19 December 2023. www.wilsoncenter.org/article/us-taking-important-imperfect-step-initiating-extended-continental-shelf-claims-what-are.

Tingstad, Abbie, Stephanie Pezard, and Yuliya Shokh. 2024a. '*China-Russia Relations in the Arctic: What Are the Northern Limits of Their Partnership?*' Washington, DC.

2024b. 'China-Russia Relations in the Arctic: What Are the Northern Limits of Their Partnership?' Washington, DC. www.rand.org/pubs/perspectives/PEA2823-1.html.

Tir, Jaroslav, and Paul F. Diehl. 2002. 'Geographic Dimensions of Enduring Rivalries'. *Political Geography* 21 (2): 263–86.

Tjønn, Halvor. 2005. 'Voldsomme Utfall Mot Norge (Strong Reactions against Norway)'. *Aftenposten*, 22 October 2005.

Toal, Gerard. 2024. *Oceans Rise Empires Fall: Why Geopolitics Hastens Climate Catastrophe*. Oxford: Oxford University Press.

Todorov, Andrey. 2020a. 'Russia in Maritime Areas off Spitsbergen (Svalbard): Is It Worth Opening the Pandora's Box?' *Marine Policy* 122 (December). https://doi.org/10.1016/j.marpol.2020.104264.

2020b. 'Svalbard in the Context of Arctic Security'. *Arktika i Sever* 39 (2020): 127–43.

2024. 'Russia's Reaction to the US Continental Shelf Announcement: Political Posturing or Setting the Stage for a Big Move?' *The Arctic Institute*, 9 April 2024. www.thearcticinstitute.org/russias-reaction-us-continental-shelf-announcement-political-posturing-setting-stage-big-move/.

Toje, Asle. 2015. 'Arktisk Nachspiel'. *Dagens Næringsliv*, 1 March 2015. www.dn.no/meninger/debatt/2015/03/01/2057/Globalt/arktisk-nachspiel.

Topalidis, Georgios Tr, Nick N. Kartalis, John R. Velentzas, and Charalampia Gr Sidiropoulou. 2024. 'New Developments in Geopolitics: A Reassessment of Theories after 2023'. *Social Sciences* 13 (2). https://doi.org/10.3390/socsci13020109.

Trevithick, Joseph. 2020. 'Navy Makes Unusual Public Display of Its Secretive Seawolf Submarine's Presence Off Norway'. *The War Zone*, 25 August 2020. www.twz.com/35997/navy-makes-unusual-public-display-of-its-secretive-seawolf-submarines-presence-off-norway.

Tripp, Emily. 2013. 'Realism: The Domination of Security Studies'. *E-International Relations*, 14 June. www.e-ir.info/2013/06/14/realism-the-domination-of-security-studies/.

Trump, Donal J. 2024. 'Truth Details'. *Truth Social*. 2024. https://truthsocial.com/@realDonaldTrump/posts/113698764270730405.

Tukker, Paul. 2022. 'The DEW Line at 65: Future Unclear for the North's Aging Radar Sites'. *CBC News: North*, 10 May 2022. www.cbc.ca/news/canada/north/dew-line-65-years-norad-1.6446875.

Tunsjø, Øystein. 2018. *The Return of Bipolarity in World Politics: China, the United States, and Geostructural Realism*. New York: Columbia University Press.

Tüysüzoğlu, Göktürk. 2021. 'Evaluation of the Systemic Balance in the Arctic within the Context of Neorealism and Regional Security Complex'. *International Journal of Politics and Security* 3 (1): 263–301.

UK Government. 2013. 'Adapting To Change: UK Policy towards the Arctic'. *London*. https://assets.publishing.service.gov.uk/government/uploads/system/uploads/attachment_data/file/251216/Adapting_To_Change_UK_policy_towards_the_Arctic.pdf.

UK Parliament. 2015. 'Chapter 6: The UK and the Arctic'. *Select Committee on the Arctic – Report of Session 2014–15*. 2015.

Ulfstein, Geir. 1995. *The Svalbard Treaty: From Terra Nullius to Norwegian Sovereignty*. Oslo: Aschehoug.

2021. 'The Svalbard Treaty and Research: Comment to Pedersen and Molenaar'. *The Polar Journal* 11 (2): 433–37.

Underwood, Kimberly. 2022. 'Russia's War Also Impacts Arctic Forum'. *SIGNAL*. 2022. www.afcea.org/signal-media/international/russias-war-also-impacts-arctic-forum.

United Nations. 1982. '*United Nations Convention on the Law of the Sea (LOSC)*'. *1833 U.N.T.S. 397 (in Force 16 November 1994), Publication No. E97.V10*. Montego Bay.

United States–Union of Soviet Socialist Republics. 1990. 'Agreement between the United States of America and The Union of the Soviet Socialist Republics on the Maritime Boundary'. 1 June 1990, provisionally in force 15 June 1990.

US Coast Guard. 2019. 'United States Coast Guard Arctic Strategic Outlook'. Washington, DC: Coast Guard. www.uscg.mil/Portals/0/Images/arctic/Arctic_Strategy_Book_APR_2019.pdf.

2024. 'Polar Security Cutter.' www.dcms.uscg.mil/Our-Organization/Assistant-Commandant-for-Acquisitions-CG-9/Programs/Surface-Programs/Polar-Icebreaker/.

US Department of Defense. 2016. 'Joint Media Availability with Secretary Carter and Minister Eriksen Søreide in Oslo, Norway.' Transcript.

2024a. '2024 Arctic Strategy.' Washington, DC. https://media.defense.gov/2024/Jul/22/2003507411/-1/-1/0/DOD-ARCTIC-STRATEGY-2024.PDF.

2024b. '*2024 Arctic Strategy*.' Washington, DC.

US Department of State. 2019. 'Looking North: Sharpening America's Arctic Focus'. *Remarks*. 2019. https://oaarchive.arctic-council.org/items/e68a8fd9-a822-41a0-b372-b5d075f4844f.

2023. 'Announcement of U.S. Extended Continental Shelf Outer Limits.' *Press Releases*. 2023. www.state.gov/announcement-of-u-s-extended-continental-shelf-outer-limits/.

US Department of War. 2025. 'Living Life on the Arctic Edge.' Multimedia. 2025. www.war.gov/Multimedia/Photo-Collections/19777/living-life-on-the-arctic-edge/.

US Navy. 2019. 'Strategic Outlook for the Arctic.' Washington, DC. https://media.defense.gov/2020/May/18/2002302034/-1/-1/1/NAVY_STRATEGIC_OUTLOOK_ARCTIC_JAN2019.PDF.

USA Today. 2023. 'Exploring Two of the Northernmost Places in the World: Norway's Svalbard Archipelago, Greenland'. *USA Today, March 14, 2023*. https://eu.usatoday.com/picture-gallery/travel/destinations/2022/12/03/svalbard-norway-look-archipelago/10728179002/.

USGS. 2008. 'Circum-Arctic Resource Appraisal: Estimates of Undiscovered Oil and Gas North of the Arctic Circle'. Washington, DC: USGS World Energy Assessment Team. http://pubs.usgs.gov/fs/2008/3049/fs2008-3049.pdf.

Utenriksdepartementet. 2020. 'Meld. St. 9 (2020–2021) Mennesker, Muligheter Og Norske Interesser i Nord'. Oslo.

Valasek, Tomas. 2011. 'The Case for a New Approach to about the CER Surviving Austerity The Case for a New'. *Centre for European Reform*, no. April 2011.

Vandiver, John. 2012. 'Nations Seek Ways to Cooperate on a Range of Arctic Issues'. *Stars and Stripes*, 30 August 2012. www.stripes.com/migration/nations-seek-ways-to-cooperate-on-a-range-of-arctic-issues-1.187454.

Varteressian, Maria. 2024. 'State Secretary Maria Varteressian's Opening Remarks at the Arctic Congress'. *Norwegian Government, Speeches and Statements*. 2024. www.regjeringen.no/en/aktuelt/state-secretary-maria-varteressians-opening-remarks-at-the-arctic-congress-in-bodo-30-may-2024/id3041361/.

Vasquez, John A. 1993. *The War Puzzle*. Cambridge: Cambridge University Press.

1995. 'Why Do Neighbors Fight? Proximity, Interaction, or Territoriality'. *Journal of Peace Research* 32 (3): 277–93.

Väyrynen, Raimo. 2003. 'Regionalism: Old and New'. *International Studies Review* 5: 25–51.

Verville, Elizabeth G. 1993. 'United States–Soviet Union'. In *International Maritime Boundaries, Vol. 1*, edited by Jonathan I. Charney and Lewis M. Alexander, 447–60. Dordrecht: Martinus Nijhoff Publishers.

Vock, Ido. 2025. 'Why Does Trump Want Greenland and What Do Its People Think?' *BBC, 2025*. www.bbc.com/news/articles/c74x4m71pmjo.

Vold Hansen, Tiril. 2024. 'Geopolitics, Diplomacy, or Idealistic Research? Framing the Research Community in Ny-Ålesund'. *Polar Journal* 14 (1). https://doi.org/10.1080/2154896X.2024.2342110.

Vu, Tuong. 2010. 'Studying the State through State Formation'. *World Politics* 62 (1): 148–75.

Wallin, Pamela, and Roméo Dallaire. 2011. 'Sovereignty & Security In Canada's Arctic'. *Special Study on the National Security and Defence Policies of Canada*. Vol. March. Ottawa.

Waltz, Kenneth N. 1959. *Man, the State, and War*. New York: Columbia University Press.

1979. *Theory of International Politics*. Boston, MA: McGraw-Hill.

Weber, Bob. 2015. 'Security Fears Growing in the Arctic Region: Survey'. *CTV News*, 22 April 2015.

Weber, Joachim, ed. 2020. *Handbook on Geopolitics and Security in the Arctic: The High North between Cooperation and Confrontation*. Cham: Springer.

Wegge, Njord. 2012. 'The EU and the Arctic: European Foreign Policy in the Making'. *Arctic Review on Law and Politics* 3 (1): 6–29.

2015. 'The Emerging Politics of the Arctic Ocean. Future Management of the Living Marine Resources'. *Marine Policy* 51: 331–38.

ed. 2023. *Sikkerhetspolitikk Og Militærmakt i Arktis*. Oslo: Cappelen Damm.

Wehrmann, Dorothea. 2019. *Critical Geopolitics of the Polar Regions: An Inter-American Perspective*. Abingdon: Routledge.

Weil, Prosper. 1989. *The Law of Maritime Delimitation – Reflections*. London: Grotius Publications Limited.

Weir, Gary E. 2002. 'Virtual War in the Ice Jungle: "We Don't Know How to Do This"'. *Journal of Strategic Studies* 28 (2): 411–27.

Weitz, Richard. 2021. 'Assessing Chinese-Russian Military Exercises: Past Progress and Future Trends'. Washington, DC. www.csis.org/analysis/assessing-chinese-russian-military-exercises-past-progress-and-future-trends.

Wendt, Alexander E. 1999. 'Social Theory of International Politics'. *American Political Science Review* 94: 429. http://ebooks.cambridge.org/ref/id/CBO9780511612183.

Wenger, Michael. 2022. 'Study Points to Responsibilities for IUU Fishing in the Bering Sea'. *Polar Journal*. https://polarjournal.ch/en/2022/11/14/study-points-to-responsibilities-for-iuu-fishing-in-the-bering-sea/.

Wezeman, S. 2016. 'Military Capabilities in the Arctic: A New Cold War in the High North?' *Stockholm*. www.sipri.org/sites/%0Adefault/files/Military-capabilities-in-the-Arctic.pdf%0A.

White, Josh. 2006. 'U.S. to Remove Military Forces and Aircraft from Iceland Base'. *The Washington Post*, 17 March 2006. www.washingtonpost.com/wp-dyn/content/article/2006/03/16/AR2006031601846.html.

The White House. 2009. 'National Security Presidential Directive 66 and Homeland Security Presidential Directive 25'. Washington, DC.

2022a. 'Guidance on Scientific and Technological Cooperation with the Russian Federation for U.S. Government and U.S. Government Affiliated Organizations'. *Press Releases*. 2022. www.whitehouse.gov/ostp/news-updates/2022/06/11/guidance-on-scientific-and-technological-cooperation-with-the-russian-federation-for-u-s-government-and-u-s-government-affiliated-organizations/.

2022b. 'National Strategy for the Arctic Region'. Washington, DC. www.whitehouse.gov/wp-content/uploads/2022/10/National-Strategy-for-the-Arctic-Region.pdf.

2022c. '*National Strategy for the Arctic Region.*' Washington, DC.

Wiegand, Krista E. 2011. *Enduring Territorial Disputes: Strategies of Bargaining, Coercive Diplomacy, and Settlement. Studies in Security and International Affairs*. Athens, GA: University of Georgia Press.

Wig, Ståle. 2014. *Seierherren – En Uautorisert Biografi Om Jonas Gahr Støre*. Oslo: Kagge forlag.

Wigell, Mikael, and Antto Vihma. 2016. 'Geopolitics versus Geoeconomics: The Case of Russia's Geostrategy and Its Effects on the EU'. *International Affairs* 92 (3): 605–27.

Wilhelmsen, Julie, and Kristian Lundby Gjerde. 2018. 'Norway and Russia in the Arctic: New Cold War Contamination?' *Arctic Review on Law and Politics* 9: 382–407.

Wilhelmsen, Julie, and Anni Roth Hjermann. 2022. 'Russian Certainty of NATO Hostility: Repercussions in the Arctic'. *Arctic Review on Law and Politics* 13: 114–42.

Willett, Robert L. 2003. *Russian Sideshow: America's Undeclared War, 1918–1920*. Washington, DC: Potomac Books.

Williams, Heather, Kari A. Bingen, and Lachlan MacKenzie. 2024. 'Why Did China and Russia Stage a Joint Bomber Exercise near Alaska?' *CSIS*, 30 July 2024. www.csis.org/analysis/why-did-china-and-russia-stage-joint-bomber-exercise-near-alaska#:~:text=WhyDidChinaandRussiaStageaJointBomberExerciseearAlaska%3F,-Photo%3ANorthAmerican&text=OnJuly24%2CtheNorth,ofbombersarenuclearcapab.

Williams, Michael C., and Iver B. Neumann. 2000. 'From Alliance to Security Community: NATO, Russia, and the Power of Identity'. *Millennium – Journal of International Studies* 29 (2): 357–87.

Willis, Matthew. 2013. 'The Arctic Council: Underpinning Stability in the Arctic'. *The Arctic Institute, 2013*. www.thearcticinstitute.org/2013/03/the-arctic-council-underpinning.html.

Wilson Rowe, Elana, ed. 2009. *Russia and the North*. Ottawa: University of Ottawa Press.

2018. *Arctic Governance: Power in Cross-Border Cooperation*. Manchester: Manchester University Press.

2020. 'Analyzing Frenemies: An Arctic Repertoire of Cooperation and Rivalry'. *Political Geography* 76 (January).

Wilson Rowe, Elana, and Helge Blakkisrud. 2014. 'A New Kind of Arctic Power? Russia's Policy Discourses and Diplomatic Practices in the Circumpolar North'. *Geopolitics* 19 (1): 66–85.

Wither, James K. 2018. 'Svalbard: NATO's Arctic "Achilles" Heel'. *The RUSI Journal* 163 (5): 28–37.

Woker, Hilde, Bernhard Schartmüller, Knut Ola Dølven, and Katalin Blix. 2020. 'The Law of the Sea and Current Practices of Marine Scientific Research in the Arctic'. *Marine Policy* 115 (May 2020): 103850.

Wormdal, Bård. 2022. 'USA Og Tyrkia Nektes å Bruke Satellittstasjon På Svalbard (US and Turkey Denied Usage of Satellite Station on Svalbard)'. *NRK*, 25 August 2022.

Xiao, Josh. 2024. 'China's Coast Guard Joined by Russians on First Patrol in Arctic'. *BNN Bloomberg*, 2 October 2024. www.bnnbloomberg.ca/business/international/2024/10/02/chinas-coast-guard-joined-by-russians-on-first-patrol-in-arctic/.

Yabuki, Susumu, and Mark Selden. 2014. 'The Origins of the Senkaku/Diaoyu Dispute between China, Taiwan, and Japan'. *The Asia-Pacific Journal* 12 (2).

Ye, Jiang. 2014. 'China's Role in Arctic Affairs in the Context of Global Governance'. *Strategic Analysis* 38 (6): 913–16. http://dx.doi.org/10.1080/09700161.2014.952938.

Ylvisåker, Line Nagell. 2020. 'Researching Fish Landing Facilities in Svalbard: "Raises A Series of Questions on Different Levels"'. *High North News*, 17 November 2020. www.highnorthnews.com/en/researching-fish-landing-facilities-svalbard-raises-series-questions-different-levels.

Young, Oran R. 1992. *Arctic Politics: Conflict and Cooperation in the Circumpolar North*. Hanover, NH: University Press of New England.

2005. 'Governing the Arctic: From Cold War Theater to Mosaic of Cooperation'. *Global Governance* 11 (1): 9–15.

2009. 'Whither the Arctic? Conflict or Cooperation in the Circumpolar North'. *Polar Record* 45 (1): 73–82.

2010. 'Arctic Governance – Pathways to the Future'. *Arctic Review on Law and Politics* 1 (2): 164–85.

Zacher, Mark W. 2001. 'The Territorial Integrity Norm: International Boundaries and the Use of Force'. *International Organization* 55 (2): 215–50.

Zellen, Barry Scott, ed. 2013. *The Fast-Changing Arctic: Rethinking Arctic Security for a Warmer World*. Calgary: Calgary Unversity Press.

Zysk, Katarzyna. 2013. 'Russia's Arctic Strategy: Ambitions and Restraints'. In *The Fast-Changing Arctic: Rethinking Arctic Security for a Warmer World*, edited by Barry Scott Zellen, 281–96. Calgary, AB: Calgary University Press.

ИноСМИ. 2017. 'Россия Нацелилась На Шпицберген (Russia Is Aiming at Svalbard)'. *Geo-Politica*, 17 November 2017. http://geo-politica.info/rossiya-natselilas-na-shpitsbergen.html.

Index

For EU product safety concerns, contact us at Calle de José Abascal, 56–1°, 28003 Madrid, Spain or eugpsr@cambridge.org.

www.ingramcontent.com/pod-product-compliance
Lightning Source LLC
LaVergne TN
LVHW020341240926
848761LV00033B/274
* 9 7 8 1 0 0 9 7 4 3 3 1 0 *